Innovation in Small-Farm Agriculture

Innovation in Small-Farm Agriculture

Improving Livelihoods and Sustainability

Edited by

Amitava Rakshit, Somsubhra Chakraborty, Manoj Parihar,
Vijay Singh Meena, P. K. Mishra, H. B. Singh

CRC Press
Taylor & Francis Group
Boca Raton London New York

CRC Press is an imprint of the
Taylor & Francis Group, an **informa** business

First edition published 2022
by CRC Press
6000 Broken Sound Parkway NW, Suite 300, Boca Raton, FL 33487-2742

and by CRC Press
4 Park Square, Milton Park, Abingdon, Oxon, OX14 4RN

CRC Press is an imprint of Taylor & Francis Group, LLC

Library of Congress Cataloging-in-Publication Data

Names: Rakshit, Amitava, editor. | Chakraborty, Somsubhra, 1984- editor. |
Parihar, Manoj, editor. | Singh Meena, Vijay, editor. |
Mishra, P. K. (Pradeep Kumar), editor. | Singh, H. B., Dr., editor.
Title: Innovation in small-farm agriculture : improving livelihoods and
sustainability / Amitava Rakshit, Somsubhra Chakraborty, Manoj Parihar,
Vijay Singh Meena, Pradeep Kumar Mishra, Harikesh Bahadur Singh.
Description: First edition | Boca Raton, FL : CRC Press, 2022. |
Includes bibliographical references and index.
Identifiers: LCCN 2021053513 | ISBN 9780367759766 (hardback) |
ISBN 9780367759773 (paperback) | ISBN 9781003164968 (ebook)
Subjects: LCSH: Farms, Small. | Sustainable agriculture. | Agricultural innovations.
Classification: LCC HD1476.A3 I563 2022 | DDC 338.1--dc23/eng/20211209
LC record available at https://lccn.loc.gov/2021053513

ISBN: 9780367759766 (hbk)
ISBN: 9780367759773 (pbk)
ISBN: 9781003164968 (ebk)

DOI: 10.1201/9781003164968

Typeset in Times LT Std
by KnowledgeWorks Global Ltd.

To all smallholders in developing countries who account for food security, poverty reduction, and sustainable development.

Contents

Preface

For achieving the Sustainable Development Goals to wipe out poverty (Goal 1), hunger (Goal 2), and to improve human health and well-being (Goal 3), a 60–110% increase in global agricultural production is needed. FAO's State of the World series, and International Food Policy Research Institute (IFPRI) visionary 2050 policy documents have also identified food security as the global concern of the 21st century. However, huge yield gaps exist in smallholder farms of developing countries with significant regional and interpersonal variations that ask for diverse strategies to enhance food production. A significant part of the universe is intensively cultivated and farming is primarily characterized by smallholder farms. Inherently, this smallholder farming system functions under a broad array of bio-physical, climate, and socioeconomic settings and the improvement of these systems is hindered by inadequate accessibility to land, fertile soil, capital, and labor. Additionally, interactions among these factors intensely affect the resource use efficacy that leads to the achievement of optimal yield. Many studies have shown that both biophysical and socioeconomic factors govern the soil fertility and crop yield variation, which in turn, are linked to diverse local climate, soil types, access to market, sociocultural, and ethnic characteristics. In fact, yield-gap analyses have recently taken adequate care of smallholder heterogeneity to find out local/regional factors of yield variation. Understanding these determinants of yield variability in smallholder systems is important to formulate informed policies to close the yield gap in major food crops. This book aims to focus on the current state of knowledge and scientific advances about the innovations in agriculture that will equip agriculture to cope with the competing challenges of addressing food and nutrition security, improving livelihoods, combatting climate change, and sustainably managing natural resources.

Amitava Rakshit
Somsubhra Chakraborty
Manoj Parihar
Vijay Singh Meena
P. K. Mishra
H. B. Singh

Acknowledgments

Much of the work reported in this book was based on the fact that innovation is not only driven by technological advances but also through novel ways of organizing farmers and connecting them to the information they need. This book is a useful source of information for researchers, trainers, progressive farmers, and other interested persons to advance not only the agriculture but also the livelihood of the farming community. Appreciation is expressed to the experts around the globe who have made major contributions. We gratefully acknowledge the valuable and generous technical support of our partners and production staff members of CRC Press and Taylor & Francis Group. Special appreciation and gratitude are expressed to our family members for their continuous and unparalleled love, help, and support.

Amitava Rakshit
Somsubhra Chakraborty
Manoj Parihar
Vijay Singh Meena
P. K. Mishra
H. B. Singh

Editors

Dr. Amitava Rakshit, an IIT-Kharagpur alumnus, is a faculty member in the Department of Soil Science and Agricultural Chemistry at Institute of Agricultural Sciences, Banaras Hindu University (BHU), Varanasi, UP, India. His research areas include nutrient use efficiency, simulations modeling, organic farming, integrated nutrient management, and bioremediation. His consulting capabilities are composting techniques, soil health management, and input quality control. He was involved in "Participatory Research" and "Lab to Land" Programmes of ICAR; Department of Agricultural Cooperation, Government of India; Department of Agriculture, Government of West Bengal; NHB, New Delhi and NHM for on farm demonstrations of agro-technologies in cereals/pulses/oilseeds/cash crops/vegetables/fruits. He is actively involved in imparting training and dissemination of technical knowledge and information to diversified end users. He has supervised approximately seven research projects, many in partnership with industry. He is widely acknowledged for his skills in linking research with the broader community in regional languages. He has been working closely with undergraduate and postgraduate students in BHU presently. He has visited Norway, Finland, Denmark, France, Austria, Russia, Thailand, Egypt, Turkey, UAE, and Bangladesh number of occasions pertaining to his research work and presentation. Dr. Rakshit has previously worked at Department of Agriculture, Government of West Bengal in research extension and implementation roles. He is the fellow of TWAS Nxt (Italy), Biovision Nxt (Freance), Society of Earth Scientists, and Crop and Weed Science Society. He is presently the chief editor of International Journal of Agriculture Environment and Biotechnology (NAAS: 4.55). He is serving as review college member of British Ecological Society, London since 2011–2012. He was awarded with Best Teacher's Award by BHU (ICAR, New Delhi) both at under graduate and postgraduate layers in 2012 and 2014, respectively. He is a member of Global Forum on Food Security and Nutrition of FAO, Rome and Commission on Ecosystem Management of International Union for Conservation of Nature. He is the author of 22 books (Springer, CRC Press, Elsevier, CBS, ATINER, ICFAI, Kalyani, Jain Publishers, IBDC, Scientific Publishers, and DPS). He has published 100 research papers, 40 book chapters, 28 popular articles, and 3 manuals.

Dr. Somsubhra Chakraborty, a Louisiana State University (United States) alumnus, is presently the faculty member in the Agricultural and Food Engineering Department at the Indian Institute of Technology (IIT) Kharagpur, India. He has worked as a postdoctoral research fellow at West Virginia University, United States. His research areas include portable proximal soil sensor measurement, digital soil mapping, mobile image-based soil analysis, and app development. Moreover, he has proved the applicability of portable XRF to quantify several soil properties and soil heavy metal contents. Dr. Chakraborty has developed a patented methodology for combining multiple proximal soil sensors to improve the predictive accuracy relative to the individual technique in isolation on the analyte of interest (US patent US-2017-0122889-A1) titled "Portable Apparatus for Soil Chemical Characterization." He has on-going collaborations with scientists from several countries including the United States, Canada, Romania, Brazil, China, and Morocco. He has received the Australia Awards Fellowship (2016) and Shastri Scholar Travel Subsidy Grant (2019–2020). With more than 50 peer-reviewed publications in reputed international journals, Dr. Chakraborty has an h-index of 23 and i10 index of 41. He is an editorial advisory board member of Geoderma (Elsevier), the global journal of soil science. He is currently supervising five PhD students and co-supervising two PhD students.

Dr. Manoj Parihar is currently working at ICAR-Vivekananda Parvatiya Krishi Anusandhan Sansthan, Almora, India as a soil scientist in Crop Production Division. He did his graduation from SKRAU, Bikaner and selected as ICAR-JRF fellow for postgraduation in Banaras Hindu University (BHU), Varanasi. He has been awarded doctorate from the same university in the year 2018. He has received various recognitions such as ICAR-SRF, UGC-BSR, and UGC-RGNF. His research works extend to arbuscular mycorrhizal fungal diversity, ecological functions, and inoculum production aspects for improving soil health and sustainable agricultural production. He has edited 2 books, and published >10 book chapters and >25 peer-reviewed journal papers.

Dr. Vijay Singh Meena is working as a project coordinator at International Maize and Wheat Improvement Center (CIMMYT), Borlaug Institute for South Asia (BISA). He worked as scientist at ICAR-Vivekananda Parvatiya Krishi Anusandhan Sansthan, Almora, India. His research areas include various aspects of soil aggregation, carbon management index, and carbon and nitrogen sequestration potential under different land types and cropping systems of northwestern Indian Himalayas. He identified carbon management index as the key indicator to measure soil degradation in different agroecosystems. His research revealed that the application of FYM and vermicompost along with vegetative barrier across the slope are highly effective in sustaining the soil quality. He reported that potassium solubilizing rhizobacteria (KSR) enhances 25–40% potassium (K) availability and help plants to uptake K from the soils. Dr. Meena identified that the combined application of organic and inorganic sources is important in sustaining the productivity of Himalayan soils and prevent soil erosion. Combined use of FYM and inorganic fertilizers on equal N basis (50 + 50 FYM) resulted in higher productivity of maize and wheat crops than an individual source. However, in-situ green manuring and inorganic fertilizers on equal N basis (50 + 50 GM) resulted in reduction of runoff and soil loss, maintained system productivity, leading to the conservation of natural resources in soils of maize-wheat cropping system. He is instrumental in the preparation and distribution of >4000 soil health cards to different hill farmers at the current institute. He recently reported the carbon and nitrogen sequestration potential of different land use and cropping systems in Indian Himalayas. He also edited seven Springer books on microbes and agricultural sustainability. He has received several scholarship and awards during his academic and professional career. In a nutshell, Dr. Meena is working in the field of natural resources management for sustainable agricultural production. He has an h-index 43, i10-index 77 with more than 5200 citations in international literature.

Dr. Pradeep Kumar Mishra is presently working as vice chancellor of Jharkhand Technical University. Previously he was professor (HAG) Chemical Engineering, IIT (BHU), and former head of the Department of Chemical Engineering at IIT (BHU) Varanasi. Professor Mishra has published more than 100 papers in reputed journals and filed 5 patents. He has authored/edited more than dozen books and contributed scores of chapters in books/encyclopedias. In addition to his technical contribution, Professor Mishra has been pioneer in the field of innovation, incubation, and entrepreneurship in eastern UP and established the first incubator (Malviya Centre for Innovation Incubation and Entrepreneurship) at BHU in 2008, which is the part of IIT (BHU) Varanasi now. He is also coordinating the activities of CISCO ThinQubator in association with CISCO and NASSCOM in the area of artificial intelligence, Internet of things, and machine learning. Professor Mishra has been decorated with various awards/honors, prominent of them are Dr. Shirin Gadhia Memorial Sustainability Award, 2017, 2nd National Prize for book writing in Hindi by Ministry of Renewable Energy, GOI and DST Lockheed Nomination for Technology Commercialization for 2010 and 2012. He is member board of governors of NISBUD, Ministry of Skill Development and Children Emancipation Society, the United Kingdom. He has also represented a delegation to Germany on the invitation of German Biogas Association on behalf of Indian Biogas Association. Professor Mishra has also been selected for prestigious Leadership of Academicians Program and successfully completed it at IIT (BHU) and Pen State University, the United States in March–April 2019. He has also been appointed as co-coordinator for 2nd LEAP being organized by the institute in collaboration with Cambridge University, United Kingdom. Professor Mishra has delivered over 100 invited talks in the area of Environment, Energy, Entrepreneurship and Motivation in addition to various field of Chemical Engineering at various forums including Institutions of eminence.

Dr. H B Singh is Professor of Excellence, Department of Biotechnology, GLA University, Mathura, India. Prior to this assignment, he served as a professor at Banaras Hindu University (2006–2019) and head from October 2014–September 2017. He spent 37 years of committed service in the field of plant pathology at different institutions in different capacities. He also guided 24 students for PhD, 6 postdoctoral, 2 women scientists, and 29 MSc(Ag.) students. He also established fruitful research collaborations with academic and industry researchers and published jointly with national and international collaborators in high impact journals and obtained 19 patents (the United States, Canada, PCT) and 5 technologies were commercialized. He published >270 research papers, 35 review articles, 83 book chapters, 27 edited books. He is a member of DBT Task Force on "Biofertilizers and Biopesticides"; DST WOS-A Task Force; DST WOS-B Task Force; DBT Task Force for Biotechnology Based Programme for SC/ST Population and Rural Development; DST NSTMIS Task Force; faculty expert member of Department Council in several Universities; served as a member of selection committees at different Universities to select faculty in Plant Pathology and Biotechnology, and Agricultural Scientists Recruitment Board (ASRB) ICAR, New Delhi; member of Research Advisory Committee (RAC) of ICAR-NBAIM, Mau and ICAR-NCIPM, New Delhi, QRT member ICAR-DMR, Solan, ICAR-NBAIR, Bangalore. Chairman: Agricultural Sciences, UP Council of Science & Technology, Lucknow (March 2020–March 2023). Professor Singh is fellow of National Academy of Agricultural Sciences (FNAAS) and has been honored with several prestigious awards. Google Citation: 7983 h-index: 46, i10-index: 163.

Contributors

Umarajashekar A
Department of Agricultural Microbiology, Agricultural College
Professor Jayashankar Telangana State Agricultural University
Jagtial, Telangana, India

O. A. Acevedo-Sandoval
Popular Autonomous University of the State of Puebla
Puebla, Mexico

Pravat Utpal Acharjee
Department of Agricultural Chemistry and Soil Science
Faculty of Agriculture
Bidhan Chandra Krishi Viswavidyalaya
Kalyani, West Bengal, India

Sankar Kumar Acharya
Department of Agricultural Extension
Faculty of Agriculture, Bidhan Chandra Krishi Viswavidyalaya
Kalyani, West Bengal, India

A. Aguiar
GreenUPorto & DGAOT
Faculty of Sciences, University of Porto
Porto, Portugal

P. Anandajayasekeram
Independent International Service Provider
Agricultural and Rural Innovation Support
Australia

S. Auci
Università degli Studi di Palermo
Palermo, Italy

Satya Prakash Barik
C V Raman Global University
Bhubaneswar, Odisha, India

Alejandro Barragán-Ocaña
Instituto Politécnico Nacional (National Polytechnic Institute)
Mexico City, Mexico

Kaushik Batabyal
Department of Agricultural Chemistry and Soil Science
Bidhan Chandra Krishi Viswavidyalaya
Nadia, West Bengal, India

Kallol Bhattacharyya
Department of Agricultural Chemistry and Soil Science
Faculty of Agriculture, Bidhan Chandra Krishi Viswavidyalaya
Nadia, West Bengal, India

Malay K. Bhowmick
Department of Agriculture (Government of West Bengal)
Kolkata, West Bengal, India
and
International Rice Research Institute (IRRI) – South Asia Regional Centre (ISARC)
Varanasi, India
and
Directorate of Agriculture (Government of West Bengal)
Kolkata, West Bengal, India

Sunanda Biswas
ICAR – Indian Agricultural Research Institute
Pusa Campus
New Delhi, India

Benjamin Ezekiel Bwadi
Department of Geography
Taraba State University
Jalingo, Nigeria

Somsubhra Chakraborty
Agricultural and Food Engineering Department
IIT Kharagpur
Kharagpur, West Bengal, India

Riti Chatterjee
Department of Agricultural Extension
Faculty of Agriculture
Bidhan Chandra Krishi Viswavidyalaya
Nadia, West Bengal, India

Snigdha Chatterjee
Department of Agricultural Chemistry and Soil Science
Bidhan Chandra Krishi Viswavidyalaya
Mohanpur, West Bengal

M. Clavel-Maqueda
Autonomous University of the State of Puebla
Puebla, Mexico

E. Cornejo-Velazquez
Autonomous University of Hidalgo State
Pachuca, Mexico

M. Coromaldi
Università degli Studi Niccolò Cusano
Rome, Italy

C. A. Costa
CERNAS – IPV Research Centre
Polytechnic of Viseu
Viseu, Portugal

Anil D
Department of Agronomy
Agricultural Research Station, Professor Jayashankar Telangana State Agricultural University
Peddapally, Telangana, India

Alexandre Marco da Silva
Department of Environmental Engineering
Institute of Science and Technology of Sorocaba – São Paulo State University (UNESP)
Sorocaba, Brazil

Ruma Das
ICAR – Indian Agricultural Research Institute
Pusa Campus
New Delhi, India

Suddhasuchi Das
Malda Krishi Vigyan Kendra
Uttar Banga Krishi Viswavidyalaya
B.S. Farm. Ratua
Malda, West Bengal, India

Vipin Das
M. S. Swaminathan Research Foundation
Community Agrobiodiversity Centre
Wayanad, India

Shubhadip Dasgupta
Department of Agricultural Chemistry and Soil Science
Faculty of Agriculture, Bidhan Chandra Krishi Viswavidyalaya
Mohanpur, West Bengal, India
and
Agricultural and Food Engineering Department
IIT Kharagpur
Kharagpur, West Bengal, India

Debashis Dash
C V Raman Global University
Bhubaneswar, Odisha, India

Pradip Dey
ICAR – Indian Institute of Soil Science
Bhopal, Madhya Pradesh, India

S. Dhillon
Punjab Agricultural University
Ludhiana, Punjab, India

S. K. Dwivedi
Indira Gandhi Krishi Vishwavidyalaya (IGKV)
Raipur, India

E. Figueiredo
Linking Landscape
Environment, Agriculture and Food (LEAF), Instituto Superior de Agronomia
Universidade de Lisboa
Lisbon, Portugal

Ranjith Kumar G
Department of Soil Science and Agricultural Chemistry
Dr. Panjabrao Deshmukh Krishi Vidyapeeth
Akola, Maharashtra, India

Deblina Ghosh
Department of Agricultural Chemistry and Soil Science
Bidhan Chandra Krishi Viswavidyalaya
Mohanpur, West Bengal, India

M. C. Godinho
Departamento of Agrarian Sciences and Environment
Agrarian School, Polytechnic Institute of Santarém
Santarem, Portugal

Himansu Sekhar Gouda
C V Raman Global University
Bhubaneswar, Odisha, India

Adalberto de Hoyos-Bermea
Instituto Politécnico Nacional (National Polytechnic Institute)
Mexico City, Mexico

Somasundaram Jayaraman
ICAR – Indian Institute of Soil Science
Bhopal, Madhya Pradesh, India

Renu Jethi
Division of Agricultural Extension
ICAR – Indian Agricultural Research Institute
New Delhi, India
and
ICAR – Vivekananda Parvatiya Krishi Anusandhan Sansthan
Almora, Uttarakhand, India

Kamalakar Jogula
Department of Soil Science and Agricultural Chemistry
Agricultural College, Professor Jayashankar Telangana State Agricultural University
Warangal, Telangana, India

Kushagra Joshi
ICAR – Vivekananda Parvatiya Krishi Anusandhan Sansthan
Almora, Uttarakhand, India

Pratibha Joshi
CATAT, ICAR – Indian Agricultural Research Institute
New Delhi, India

Ravinder Juttu
Department of Soil Science and Agricultural Chemistry
Regional Sugarcane and Rice Research Station
Rudrur, Nizamabad, Telangana, India
and
Professor Jayashankar Telangana State Agricultural University
Hyderabad, Telangana, India

Rubina Khanam
ICAR – National Rice Research Institute
Cuttack, Odisha, India

Suryakanta Khandai
International Rice Research Institute (IRRI)
Guwahati, Assam, India

Dominik Klauser
Syngenta Foundation for Sustainable Agriculture
Basel, Switzerland

Jayasree Krishnankutty
Communication Centre
Kerala Agricultural University
Thrissur, Kerala, India

Pedda Ghouse Peera Sheikh Kulsum
C V Raman Global University
Bhubaneswar, Odisha, India

Ashok Kumar
International Rice Research Institute (IRRI)
Bhubaneshwar, Odisha, India

N. Anil Kumar
M. S. Swaminathan Research Foundation
Community Agrobiodiversity Centre
Wayanad, India

V. Kumar
International Rice Research Institute (IRRI)
Guwahati, Assam, India

Virender Kumar
International Rice Research Institute (IRRI)
Los Baños, Philippines

Gunja Kumari
Regional Research Station (Terai Zone)
Uttar Banga Krishi Viswavidyalaya
Coochbehar, West Bengal, India

Merlin Lopus
M. S. Swaminathan Research Foundation
Community Agrobiodiversity Centre
Wayanad, India

Katya Amparo Luna-López
Instituto Politécnico Nacional (National Polytechnic Institute)
Mexico City, Mexico

Girijesh Singh Mahra
Division of Agricultural Extension
ICAR – Indian Agricultural Research Institute
New Delhi, India

Vijay Singh Meena
ICAR – Vivekananda Parvatiya Krishi Anusandhan Sansthan
Almora, Uttarakhand, India
and
Borlaug Institute for South Asia (BISA)
CIMMYT
Samastipur, Bihar, India

Sweta Shikta Mohapatra
C V Raman Global University
Bhubaneswar, Odisha, India

Prithusayak Mondal
Regional Research Station (Terai Zone)
Uttar Banga Krishi Viswavidyalaya
Coochbehar, West Bengal, India

Dhanya MS
Centre for Environmental Science and Technology
Central University of Punjab
Bathinda, Punjab, India

Sidhu Murmu
Department of Agricultural Chemistry and Soil Science
Bidhan Chandra Krishi Viswavidyalaya
Mohanpur, West Bengal, India

Firuza Begham Mustafa
Department of Geography
University of Malaya
Kuala Lumpur, Malaysia

Anandkumar Naorem
ICAR – Central Arid Zone Research Institute
Regional Research Station
Bhuj, Gujarat, India

J. Neto
CERNAS – IPV Research Centre, Polytechnic of Viseu
Viseu, Portugal

C. Parente
Department of Sociology and Institute of Sociology
Faculty of Arts and Humanities, University of Porto
Porto, Portugal

Sampad R. Patra
Department of Agriculture (Government of West Bengal)
Kolkata, West Bengal, India

Santanu Kumar Patra
C V Raman Global University
Bhubaneswar, Odisha, India

Arunava Pattanayak
ICAR – Vivekananda Parvatiya Krishi Anusandhan Sansthan
Almora, Uttarakhand, India
and
ICAR – Indian Institute of Agricultural Biotechnology
Ranchi, Jharkhand, India

R. Pires
UNIDEMI – New University of Lisbon and Polytechnic Institute of Setúbal
Setúbal, Portugal

Jogarao Poiba
Regional Agricultural Research Station
Visakhapatnam, Andhra Pradesh, India

Shiv Prasad
Division of Environment Science
ICAR – Indian Agricultural Research Institute
New Delhi, India

Rajesh K. Raju
Communication Centre
Kerala Agricultural University
Thrissur, Kerala, India

Amitava Rakshit
Department of Soil Science and Agricultural Chemistry
Institute of Agricultural Sciences, Banaras Hindu University
Varanasi, India

Dheeraj Rathore
School of Environment and Sustainable Development
Central University of Gujarat
Gandhinagar, Gujarat, India

S. Rawal
Chaudhary Charan Singh Haryana Agricultural University
Hisar, Haryana, India

Sharan Bhoopal Reddy
Department of Soil Science and Agricultural Chemistry
University of Agriculture Sciences
Raichur, Karnataka, India

H. Romero-Trejo
Popular Autonomous University of the State of Puebla
Puebla, Mexico

Á. Rosa
ISCTE – University Institute of Lisbon and BRU-UNIDE/ISCTE-IUL
Lisboa, Portugal

Deepa Roy
Department of Agricultural Extension
Faculty of Agriculture, Uttar Banga Krishi Viswavidyalaya
Coochbehar, West Bengal, India

Rakesh S
Department of Soil Science and Agricultural Chemistry
Uttar Banga Krishi Viswavidyalaya
Coochbehar, West Bengal, India

Pardeep Sagwal
International Rice Research Institute (IRRI) – South Asia Regional Centre (ISARC)
Varanasi, Uttar Pradesh, India

V. Sangeetha
Division of Agricultural Extension
ICAR – Indian Agricultural Research Institute
New Delhi, India

M. Saraiva
Management Department of the School of Social Sciences
University of Évora and BRU-UNIDE/ISCTE-IUL
Évora, Portugal

Amit Baran Sarangi
Department of PSMA
Faculty of Horticulture, Bidhan Chandra Krishi Viswavidyalaya
Nadia, West Bengal, India

Sujit Sarkar
CAR – Indian Agricultural Research Institute New Delhi
Regional Station
Kalimpong, West Bengal, India

Sudip Sengupta
Department of Agricultural Chemistry and Soil Science
Faculty of Agriculture, Bidhan Chandra Krishi Viswavidyalaya
Nadia, West Bengal, India

Kadambot H. M. Siddique
The UWA Institute of Agriculture
University of Western Australia
Crawley, Western Australia, Australia

E. Simões
School of Agricultural and Environmental Sciences at the University of Cape Verde
Praia, Cabo Verde

Anoop Singh
Department of Scientific and Industrial Research (DSIR)
Ministry of Science and Technology, Government of India
Technology Bhawan
New Delhi, India

Kanwar Singh
International Rice Research Institute (IRRI)
Guwahati, Assam, India

O. P. Singh
Department of Agricultural Economics
Institute of Agricultural Sciences, Banaras Hindu University
Varanasi, Uttar Pradesh, India

P. K. Singh
Department of Agricultural Economics
Institute of Agricultural Sciences, Banaras Hindu University
Varanasi, Uttar Pradesh, India

Sudhanshu Singh
International Rice Research Institute (IRRI) – South Asia Regional Centre (ISARC)
Varanasi, Uttar Pradesh, India

Arun Sreekumar
Institute of Management
Ahmedabad, Gujarat, India

Amit K. Srivastava
International Rice Research Institute (IRRI) – South Asia Regional Centre (ISARC)
Varanasi, Uttar Pradesh, India

K. Srivastava
International Rice Research Institute (IRRI) – South Asia Regional Centre (ISARC)
Varanasi, Uttar Pradesh, India

Raveendran Telapurath
M. S. Swaminathan Research Foundation
Community Agrobiodiversity Centre
Wayanad, Kerela, India

N. Tripathi
Division of Crop Protection
ICAR – Indian Vegetable Research Institute
Varanasi, Uttar Pradesh, India

Shiva Kumar Udayana
MSSSOA
Centurion University of Technology and Management
Paralakhemundi, Odisha, India

E. Valério
Departamento of Agrarian Sciences and Environment
Agrarian School, Polytechnic Institute of Santarém
Santarem, Portugal

Tamara Arizbe Virgilio-Virgilio
Instituto Politécnico Nacional (National Polytechnic Institute)
Mexico City, Mexico

Section I

Issues, Ideas, and Challenges in Agricultural Sector

1

Factors that Ensure a Sustained and Scaled-Up Delivery of Innovation

Dominik Klauser
Syngenta Foundation for Sustainable Agriculture, Basel, Switzerland

CONTENTS

1.1 Introduction

> *"If you have a hammer, all you see is nails". This old saying applies well to countless initiatives in agricultural research for development where numerous donors, NGOs and implementers have a strong tendency to focus on the tool they want to promote (the hammer) rather than the problem they want to solve. This often translates into a poor or biased understanding of what is really needed to overcome the most limiting problems of smallholder farming systems (the task) and hence the tools needed to tackle them. So once you understand the problem, you might well find out that it's not the nails you have to fix, but the screws. So ideally, you want a different tool to do so.*

Investment in agricultural research is widely regarded as the most effective way to sustainably reduce poverty and food insecurity in developing countries (de Janvry and Sadoulet 2010, Christiaensen et al. 2011). However, particularly smallholder farming systems have often not benefitted much from innovation and technological progress (Pretty et al. 2011). Recent reviews indicate that many potentially beneficial innovations are not adopted at scale, limiting their impact on the productivity and sustainability of smallholder farming systems (Stevenson and Vlek 2018, Stevenson et al. 2019). However, particularly in the context of climate change and the rapid degeneration of our environmental resource base, there is an urgent need for more innovation "to reach the farm" in order to meet the food demand of future generations in a sustainable way. It is widely agreed that only continuous innovation and its adoption at scale will help farmers to address these challenges.

Several factors have been identified that limit the scaled-up adoption of innovation. Among others, they include short-sighted project management in the agricultural research for development space, with a focus on "quick wins" and "cheap numbers" rather than having an ambition of sustained, systemic impact (Leeuwis et al. 2018, Hall and Dijkman 2019). In some regard, this development happens in response to the growing impatience of donors to see fast and visible impact and desire more thematic prescription for their investments. However, this can lead to a diluted focus on and understanding of the actual problem, hence limiting the chances of success of sustained adoption. This disconnect between

DOI: 10.1201/9781003164968-2

donor priorities, short-sighted funding cycles, and the reality on the ground has been widely discussed in recent publications (Glover et al. 2016, Cooley and Howard 2019, Hall and Dijkman 2019, Lobell 2020).

This chapter will focus on how to better connect innovation to its users and suppliers through a stronger focus on customers and the context specific realities that influence their adoption. I propose that these factors need to be included more prominently into the design of many innovation processes. They can be clustered into four major categories, which are discussed in separate subchapters below.

1.2 Focus on the Problem, Not the Solution

Agricultural research for development is fueled with ideology. We all have our bias to what we believe is a sustainable and remunerative farming system. To me, this is not necessarily a bad thing. A strong vision on what we consider "good and sustainable" can help us stay motivated and committed to our cause, which, at the end, is a better world. However, it can also lead to a bias toward the solution (the hammer or tool) rather than a desire to fully understand the problem. A full understanding of the major limiting challenges of smallholder farming systems is imminently important to select the appropriate solutions to tackle problem and often involves the assessment of several solution scenarios (the tools [Sartas et al. 2020]). It can also dilute our customer focus as we (the researchers, donors, implementers) claim to already know what will be best for "them" (smallholder communities), oftentimes creating misleading technology-centric narratives of transformation (Hall and Dijkman 2019). Counteracting this tendency implies a thorough understanding of the farming systems we work in. We have to better understand the challenges they face, the opportunities they offer, and the context-specific factors that will drive the adoption of innovation. It is remarkable how many initiatives in Agricultural Research for Development continue to cling onto the solution despite of an obvious disconnect to the problem. Another factor that needs consideration is to differentiate between the symptoms and the root causes of the problems we want to solve. Arguably, only the latter approach can deliver the sustained, systemic impact we are striving for. More chemical inputs may reduce pest pressure in monocropping systems in the short-term but are unlikely to solve the problem of growing, specialized pest populations. More efficient pumps for groundwater extraction might provide water at a lower cost but will not help refill depleted groundwater reservoirs. Providing free seed might secure a season's production but will not help establish functional, reliable, and self-sustainable seed supply systems in the future. Instead of "just" fixing the symptoms, there is hence a need for longer-term programmatic interventions that address the root causes of problems in a sustainable way (Schut et al. 2020). This has to be built on a thorough understanding and better appreciation of the systems we work in.

1.3 Focus on Your Customer

This sounds obvious. However, in reality, it is often quite difficult to get it right. Why? The first difficulty is knowing your customer. In many cases, customers are not only the users of the solutions that are developed (i.e., smallholder farmers) but also includes their suppliers, producers, regulators, or other stakeholders that influence their adoption. The latter for instance includes consumers, downstream value chain partners (processors and traders), or governments that in one way or another influence the value proposition of the solution (such as through certification, subsidies or other (dis-)incentives) (Klauser and Negra 2020). In the last two decades, the Syngenta Foundation for Sustainable Agriculture has built a substantial experience from partnerships with the private and public sector for crop breeding. Common observations from these partnerships included a narrow focus on specific, scientifically interesting traits (such as biotic and abiotic stress tolerance or yield potential), while other traits and attributes that have a similar importance in determining the adoption of improved varieties were widely disregarded (Persley and Anthony 2017). These for instance include the nutritional composition of crops (processor preferences), shelf-life (consumers and traders), a crop's multiplication rate (which determines the business case for seed producers), or simply the fact that investment went into the development of genetically

Creating a Product Profile

Trait category	No	Trait	Trait description	Variety benchmark	Performance required (=, >, >=, >>)
Crop yield	1				
	2				
Plant architecture	1				
Seed production	2				
Biotic stress	1				
	2				
Abiotic stress	1				
	2				
Crop handling, harvest, storage, transport	1				
	2				
Value chain clients, consumers, processors	1				
	2				

FIGURE 1.1 Template for a target product profile (TPP). Performance aspects (traits) are identified, prioritized, and benchmarked against an existing variety in a given farming context. (**Courtesy of Persley and Anthony (2017).**)

modified solutions in the absence of any conducive regulatory systems. This has led to a major adoption gap for new varieties of many smallholder-relevant crops. As a consequence, many smallholders are still mostly using old varieties, not benefitting from the donor investment in genetic gains from new, improved cultivars.

We addressed this challenge by introducing the concept of target product profiles to many breeding programs. Target product profiles highlight the minimum requirement for candidate varieties to be met for relevant traits (Figure 1.1) to ensure the endorsement of the various stakeholders that will determine a scaled-up adoption. The need for stronger emphasis on such considerations has also led us to launch an initiative on "demand-led breeding"[1] to foster more customer-centric approaches for variety development. This initiative is described in more detail in Case Study 1.

CASE STUDY 1: DEMAND-LED BREEDING

Many breeding programs for smallholder relevant crops have struggled to generate a significant impact at farm-level, despite of substantial donor investment. Among other reasons discussed in this chapter, one aspect is an often too narrow focus on few breeding targets (traits), such as abiotic and biotic stress tolerance. However, in most cases, a variety of different traits is relevant for the various stakeholders in a crop's value chain – from seed producers to consumers. Together with the Australian Centre for International Agricultural Research (ACIAR), the Syngenta Foundation for Sustainable Agriculture has led the development of an initiative to embed more demand-led approaches into breeding processes. This with the aim to promote the inclusion of preferences and prerequisites of the entire crop value chain to create sufficient "pull" to drive the adoption

[1] https://www.demandledbreeding.org/

of innovation (Persley and Anthony 2017). This initiative is built on ten key points to consider in breeding programs as mentioned below:

1. Understanding clients, which is central to demand-led variety design and increasing adoption of new varieties: clarity is required on: Who are the clients? What factors influence their buying decisions? What are the needs, preferences, and problems of each client?
2. Farmer adoption: demand-led approaches should increase the likelihood of new varieties being adopted by farmers.
3. Value chains: demand-led approaches build on and go beyond farmer participatory breeding. They include consultations not only with farmers but with all clients and stakeholders along the whole crop value chain.
4. Urban and rural consumers: breeders must consider needs and preferences of consumers living in both rural and urban environments. Rapid (rural and urban) appraisals can be extended to gathering information not only from farmers but also from consumers and clients who live in towns and cities.
5. Markets and client segmentation: breeders need to understand markets and client segmentation to be able to prioritize their breeding targets.
6. Market research and intelligence gathering: market research at the start of a breeding program needs to be complemented with continuing consultations with stakeholders at key decision points along with the development stage plan from new variety design to post-market release.
7. Breeding entrepreneurship: this can contribute to economic growth, better livelihoods for smallholder farmers and increased food security. Improved varieties can change lives.
8. Market creation: to maximize market creation and nurture innovation, a balance is required between using demand-led approaches and enabling new technologies to drive innovations. Both approaches have value and complement one another.
9. Role of the plant breeder: plant breeders do much more than making crosses and leading selection programs. A breeder must also be an integrator of inputs and be able to assimilate information and incorporate a broad range of views, including those of nontechnical experts. This requires assimilating data, looking at its implications, and making decisions based on information from diverse areas such as agricultural economics, markets, and market research as well as the core scientific functions for breeding.
10. Breeding experience: demand-led approaches retain emphasis and put a value on the breeders' eyes and experience in assessing germplasm.

The same observations have also led us to develop a basic framework on aspects to consider when introducing new solutions to smallholder farming systems (Figure 1.2). This framework is adapted from previous work on introducing new crop varieties[2] and can be clustered into four main categories that determine adoption, namely: (i) compatibility with existing farming systems in terms of their capacity to adopt innovation (such as knowledge, financial capacity, existing infrastructure); (ii) the potential benefit a new solution offers against the status quo, particularly in the context of environmental and resilience aspects; (iii) the market potential of new solutions that will determine the business case for their suppliers; (iv) the potential to create value for farmers, i.e., the overall business case but also less tangible value including societal and cultural aspects.

Addressing these aspects more prominently in the design phase of innovation processes will greatly help getting more innovation to farmers in a sustainable and scalable way. This, in turn, will help creating substantial economic, environmental, and societal impact.

---

[2] https://www.syngentafoundation.org/sites/g/files/zhg576/f/2020/04/08/rotation_crops_for_cereal_farmers_0.pdf

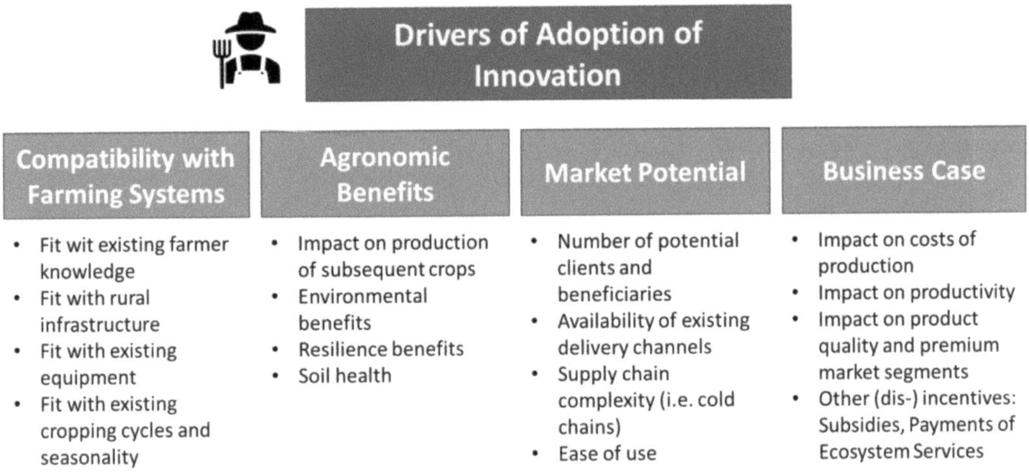

FIGURE 1.2 Factors to consider when introducing new solutions to smallholder farmers.

1.4 Partner to Ensure Sustained and Scalable Delivery Mechanisms

Many projects in Agricultural Research for Development have the ambition to both develop and deliver new solutions to as many beneficiaries as possible and within a project's lifetime. Whereas this can be perceived as a noble ambition (but in fact is mostly a response to donors' desire for more immediate impact), in most cases, counting adoption at the end of a project is a poor metric for success (Woltering et al. 2019). Most projects that are designed this way do not achieve to sustain impact after the project phase itself (Cooley and Howard 2019). Why? For a variety of reasons. Firstly, the use of project funds to deliver solutions can be unsustainable. Whereas it can help catalyze early adoption to a critical threshold and de-risk the business case for early adopters and suppliers, it, in most cases, just inflates early adoption numbers at the sake of long-term business sustainability built on a solid value proposition and functional market systems which will ensure the sustained and scaled-up adoption of most products post a project's lifetime. Secondly, the development and the delivery of innovation require different mindsets. Whereas, as argued in the previous subchapter, certain aspects of market information will benefit from being embedded into innovation processes, taking a product to market require different skills. Among others, they include business development, entrepreneurial skills, marketing, IP management, and licensing. By nature, these skills are more embedded in private sector organizations than academic institutions, public sector research bodies, and not-for-profit entities.

Scale is not sustainable in the absence of local delivery mechanisms with self-generating financing (Woltering et al. 2019). Hence, rather than delivering innovation as part of a research or innovation project, it is often best to focus on creating an optimized product against defined target product profiles codeveloped with delivery partners. This can of course be done as an iterative process that involves continuous customer feedback. However, it best leaves out any ambition of product delivery during a project's lifetime but includes a clear plan and appropriate partnerships in place that ensure a sustained delivery of a finished product. In the context of developing new crop varieties for smallholder farming systems, we have noticed a certain disconnect between variety development and their delivery, resulting in many good varieties remaining in research stations rather than reaching farmers' fields. We have hence created a platform to link the creators of new varieties (breeders) with multipliers and distributors (oftentimes local seed companies), facilitating aspects of business development, technology transfer, IP management, and supply of early generation seed. This program is branded Seeds2B[3].

[3] www.seeds2b.org

CASE STUDY 2: SEEDS2B

Billions of dollars of donor investment have been put into crop improvement programs for smallholder relevant crops in the last two decades. Whereas the varietal output from these investments offer opportunities for improvement themselves (see Case Study 1), even good varieties with a holistic target product profile often struggle to create the impact they could deliver (Walker and Alwang 2015). Our analysis led to the conclusion that the lack of sustainable seed delivery mechanisms poses a major bottleneck, with many breeders not having sufficient capacity, resources, and expertise to produce and deliver planting material of improved varieties at scale and in a sustained way. As a response to this, we created the Seeds2B program. It connects (mostly public sector) breeders with (mostly private) local seed producers. Doing so, it provides a variety of services to both parties, including licensing and profit-sharing mechanisms, local adaptation trials to de-risk market building, the establishment of value chain partnerships to ensure future seed demand, and the capacity building through funding vehicles and technical assistance. Since its inception in 2014, Seeds2B has brokered major partnerships to deliver improved varieties of clonal and open-pollinated crops across the developing world, with a particular focus on East and West Africa.

1.5 Have a Plan and an Exit Strategy!

The innovation of new products and solutions is a sequenced process. You need to have a clear vision of what success will look like (a viable, attractive solution with a clear value proposition) and what your progress is against achieving this. You will also have to make sure that you observe the right indicators that highlight in case success will become unlikely or even impossible. In order to ensure the impact of new solutions, it is important that a vision of success will not only include the solution itself but also a vision on how it will be delivered, stewarded, supported, and regulated. In private sector innovation, most of these aspects are covered by a process referred to as a "product life cycle". A product life cycle sequences and maps the necessary steps to get from an idea to a product and to then get the product to the market. A stronger endorsement of this approach across public and private innovation will likely benefit agricultural research for development. Highlights how such a process can work in practices by showing evidence from a recent partnership between CIMMYT, Syngenta, and the Syngenta Foundation for Sustainable Agriculture to develop and deliver improved, drought-tolerant maize varieties to resource-poor smallholder farmers in drought-prone areas across South Asia. Doing so, it addressed these aspects from product design, development, and delivery, involving different partners throughout the several stages of a product life cycle.

Around ten Indian seed partners are now promoting the hybrid in the three targeted states. Out of a typical 2.2 ha total area per smallholding there, farmers devote about 0.75 ha to maize. Average AAA maize yields are 2.5 t/ha, compared to OPVs' 1.5 t/ha. This difference gives smallholders additional net income of $100 on a full hectare. The extra 75 dollars from their planted area represent a very valuable boost to total household income: smallholder families in the AAA maize area generate a net income of about $1000 per year, including government subsidies.

The current seed partners are either small companies or NGOs. To enable smallholders' informed adoption and repeat purchases of AAA maize, these partners need to conduct appropriate marketing and provide support. Syngenta Foundation training for them began before commercialization and continues in the classroom and field. Seed partners learn about technicalities of production and receive education on business opportunities and scoping, product positioning, pricing, promotional activities, sales and production planning, legal frameworks, and registration licensing agreements. This extensive training program helped the local seed partners to sell 18 tons of AAA maize in 2018, 50 tons in 2019, and, despite the very challenging circumstances, 120 tons in 2020. The AAA initiative is a rare example of a public-private partnership delivering products to smallholders in central India.

CASE STUDY 3: AAA MAIZE

In central India, hybrid maize cultivation has increased significantly over the last ten years, thanks to private sector investment into the development and delivery of improved hybrid varieties. Farmers nonetheless still plant open pollinated varieties (OPVs) on over 1.3 million hectares. That is about half of the total maize area. Very little maize is irrigated. SFSA and partners hence saw potential for developing affordable, drought-tolerant hybrid maize varieties that match local demand and can be grown on about half a million hectares. This translates into a new market for some 10,000 tons of seed. Drought and heat stress are two major factors limiting maize productivity. In Central India, rain is becoming increasingly erratic and average temperatures are rising. Seed plays a crucial role in smallholders' resilience, i.e., their ability to farm successfully under increasingly difficult climatic conditions. Enabling them to do requires two important steps: first breeding improved varieties and then producing and selling them to smallholders. CIMMYT and Syngenta are both strongly committed to maize improvement through R&D. The Syngenta Foundation funded the program that led to AAA maize. From the beginning of the partnership, it was agreed that whereas Syngenta and CIMMYT were leading the development of the varieties, seed production of these varieties would fall under the mandate of local seed companies, building on existing delivery platforms and linkages to smallholder farming communities. Contractual arrangements had hence to be developed to both remunerate the innovation partners but to also ensure a continuous supply of parental material to the local seed companies so that they can sustain and grow their production based on market demand.

1.6 Concluding Remarks and Future Perspectives

There is no question, smallholder farming systems will always be difficult ecosystems for innovation. Their fragmented and diverse nature as well as the limited resource base of many farmers and the poor rural infrastructure that serves them will limit smallholder farmers' appetite and capacity to invest in new solutions. However, as smallholder farmers are entrepreneurs, they will be open for innovation that creates value for them, be it inputs, practices, extension, or value chain linkages. Learning from past success stories and failures through a more honest and open approach, a much better understanding of limiting factors and pathways to tackle them has been developed in recent years (Wigboldus et al. 2016; Leeuwis et al. 2018; Cooley and Howard 2019; Woltering et al. 2019; Sartas et al. 2020; Schut et al. 2020). This includes a better understanding for the need to create sufficient "pull" for the adoption of innovation, such as market and public-sector signals and incentives (Klauser and Negra 2020). This chapter deals with another dimension to tackle the challenge, namely, by endorsing more realistic, context-specific, and customer-centric approaches to design, develop, and deliver innovation. It includes a better understanding of the challenge and its root causes, the context into which the solution will be delivered, and the need for sustainable, value-creating local platforms to get innovation "to the farm". Whereas a focus on these aspects will not remove all the structural challenges that limit the adoption of innovation in a smallholder context, we believe that it will greatly help to achieve a higher success rate in terms achieving impact from innovation. This, in turn, will lead to an increased return on investment into agricultural research for development, hence creating a strong(er) narrative for future investment in the sector.

REFERENCES

Christiaensen, L., L. Demery and J. Kuhl (2011). "The (evolving) role of agriculture in poverty reduction—An empirical perspective." Journal of Development Economics 96(2): 239–254.

Cooley, L. and J. Howard (2019). Scale Up Scourcebook, Purdue University.

de Janvry, A. and E. Sadoulet (2010). "Agricultural growth and poverty reduction." The World Bank Research Observer 25(1): 1–20.

Glover, D., J. Sumberg and J. A. Andersson (2016). "The adoption problem; or why we still understand so little about technological change in African agriculture." Outlook on Agriculture 45(1): 3–6.
Hall, A. and J. Dijkman (2019). Public Agricultural Research in an Era of Transformation: The Challenge of Agri-Food.
Klauser, D. and C. Negra (2020). "Getting down to Earth (and business): Focus on African smallholders' incentives for improved soil management." Frontiers in Sustainable Food Systems 4(172): 1–10.
Leeuwis, C., L. Klerkx and M. Schut (2018). "Reforming the research policy and impact culture in the CGIAR: Integrating science and systemic capacity development." Global Food Security 16: 17–21.
Lobell, D. B. (2020). "Viewpoint: Principles and priorities for one CGIAR." Food Policy 91: 101825.
Persley, G. J. and V. M. Anthony (2017). The Business of Plant Breeding: Market-Led Approaches to New Variety Design in Africa. Wallingford, CABI.
Pretty, J., C. Toulmin and S. Williams (2011). "Sustainable intensification in African agriculture." International Journal of Agricultural Sustainability 9(1): 5–24.
Sartas, M., M. Schut, C. Proietti, G. Thiele and C. Leeuwis (2020). "Scaling Readiness: Science and practice of an approach to enhance impact of research for development." Agricultural Systems 183: 102874.
Schut, M., C. Leeuwis and G. Thiele (2020). "Science of scaling: Understanding and guiding the scaling of innovation for societal outcomes." Agricultural Systems 184: 102908.
Stevenson, J., B. Vanlauwe, K. Macours, N. Johnson, L. Krishnan, F. Place, D. Spielman, K. Hughes and P. Vlek (2019). "Farmer adoption of plot- and farm-level natural resource management practices: Between rhetoric and reality." Global Food Security 20: 101–104.
Stevenson, J. R. and P. Vlek (2018). Assessing the Adoption and Diffusion of Natural Resource Management Practices: Synthesis of a New Set of Empirical Studies. Rome, Independent Science and Partnership Council (ISPC).
Walker, T. S. and J. Alwang (2015). Crop improvement, Adoption and Impact of Improved Varieties in Food Crops in Sub-Saharan Africa. CABI. Wallingford, UK.
Wigboldus, S., L. Klerkx, C. Leeuwis, M. Schut, S. Muilerman and H. Jochemsen (2016). "Systemic perspectives on scaling agricultural innovations. A review." Agronomy for Sustainable Development 36(3): 46.
Woltering, L., K. Fehlenberg, B. Gerard, J. Ubels and L. Cooley (2019). "Scaling – From "reaching many" to sustainable systems change at scale: A critical shift in mindset." Agricultural Systems 176: 102652.

2

Sustaining Smallholder Farming through Collective Action and Entrepreneurship

Jayasree Krishnankutty[1], Arun Sreekumar[2], Rajesh K. Raju[3], and Kadambot H. M. Siddique[4]
[1]*Directorate of Extension, Kerala Agricultural University, Thrissur, Kerala, India*
[2]*Indian Institute of Management, Ahmedabad, Gujarat, India*
[3]*Communication Centre, Kerala Agricultural University, Thrissur, Kerala, India*
[4]*UWA Institute of Agriculture, The University of Western Australia, Crawley, WA, Australia*

CONTENTS

2.1 Introduction

In recent years, studies on entrepreneurship in the agricultural sector have begun to focus on small and marginal farmers.[1] One of the reasons for such interest is the sheer size of this segment and the potential to make substantial improvements to rural livelihoods (Lans et al. 2017). A second reason is the adaptive nature of agriculture that lends itself to various forms of enterprise (McElwee 2006). Smallholder farmers, a term we use to include small and marginal farmers, look for better ways to organize their farms, new crops and cultivars, better animals, and alternative technologies to increase productivity. Smallholder farming, therefore, offers entrepreneurial opportunities, such as the development of new products and innovations in the business processes, distribution, and marketing in the agricultural sector (Pindado and Sánchez 2017).

Smallholder farmers generally farm for one of four reasons: exclusively for home consumption with infrequent surpluses; primarily for home consumption, with some surplus for market; partly for home

[1] Marginal farmers are those farmers who cultivate up to 1 ha of land, whereas small farmers are those who cultivate between 1 and 2 ha of land.

DOI: 10.1201/9781003164968-3

consumption and partly for market; and exclusively for market. In India, most smallholder farmers traditionally belong to the middle two categories (Haque 1996). However, an increasing number of smallholder farmers are now taking up farming as exclusive commercial enterprises, growing fruit crops, vegetables, and even cash crops targeted exclusively for market. Despite the ongoing vertical integration of agribusiness industries and asset constraints in rural areas, smallholder farmers have held their ground through entrepreneurial efforts and developing partnerships (Carter 1998). Hence, there is hope for the viability of smallholder farm enterprises and a boom in the availability of smallholder produce in markets using innovative marketing methods.

Smallholder farm entrepreneurs are defined as those individuals with the potential to generate value through the creation, expansion, or innovation of economic activity by identifying and exploring new agricultural products, agro-processes, or markets (Thindisa 2014). For smallholders to transition from subsistence farmers to entrepreneurs, having a marketable surplus is not an adequate condition. They must recognize market opportunities, organize their resources, use innovative technologies to enhance productivity, and ensure a sustainable source of revenue to operate their farm enterprise (Carter 1998). Farmers have overcome resource limitations and found ways to generate higher revenue through food production techniques, diversification of farm portfolios, and active market promotion. These activities require functional skills, such as salesmanship, risk-taking, opportunity recognition, financial management, and marketing – all of which are influenced by personal, locational, and institutional factors (Morgan et al. 2010).

Monoactive farm entrepreneurs leverage these skills and capacities from their social network when institutional support is lacking (Vik and McElwee 2011). Such entrepreneurs manage their farm enterprise individually but seek active support and resources through their cooperation and networking with social ties. At the same time, a growing number of smallholder farmers in India and other developing countries unite as a group to manage their business. In India, the government has traditionally supported individual farmer's skills and knowledge development through a variety of extension programs offered by the Indian Council for Agricultural Research (ICAR) and the Krishi Vigyan Kendra (KVK) system. However, since the 1990s, the government has deemphasized the centralized extension programs offered to individual farmers and focused on building entrepreneurial skills and capacities of grassroots farmer collectives (Anil et al. 2020). In the following sections, we discuss the forms of collective entrepreneurship, examining the opportunities and challenges associated with managing farm businesses collectively.

2.2 Collective Entrepreneurship among Smallholder Farmers

Farmer collectives are defined as formal or informal membership-based group action institutions that support their farmer members in pursuing their individual or collective interests (Bizikova et al. 2020). This definition includes farmer associations, cooperatives, producer organizations, and self-help groups (SHGs). These collectives help small-scale farmers access markets, credit, and rural extension services and manage their shared assets and natural resources. Farmer collectives can provide skills and capacities for adopting innovative farming practices and marketing agricultural commodities.

Farmer collectives are an example of collective entrepreneurship. In collective entrepreneurship, decisions about the deployment of assets are made jointly by a group of people. However, in a strict sense, farmer organizations (FOs) in their various forms may not conform to this model of joint decision-making on asset use. Miles et al. (2006) presented a broader view, describing FOs as examples of collaborative entrepreneurship in which firms (i.e., individual farmers) form a network to match their limited resources with unexplored market opportunities, constantly leveraging innovations in agriculture to better use their available resources. This resource-based view of farmer collectives fits the situation for smallholder farmers, as their business growth is generally curtailed by the availability of knowledge and financial resources to exploit new technologies and market opportunities. Further, within these collectives, there are usually three forms of cooperation among farmers: (1) farmers sharing economic resources, such as agricultural inputs, knowledge, experience, labor, and credit; (2) mutual sharing of organizational resources, such as leadership and governance mechanisms; (3) using existing social resources, such as interpersonal ties with other farmers, marketing agents, and other community members.

The collectivization of farmers to share resources presents opportunities for overcoming significant barriers that limit the growth of farm entrepreneurship. The two major problems faced by smallholder entrepreneurship are related to high transaction costs (Trebbin and Hassler 2012) and vulnerability to risks in agricultural production (Fischer and Qaim 2012). High transaction costs emerge from procuring inputs, accessing supply chains, accessing credit, and obtaining market information. Risk vulnerability can be attributed to limited resource buffers for crop and market failures. In these situations, farm collectives can help to reduce costs by increasing the volume of transactions, providing more bargaining power, and increasing resilience by creating a buffer of shared resources.

In summary, farmers can collectively create business-like organizations with better access to resources and markets than farmers operating individually. Farmers need to cooperate with other farmers to acquire, bundle, and deploy resources required for farming and marketing their commodities. The resources bundled by farmer collectives can provide economies of scope to collective farmer enterprises. The homogeneity of commodities can provide economies of scale to market agricultural commodities. In succeeding sections, we discuss the various types of smallholder collectives operating in India.

2.2.1 Farmer Cooperatives

In India, Primary Agricultural Credit Societies (PACS) are prototype farmer cooperatives that began after enacting the Cooperative Credit Societies Act, 1904. Following successive amendments, the government started providing financial, technical, and administrative support to cooperatives. Cooperatives are based on the premise of collaborative entrepreneurship – they were envisioned to function as self-sustained, member-controlled businesses where farmers share knowledge and material resources for production and marketing. Farmer cooperatives are small-scale operations embedded within local contexts and well placed to go beyond classic extension services to help to develop more contextualized technologies through cooperation (Yang et al. 2014).

In the last five decades, a network of farmer cooperatives in India has been promoted, including national-level cooperatives (e.g., National Agricultural Cooperative Marketing Federation of India Ltd. [NAFED], Tribal Cooperative Marketing Development Federation of India [TRIFED]), state-level general and commodity-specific organizations, and primary-level marketing and credit societies. Primary-level marketing cooperatives are largely preoccupied with input supply rather than output marketing. The PACS at the village or cluster level mainly handle credit and inputs rather than outputs. For some states (Gujarat, Maharashtra) and commodities (milk, oilseeds, sugarcane), cooperatives have played an important role in output marketing. At present, the cooperative marketing structure comprises 2,633 general-purpose primary cooperative marketing societies at the Mandi (major towns) level, 3,290 specialized primary marketing societies for oilseeds, 172 district Central Federations, and the NAFED (Singh 2015).

Farmer cooperatives play an important role in organizing production and marketing functions in developed countries. For example, more than 30,000 cooperatives have more than nine million members within the European Union alone. These cooperatives account for 50% of the market share for delivering inputs and 60% for agricultural produce (World Bank 2008). Farmer cooperatives were historically introduced in sub-Saharan Africa during the colonial period to promote cash crop production by peasant farmers (Hussi et al. 1993). After independence, many governments and donors continued promoting cooperatives and other rural organizations as a potential source of decentralized grassroots participation in agricultural credit, inputs, and commodity markets (Lele and Christiansen 1989; Hussi et al. 1993). The performance of these cooperatives was mixed. For example, in Kenya, semiautonomous agencies such as the Kenya Tea Development Authority (KTDA) and the coffee and dairy cooperatives were important for the growth of smallholder production, while some parastatals and cooperatives had mediocre records.

In India, cooperatives have been generally successful, particularly among sugarcane farmers and dairy farmers in western India. However, examples of robust and self-sustaining cooperatives are limited, despite large investments from successive governments and public corporations. Unsatisfactory performance is often attributed to technological problems and poor management (Wolf 1986; Lele and Christiansen 1989). Others have pointed to corruption, inefficient governance, lack of market orientation, and bureaucratic lethargy as reasons for their failure (Shah 2016). In a globalized world, most farmer cooperatives in India have struggled to maintain market competitiveness with the rapid entry of large private corporations into marketing. At the same time, there are notable exceptions of cooperatives such as AMUL and cooperative federations such as NAFED that have successfully established viable businesses by maintaining cohesiveness, tenacity in governance processes, and operating effectiveness (Shah 2016).

In summary, farmer cooperatives were instrumental in kick starting collective entrepreneurship in India, with various types of support from state agencies. The cooperative movement has seen some successful examples of input aggregation and output marketing. However, these examples are limited, and many farmer cooperatives have failed to replicate their profitability. Since the 1990s, the focus has shifted from cooperatives to farmer clusters and producer companies, which we describe next.

2.2.2 Clusters

Clusters represent a bottom-up approach for forming farmer collectives rather than the top-down approach adopted for forming cooperatives through government initiatives. Clusters are aggregations of farmers in a localized geographical region with commonalities in resource availability, crops grown, or markets served. Clusters usually have a more democratic governance system than cooperatives, with leaders chosen internally with minimal external intervention.

There are several examples of successful clusters in India, particularly for the cultivation and marketing of horticultural crops. Despite being the world's third-largest producer of fruits and the second-largest producer of vegetables, India has had little success exporting these products, with only 2.3% of world horticultural trade in 2004 (Roy and Thorat 2008). A major factor limiting the growth of its agro-export industries, particularly given the fall in trade barriers, is the inability of India's smallholder-dominated production systems to meet the food safety requirements of export markets. In addition, smallholders have little knowledge of the feasibility of export production and the associated processes involved. However, the Maharashtra grape cluster is a successful case study highlighting how smallholders can overcome these constraints (FAO 2007).

The scheme for agro-processing clusters by the Ministry of Food Processing Industries aims to develop modern infrastructure and common facilities so that groups of entrepreneurs can set up food-processing units using the cluster approach by linking groups of producers/farmers to processors and markets through a well-equipped supply chain. Each agro-processing cluster has two basic components: SHG and FO.

2.2.3 Self-Help Groups

The formation of SHGs represents a major shift in agricultural policy toward more participatory and market-led approaches for agricultural development. SHGs are internally governed groups of 8–12 members, usually localized within a village. Anil et al. (2020) reviewed the role of these groups in local agricultural development through a case study approach. They found that SHGs are formed by a shared social and development need for skills, knowledge, or finance. As groups gain knowledge and financial capital, they begin to explore new possibilities in farming and new opportunities in the market. Multiple SHGs can also aggregate to form federations, amplifying their economies of scale and bargaining power in the market.

The SHG movement in India gained momentum in the last two decades, resulting in some changes to the marketing behavior of small and marginal farmers, who are now slowly shifting from being production-oriented to market-oriented. The experience and exposure obtained by mainly rural women and households through the SHG movement have enhanced marketing skills and improved entrepreneurship orientation (Krishnankutty and Shinoji 2014). A nationwide program linking SHGs to the banking system was launched in 1992. There are three types of SHGs: (1) formed and financed by banks; (2) formed by other agencies but financed by banks; (3) financed by banks using nongovernment organizations (NGOs). By March 2004, 10.8 lakh SHGs were linked to banks, with 90% of them women's groups. However, the microfinance program did not explicitly target the agricultural sector. Extending the SHG program to farmers will require internalization with PACS, which may not be easy (Figure 2.1).

FIGURE 2.1 Bio compost unit operated by a Self-Help Group in Kerala, India. (**Photograph by Jayasree Krishnankutty.**)

2.2.4 Farmer Organizations

FOs have a central role in smallholder development but need specific attention and support to bolster their effectiveness and sustainability.

When farmers unite, they can articulate their needs and organize services like inputs, credit, implements, transport, and service providers; if necessary, they can pressure groups and demand services (Singh 2012). FOs strengthen the political power of farmers by increasing the likelihood that policymakers and the public hear their needs and opinions. Economically, FOs help farmers form enterprises and process, and market their products more effectively to increase their incomes. FOs can achieve economies of scale, reduce costs, and facilitate the processing and marketing of agricultural commodities for individual farmers (Penunia 2011). The strength and unity of cooperating and working together empower small-scale producers, who often lack the skills, knowledge, and experience to participate in the market successfully. By working as a group, farmers can take risks and gain confidence.

For instance, farmer groups in Maharashtra tapped into nearby urban markets through the collective marketing of flowers (Kailas and Wagle 2015). Shrimp farmers in Tamil Nadu have organized themselves to profitably cultivate and market their produce for over two decades (Kumaran et al. 2012).

In recent years, KVKs and other organizations have formed commodity-based farmers' clubs. By March 31, 2005, NABARD had organized 13,664 farmers' clubs. The promotion of these clubs helps to create basic infrastructure for their effective functioning, including assistance for professional management.

2.2.5 Farmer Producer Companies

In contrast to large farmers, the core of the problems faced by small and marginal farmers can be attributed to their limited negotiating power and inability to profit from economies of scale. Among the many techniques developed and tried worldwide, the community strategy has proven to increase negotiating power. As a result, a consortium of growers committed to social responsibility and working to the standards – Farmer Producer Companies (FPCs) – has been suggested to address the challenges faced by small and marginal farmers (Figure 2.2).

FPCs are stand-alone, self-reliant institutions with all the essential characteristics of a private enterprise, while simultaneously incorporating features of mutual assistance in their mandate. Hence, FPCs can be considered a hybrid of farmer cooperatives and private corporations. The core purpose of FPCs is to collectivize farmers for backward linkages for inputs, such as fertilizers, training, seeds, and credit, and forward linkages, such as group marketing, marketing contracts, and agricultural processing.

FIGURE 2.2 Farmer producer company outlet for geographical indicator – organic jaggery, Kerala, India. (**Photograph by Akhil Ajith.**)

Producer companies, therefore, have all the qualities of free enterprise, with the ability to make decisions on how and where to market their commodities or products. Farmers are shareholders of producer companies, as opposed to being 'members' in cooperatives. This feature democratizes the governance of FPCs and ensures greater accountability for its leadership. The profits earned by each shareholder farmer are commensurate with the volume of commodities traded and the shareholder's contribution to the business. Importantly, FPCs can raise capital from external sources and have the freedom to form relationships with external entities in the marketplace. As a result, FPCs can craft their own entrepreneurial vision, freely explore business opportunities, and develop new value chains for taking their products to the market.

FPCs have a high degree of shared accountability among their members, resulting in high-quality, reliable goods. They have had a notable effect on manufacturing value chains, ensuring a minimum price for goods and reducing related risks; however, empirical evidence shows that the business models and overall output of these generating companies vary significantly. To tackle the problems associated with advertisement-based smart marketing techniques, strategies aligned with 'smart supply chain operations' are needed so that FPC goods meet both national and foreign demands. Perhaps smart marketing using farm labels will be the next agricultural trend. In this regard, FPCs play an important role in assisting smallholder farmers to improve their position within developing value chains. The potential benefits of producer organizations are commodity- and context-specific and dependent on concrete collective activities (Fischer and Qaim 2012). Indeed, there is much discussion on whether effective producer organizations, for a given context, offer benefits to all members, regardless of farm size. Most studies on the economic effects of FPCs focus on their impact on overall household welfare, farm revenue, and farm profit, not the FPC value chain and traceability. Establishing a reputation for quality and traceability for products manufactured and sold by FPCs will quickly gain customer acceptance and help our resource-poor farmers command premium prices.

There is evidence that FPCs in India have potential for integrating small farmers with the agricultural supply chain through innovative business models (Trebbin and Hassler 2012). Although there is high variability in the performance and profitability of FPCs, many have succeeded in tapping market opportunities despite market dominance by large corporations (Trebbin 2014). Buoyed by the initial success, the Indian government is promoting the formation of FPCs across the country. In 2013, the Department of Agriculture and Cooperation issued indicative guidelines for the formation, operation, and management of producer companies, providing an impetus for the formation of FPCs in several districts of the country (Table 2.1).

TABLE 2.1

Number of Farmer Producer Organizations by State in India (NABARD 2021)

State	Districts	Farmer Producer Organizations	Shareholders
Andaman and Nicobar	2	3	307
Andhra Pradesh	13	95	47,340
Arunachal Pradesh	5	8	1,813
Assam	17	41	14,380
Bihar	32	118	41,604
Chhattisgarh	15	57	24,797
Delhi	1	1	10
Goa	1	2	104
Gujarat	22	117	42,530
Haryana	17	50	26,297
Himachal Pradesh	9	51	11,004
Jammu and Kashmir	10	13	1,548
Jharkhand	20	60	30,950
Karnataka	27	159	78,618
Kerala	14	105	49,256
Lakshadweep	1	1	50
Madhya Pradesh	36	160	71,404
Maharashtra	21	120	39,927
Manipur	7	10	3,717
Meghalaya	7	11	1,798
Mizoram	7	16	3,266
Odisha	28	100	55,905
Punjab	20	70	8,734
Rajasthan	28	143	57,945
Sikkim	1	4	856
Tamil Nadu	31	170	122,858
Telangana	19	68	23,349
Tripura	1	1	210
Uttar Pradesh	43	116	48,118
Uttarakhand	13	50	18,751
West Bengal	18	150	79,991
Total	**486**	**2,070**	**907,437**

2.3 Challenges for Collectives

While farmer collectives provide numerous benefits for running a profitable agricultural enterprise, they are vulnerable to various challenges that threaten their success. As discussed earlier, collectives succeed through the sharing of complementary heterogeneous resources among farmers. The implicit assumption in such sharing is that the members believe that the benefits from cooperation exceed their contribution of resources. However, individuals acting opportunistically may contribute nothing to the collective and still enjoy gains from collective effort. This is particularly common as the size of the collective increases (Wincent et al. 2010). Avoiding such free-riding behavior requires strong governance mechanisms and their enforcement within the group. In addition to this challenge that impedes group coordination in various settings, some challenges pertain specifically to farmer collectives. These collectives are typically formed in rural areas with strong collectivistic norms and social bonding. These norms may contradict the business-like norms adopted by collectives (Olson 1965), particularly by producer companies. For example, imposing penalties for free-riding behavior or rewarding well-performing members at the expense of other members may become difficult in a collective embedded within strong interpersonal ties. Added to this, many farmer collectives may already have weak governance mechanisms, lacking professional leadership and managerial acumen. While forming collectives has clear benefits for farmers, they must be cognizant of the challenges faced in the operation of collective enterprises in the agricultural sector. We describe some of these pertinent challenges below.

2.3.1 Economic Viability and Inclusiveness

To ensure economic viability, farmer groups set certain restrictions on membership. While economies of scale increase with group size, this can be associated with higher transaction costs in mobilizing widely dispersed farmers and heterogeneity that can undermine group cohesion and reduce trust among members. Inclusiveness and expressed interest for tackling poverty may suggest wider and open membership, but resource-poor farmers may not be able to generate marketable surpluses or assets that foster trust and creditworthiness. Nevertheless, many FOs need a balance among economic viability, inclusiveness, and other social goals (Shiferaw and Muricho 2011).

Two case studies focusing on membership in producer marketing groups (PMGs) in Kenya (Shiferaw et al. 2009) and farmer cooperatives in Ethiopia (Bernard and Spielman 2009) that deal with the collective marketing of grain cereals and pulses, produced by small farmers in Africa, were examined. They revealed a 'middle class' effect for participation in cooperatives, with less participation by the resource-poor and wealthiest households at the bottom and top ends, respectively, of selected wealth indicators (Bernard and Spielman 2009).

2.3.2 Maintaining Competitiveness

The risks associated with globalization and increasing competition from external market forces, including supermarket chains, transnational agri-food companies, and subsidized producers in distant locations, pose a serious challenge for small-time local producers. In liberalized markets, the success of producer organizations will depend on access to new technology for reducing production costs, maintaining consistent supply, producing required volumes, and meeting the increasingly stringent food quality and food safety standards.

Smallholder collectives generally fall short of efficiently mobilizing internal and external resources, including finance and marketing assets, to carry out their functions. The functional orientation of farmer groups and their internal features are important determinants of the success of FOs. Large groups may be less successful than small groups in furthering their interests but only to a certain threshold size. This is mainly because the transaction and managerial costs of cooperation increase faster than the gains as group size increases beyond a certain level (Hussi et al. 1993), implying that optimal group size will depend on the type of joint activities in the group.

The existing evidence suggests that producer organizations perform better when dealing with cash crops (e.g., coffee, tea, cocoa, cotton, tobacco) and food products with high demand (e.g., fruits and vegetables, milk) in the agri-food industry, especially in the high-value markets connecting small producers with processors, exporters, and retail chains. When there are no large alternative suppliers, the potentials for collective action to aggregate standardized and high-quality products from small producers to supply identified markets make it mutually beneficial for the private sector and producer organizations to establish strategic partnerships. There are several such examples of success: potatoes to supply a fast-food outlet in Uganda (Kaganzi et al. 2009), green beans in Kenya and fruits in India (Narrod et al. 2009), and horticultural products in Honduras and El Salvador (Hellin et al. 2009). The challenge is leveraging these lessons and expanding into other low-value crops and staple foods, such as cereals, cassava, and pulses.

2.3.3 Urban Migration

Small-scale farmers who undergo contractions of their natural, financial, and human resources are increasingly vulnerable to risk factors, undergoing loss of social capital and forced to liquidate their capital assets and reconfigure their livelihood strategies (Bragdon and Smith 2015). Migration to urban areas is an increasingly prevalent solution for this. Money through government doles and relief cannot solve the problem of small-farm viability. Essentially, there is no salvation for the landless rural poor, who migrate to urban areas and become confined to slums. Seasonal migration changes to permanent migration. The development agenda of governments that offer enormous areas of farmland (by displacing farmers) for special economic zones (SEZs) (Vombatkere 2009) accelerates this process. The impact of this is visible in the social fabric of both urban and rural areas. It is an issue that needs addressing systematically, as most farmers have little formal education, and farming is all they know to do for a living.

2.4 Way Forward

2.4.1 Small Farmer Innovation Systems

Relatively little attention has been given to farmers' capacities to experiment and adapt to meet their own needs (Waters-Bayer et al. 2009). Lorentzen (2013) observed that most research in agricultural innovation has focused on formal organizations rather than individuals, households, and communities as the main units of analysis (Figure 2.3).

There is growing recognition of the adaptive capacity of small farmers, which should translate in coming years into a wider appreciation for their innovative capacity. There is a substantial overlap between the concepts of innovation and adaptation, notably, the adaptive capacity of small farmers and their ability to experiment with new varieties and management practices to suit changing growing conditions (Tittonell et al. 2012). The reality of climate change offers effective novel partnerships between farmers and formal sector research and development organizations, an example of institutional innovation (Chhetri and Easterling 2010).

The push-and-pull factors influencing small farmer behavior depend on their circumstances. Tittonell (2014) presented a typology of farm livelihood strategies in smallholder agriculture across Africa: (1) 'Hanging-in': farmers in situations with poor resource potential and market opportunities who engage in subsistence farming activities; (2) 'Stepping up': farmers in situations with high agricultural potential who invest in assets to expand current production (semicommercial farming);

FIGURE 2.3 Transforming sandy soil for vegetable cultivation through the diligent use of mulches and organic manure. (**Photograph by Jayasree Krishnankutty.**)

(3) 'Stepping-out': farmers with accumulated assets who may engage in nonfarm activities. 'Hanging-in' farmers are likely to innovate in response to push factors, while those 'stepping up' and 'out' may respond to new market opportunities. However, farms typically fluctuate between the two regimes. Smallholders who 'step out' may only do so temporarily or partially.

2.5 Conclusions

With the existing and impending challenges that smallholder farmers face, they are real-time entities invested in farming for the long term, especially in the developing world. Their potential is immense as they are survivors who thrive on little and contribute much more than they take. Smallholder farmers are emerging in some parts of the world as educated young people, urban dwellers, and retired people that have started small-time farming in the wake of nutrition insecurity and increased awareness of food safety issues. Still, they are generally characterized by firm social bonding, norms, and cohesive behavior. Transforming these to entrepreneurial activities causes incongruences that sometimes adversely affect the efficient functioning of the collectives, ultimately affecting their sustainability. Streamlining the modus operandi of collectives in a participatory manner might help address this issue. Governments must recognize smallholders as key stakeholders who can steady agriculture where it is greatly threatened. Special programs must be designed to sustain smallholder farming systems, technically, financially, and socially.

REFERENCES

Anil, B., Tonts, M. and Siddique, K.H. 2020. Community-Based Self-Help Groups in Agriculture. In *Local Knowledge, Intellectual Property and Agricultural Innovation* (pp. 217–239). Springer, Singapore.

Bernard, T. and Spielman, D.J. 2009. Reaching the rural poor through rural producer organizations? A study of agricultural marketing cooperatives in Ethiopia. *Food Policy* 34(1): 60–69.

Bizikova, L., Nkonya, E., Minah, M., Hanisch, M., Turaga, R.M.R., Speranza, C.I., Karthikeyan, M., Tang, L., Ghezzi-Kopel, K., Kelly, J. and Celestin, A.C. 2020. A scoping review of the contributions of farmers' organizations to smallholder agriculture. *Nature Food* 1(10): 620–630.

Bragdon, S.H. and Smith, C. 2015. Small-scale farmer innovation. Quaker United Nations Office. 2015. http://quno.org/areas-of-work/food-sustainability.

Carter, S. 1998. Portfolio entrepreneurship in the farm sector: indigenous growth in rural areas? *Entrepreneurship & Regional Development* 10(1): 17–32.

Chhetri, N.B. and Easterling, W.E. 2010. Adapting to climate change: Retrospective analysis of climate technology interaction in the rice-based farming system of Nepal. *Annals of the Association of American Geographers* 100(5): 1156–1176.

Fischer, E. and Qaim, M. 2012. Linking smallholders to markets: determinants and impacts of farmer collective action in Kenya. *World Development* 40(6): 1255–1268.

FAO (Food and Agricultural Organization). 2007. "Agro Based Clusters in Asia." FAO, Rome. Http://Www.Fao.Org/3/I1560e/I1560e02.Pdf. 2007.

Haque, T. 1996. *Sustainability of Small Holder Agriculture in India*. New Delhi: Concept Publishing Company.

Hellin, J., Lundy, M. and Meijer, M. 2009. Farmer organization, collective action and market access in Meso-America. *Food Policy* 34(1): 16–22.

Hussi, P., Murphy, J., Lindberg, O. and Brenneman, L. 1993. *The Development of Cooperatives and Other Rural Organizations: The Role of the World Bank*. The World Bank, Washington, DC.

Kaganzi, E., Ferris, S., Barham, J., Abenakyo, A., Sanginga, P. and Njuki, J. 2009. Sustaining linkages to high value markets through collective action in Uganda. *Food Policy* 34(1): 23–30.

Kailas, A. and Wagle, S.M. 2015. "Collective Marketing for Better Incomes" LEISA INDIA. December 23, 2015. https://leisaindia.org/collective-marketing-for-better-incomes/.

Krishnankutty, J. and Shinoji, K.C. 2014. SHGs promoting better marketing behaviour among marginal vegetable farmers. *International Journal of Basic and Applied Agricultural Research* 12(1): 15–19.

Kumaran, M., Kalaimani, N., Ponnusamy, K., Chandrasekharan, V.S. and Vimala, D. 2012. "Farmer Organisations." Leisa India. September 15, 2012. https://leisaindia.org/farmer-organisations/.

Lans, T., Seuneke, P. and Laurens, K. 2017. Agricultural Entrepreneurship. In *Encyclopedia of Creativity, Invention, Innovation and Entrepreneurship* (pp. 1–7). Springer, New York, NY. http://dx.doi.org/10.1007/978-1-4614-6616-1_496-2.

Lele, U.J. and Christiansen, R.E. 1989. "Markets, Marketing Boards, and Cooperatives in Africa." United Nations Digital Library System. 1989. https://digitallibrary.un.org/record/68313?ln=en.

Lorentzen, J. 2013. Low-Income Countries and Innovation Studies: A Review of Recent Literature. In *Challenges of African Transformation: Exploring Through Innovation Approach*, by Mammo Muchie and Angathevar Baskaran (pp. 80–107). Africa Institute of South Africa, Pretoria, South Africa.

McElwee, G. 2006. Farmers as entrepreneurs: developing competitive skills. *Journal of Developmental Entrepreneurship* 11(03): 187–206. https://doi.org/10.1142/s1084946706000398.

Miles, R.E., Miles, G. and Snow, C.C. 2006. Collaborative entrepreneurship. *Organizational Dynamics* 35(1): 1–11. https://doi.org/10.1016/j.orgdyn.2005.12.004

Morgan, S.L., Marsden, T., Miele, M. and Morley, A. 2010. Agricultural multifunctionality and farmers' entrepreneurial skills: A study of Tuscan and Welsh farmers. *Journal of Rural Studies* 26(2): 116–129.

NABARD (National Bank for Agriculture and Rural Development). 2021. NABARD Portal On Farmer Producers' Organisations (Promoted under PRODUCE Fund). https://nabfpo.in/images/staticFPO.html

Narrod, C., Roy, D., Okello, J., Avendaño, B., Rich, K. and Thorat, A. 2009. Public–private partnerships and collective action in high value fruit and vegetable supply chains. *Food Policy* 34(1): 8–15.

Olson, M. 1965. *The Logic of Collective Action*. Harvard University Press, Cambridge.

Penunia, E.A. 2011. The role of farmers' organizations in empowering and promoting the leadership of rural women. *Accra, Ghana: UN Women, FAO, IFAD and WFP.*

Pindado, E. and Sánchez, M. 2017. Researching the entrepreneurial behaviour of new and existing ventures in European agriculture. *Small Business Economics* 49(2): 421–444.

Roy, D. and Thorat, A. 2008. Success in high value horticultural export markets for the small farmers: The case of Mahagrapes in India. *World Development* 36(10): 1874–1890.

Shah, T. 2016. Farmer producer companies. *Economic & Political Weekly* 51(8): 15.

Shiferaw, B. and Muricho, G. 2011. Farmer Organizations and Collective Action Institutions for Improving Market Access and Technology Adoption in Sub-Saharan Africa: Review of Experiences and Implications for Policy. *Towards Priority Actions for Market Development for African Farmers.* International Livestock Research Institute (ILRI), Nairobi, Kenya.

Shiferaw, B., Obare, G., Muricho, G. and Silim, S. 2009. Leveraging institutions for collective action to improve markets for smallholder producers in less-favored areas. *African Journal of Agricultural and Resource Economics* 3(1): 1–18.

Singh, H. 2015. "Cooperative Marketing in India: Advantages and Progress." Jagranjosh.Com. October 20, 2015. https://www.jagranjosh.com/general-knowledge/cooperative-marketing-in-india-advantages-and-progess-1445333973-1.

Singh, R.P. 2012. "Farmer Organisations: Formation and Management." Share and Discover Knowledge on SlideShare. 2012. https://www.slideshare.net/rpratap11/farmer-organisations.

Thindisa, L.M.V. 2014. *Participation by smallholder farming entrepreneurs in agro-processing activities in South Africa* (Doctoral dissertation).

Tittonell, P. 2014. Livelihood Strategies, Resilience and Transformability in African Agroecosystems. *Agricultural Systems*: 3–14. https://doi.org/10.1016/j.agsy.2013.10.010.

Tittonell, P., Scopel, E., Andrieu, N., Posthumus, H., Mapfumo, P., Corbeels, M., Van Halsema, G.E., Lahmar, R., Lugandu, S., Rakotoarisoa, J. and Mtambanengwe, F. 2012. Agroecology-based aggradation-conservation agriculture (ABACO): Targeting innovations to combat soil degradation and food insecurity in semi-arid Africa. *Field Crops Research* 132: 168–174.

Trebbin, A. and Hassler, M. 2012. Farmers' producer companies in India: a new concept for collective action? *Environment and Planning A* 44(2): 411–427.

Trebbin, A. 2014. Linking small farmers to modern retail through producer organizations–Experiences with producer companies in India. *Food Policy* 45: 35–44.

Vik, J. and McElwee, G. 2011. Diversification and the entrepreneurial motivations of farmers in Norway. *Journal of Small Business Management* 49(3): 390–410.

Vombatkere, S.G. 2009. "Why Are Farmers Moving to the Cities? *Mainstream* 47(28). https://www.mainstreamweekly.net/article1473.html

Waters-Bayer, A., van Veldhuizen, L., Wongtschowski, M., Wettasinha, C., Waters-Bayer, A. and Kaaria, S. 2009. Recognizing and enhancing processes of local innovation. In *Innovation Africa: Enriching Farmers' Livelihoods*. Earthscan, London.
Wincent, J., Anokhin, S. and Örtqvist, D. 2010. Does network board capital matter? A study of innovative performance in strategic SME networks. *Journal of Business Research* 63(3): 265–275.
Wolf, T. 1986. State intervention at the cabbage-roots: A case study from Kenya. *IDS Bulletin* 17(1): 47–50.
World Bank. 2008. "World Development Report 2008 : Agriculture for Development." Open Knowledge Repository.
Yang, H., Klerkx, L. and Leeuwis, C. 2014. Functions and limitations of farmer cooperatives as innovation intermediaries: Findings from China. *Agricultural Systems* 127: 115–125.

3

Technological Innovation Strategy to Strengthen the Competitive Advantages of Smallholder Farmers

E. Cornejo-Velazquez[1], M. Clavel-Maqueda[2], O.A. Acevedo-Sandoval[2], and H. Romero-Trejo[2]
[1]*Autonomous University of Hidalgo State, Pachuca, Mexico*
[2]*Popular Autonomous University of the State of Puebla, Puebla, Mexico*

CONTENTS

3.1 Introduction

Climate events are more frequent and intense; floods, droughts and storms have adverse impacts on agriculture and food security in terms of availability, accessibility and utilization (FAO 2015b). Scientific evidence shows that climate change and its effects are not hypothetical future events; the consequences of global warming are becoming obvious (Bayala et al. 2017; FAO 2017; Singh and Singh 2017; Valentini 2017). Sustainable food production in the 21st century faces the effects of climate change, increasing world population, degradation of natural resources and decreasing available water. The Food and Agriculture Organization of the United Nations (FAO) estimates that by 2050, the planet's population will reach 9 billion people, and the demand for food will require an increase in production of at least 60% (FAO 2015a).

Main effects of climate change are reflected in lower crop yields and producer incomes (Costinot et al. 2016). In addition, changes in climate modify the dynamics of pest, pathogen and weed populations (Sridhar et al. 2020); these imply increases in the production costs of agricultural systems (Struik and Kuyper 2017).

In coming years, given the magnitude and speed of the effects of climate change, large segments of the world's population will have to improve the effectiveness of adaptation strategies and strengthen mitigation efforts. In this scenario, planning agricultural development is a challenge that requires technological strategies and public policies focused on reducing the vulnerability of producers and increasing their capacity to adapt effectively (Zilberman et al. 2018).

Agricultural transformations that took place at the end of the last century were based on intensification using large amounts of inputs (FAO 2017). In many countries, this approach has had serious environmental consequences, reflected in massive deforestation and soil and water depletion.

DOI: 10.1201/9781003164968-4

Globally, 85% of farms are less than 2 hectares in size (FAO 2017). Smallholders provide food and sources of employment in their communities and regions, directly impacting their economy. Smallholder farmers in developing countries are characterized by having insufficient resources to cover operational, financial and technological requirements. It is therefore a challenge not to leave them out of structural and technological transformations.

Given this scenario, there is a need to design strategies for small farmers to make innovative use of information and communications technologies (ICT) as a mechanism for mitigating and adapting production systems to the effects of climate change (Kuhl 2020) and to build competitive advantages that will strengthen their participation in the markets where they sell their products.

3.2 Competitive Advantages of Smallholder Farmers

Agri-food value chains are made up of the set of activities that an agricultural product requires. From its production, distribution to consumers and the ways in which it is discarded or recycled once consumed. The basic principle is that the competitiveness of agricultural products is not determined by what happens in one link but by the performance of all the links. Thus, leading companies are interested in ensuring proper coordination between each stage, seeking the best level of systemic competitiveness.

Classical economic theory considers the comparative advantages of a region or nation in the abundant endowment of basic factors of production (land, labor and capital) and the relative abundance of natural resources. In globalization, technology is considered an innovative element, as well as consumption patterns and a greater awareness of the conservation of natural resources. In view of this, there is a conceptualization of competitiveness in which comparative advantages as drivers of development evolve into competitive advantages.

Competitive advantages are created through product differentiation and cost reduction (Porter 1990). In this context, the capacity for innovation, the application of technology and specialized aspects (infrastructure, research, human resources, public services) are important for generating competitive advantages for agricultural producers. On the other hand, smallholder producers play an important role in the social, environmental and economic well-being of their families and communities, so they must adopt an attitude of continuous improvement in search of a more competitive position for their own well-being and that of society. However, because they operate on a small scale, they have limited access to productive and financial resources. In addition, greater levels of rigidity, this places them in a situation of greater vulnerability and risk with respect to their competitors.

3.3 Systemic Approach to Agricultural Production

Agricultural production is organized to produce food through the management of available resources within a social, economic and institutional environment (Hall et al. 2001). Its components are the available natural resources: land types and water sources, climate and biodiversity as well as human, social and financial capital. Biophysical, socioeconomic and human elements are interdependent, and therefore, agricultural production systems can be analyzed from a systemic approach by considering a set of parts forming a complex whole (Behera 2013).

Agricultural system, as a concept, takes into account the components of soil, water, crops, livestock, labor, capital, energy and other resources in the farmland to manage agricultural activities (Mahapatra 1994). System can be represented as a matrix of the interrelationships of its components and other inputs controlled by farmers and influenced to varying degrees by political, economic, institutional and social forces. Behera (2013) proposes a simplified model where we put the farmer at the center. He is who makes decisions taking into consideration his priorities, knowledge, experiences and the resources at his disposal. From this perspective, the availability of markets and price levels influence the farmer's decision to purchase inputs as well as the timing of the sale of his products. On the other hand, economic and social infrastructure determines transportation costs and the availability of services for consumers. Similarly, information and education services affect the producer's strategies and decisions.

3.4 Technological Innovation in Agricultural Systems

Use of technology in the agricultural sector has historically served as a mediating tool between man and nature. Its basic function is to contribute substantially to transforming nature for the benefit of the people living in the countryside. Replacement of traditional work by technological innovation is based on a set of specific knowledge and processes. Those changes could transform and solve a problem in the operation of the agricultural sector as a linear process. This brings with it a series of economic and socio-cultural situations that many authors have worked on in terms of their social impact and implications. Technological innovation is a key element in the construction of competitive advantages in agricultural production systems and in the development of the sector at the local, national and international levels.

In the process of agricultural modernization, the generation and development of technologies is shaped by the ideological expectation of capital formation and increased productivity. Scientific literature describes how agricultural technology has been developed for better crop yields and higher economic gains (Rehman and Hussain 2016). ICT allows people and electronic devices to connect worldwide at high speeds at any time (Gartner 2016). Even remote and developing regions have increased possibilities to go online through telephony and Internet providers (Danes et al. 2014).

Use of satellites, unmanned aerial vehicles (also named drones) and remote sensors to obtain data linked to crop growth and development, soil moisture and other dynamic variables in real time expand the possibilities for crop management (Capolupo et al. 2015; Domingues Franceschini et al. 2017; Donatelli et al. 2014). High-performance computing allows us to make sense of structured and unstructured data collected through new digital sensor technologies (Elliott et al. 2013). Coupled with this, web and cloud computing technologies allow such capabilities to be made available to a large number of users at an affordable cost (Foster 2015). Together these technologies provide the possibility for higher-quality information to support crop-data-driven decision-making.

Janssen et al. (2017) describe two opportunities for applying modern ICT in agricultural systems:

1. Advances in data processing, remote sensor-based monitoring, and high-performance computing can be used to advance production systems model science.
2. New technologies can be used to transform farming system modeling practices and applications by making them more collaborative, distributed, flexible, and accessible.

Taking advantage of these opportunities requires the integration of different disciplines to exploit new data sources to produce and apply new agricultural production system models. This requires frameworks, data processing applications and tools for visualizing results.

From this perspective, the knowledge chain concept (Janssen et al. 2017; Lokers et al. 2016) postulates that raw data when combined with description and quality attributes lead to information. Information can be linked to other sources of information and placed in causal flows to produce knowledge. Ultimately, knowledge serves as an input for wisdom-based decisions, which cannot be digitize and it exists at the level of the agricultural production system for the decision-making process. For its part, the application chain concept (Janssen et al. 2017) considers that agricultural production systems can make use of software (data access, processing, analysis and visualization) and hardware (servers, processing and storage units) infrastructure to provide information and knowledge related to yield forecasts, effects of a change in crop management or estimates of disease-related damage to smallholder farmers and users of ICT infrastructure.

In recent years, robotics and automation have come together to build automated and intelligent agricultural systems that operate in the fields performing complete work cycles in planting and harvesting. This approach seeks to increase the efficiency of agricultural systems in terms of reducing costs and improving quality by developing differentiated products. Precision farming paradigm describes the management of crops with the goal of improving productivity and resource use, either through higher yields or reduced input inputs and adverse environmental effects. Smart farming technologies are required for efficient application of inputs (seeds, fertilizers, chemicals, water, fuel and labor), increased work speed, convenience and flexibility in crop management.

Use of different technologies in agricultural production systems is oriented toward intelligent operations. They seek the adaptability of systems to achieve higher levels of production and, therefore, increase efficiency, mainly by optimizing the use of water, fertilizers and phytosanitary products. Production systems are improved with software services that employ algorithms to transform data into value-added information with emphasis on optimizing products, production processes, reducing risks and limiting vulnerabilities.

Scientific literature reports the emergence of agricultural ecosystems with technological platforms. In them, it combines data from various sources in the field with external sources, allowing smallholder farmers to monitor their crops with real-time information to support decision-making. Such conditions accelerate and strengthen the incorporation of technological innovations into agricultural production systems and contribute to strengthening the competitiveness of small farmers by building competitive advantages.

3.5 Technological Innovation Strategy

Agricultural production systems depend on different conditioning factors: genetic, climatological, agroecosystem and the interventions carried out. Figure 3.1 presents the functional components of the agricultural production system: crop plot, production cycle, soil characteristics, climate conditions, management practices and management, all related to the phenological development stages of the crop.

The management and exploitation of data describing the status of each of the components of the agricultural production system allows strengthening the competitive advantages of smallholder producers

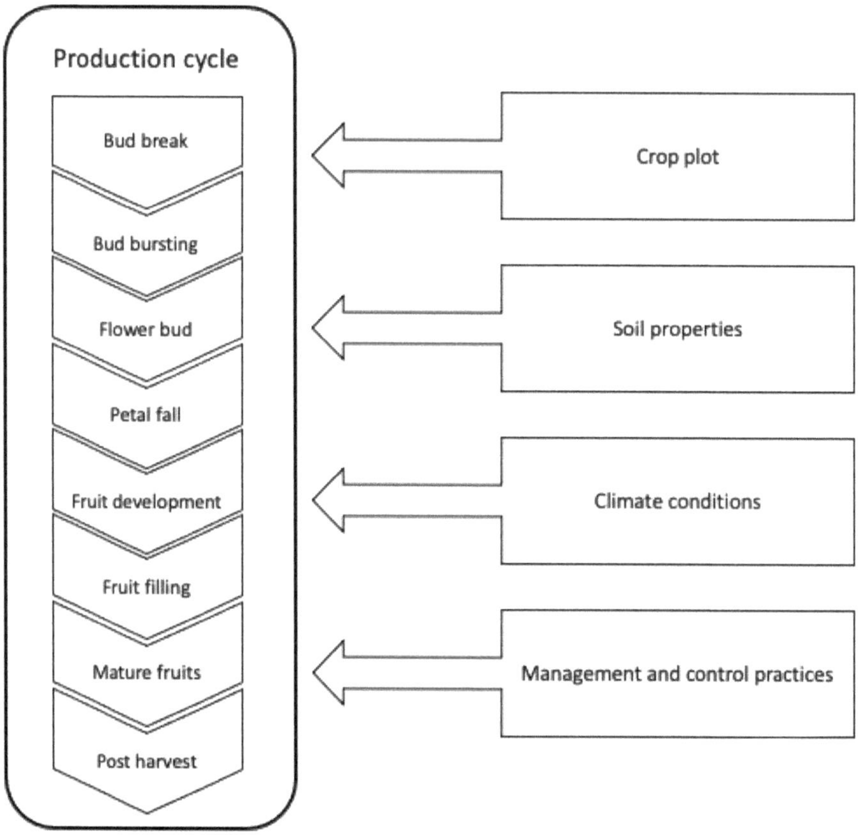

FIGURE 3.1 Components of the agricultural production system.

through the planning of management practices and plantation management focused on improving crop yields, ensuring product quality, reducing production costs and achieving product differentiation. From the systemic approach, the agricultural production system is conceptualized from the central idea that each agricultural producer, according to the characteristics of his crop plot, follows a productive logic that depends on the endowment of productive factors and the limiting factors of the production unit.

With this perspective, a descriptive model of the agricultural production system at the level of the crop plot includes structural aspects: land, water, and other natural resources; functional aspects related to the interactions and exchanges between the components; and dynamic aspects to represent the evolution of the system's conditions. Smallholder farmers can implement data-driven strategic planning for management and decision-making in the management of their crops by integrating these components. Model of technological innovation at the crop plot scale presented incorporates the horizontal and vertical heterogeneity of the agricultural production system. Horizontal characteristics refer to microclimate conditions, the condition of the plot and its planting, as well as management and control practices, while the vertical characteristics are the soil properties that vary with depth and are related to soil type, porosity and surface slope. This model is used to design the strategy for technological innovation in the production systems of smallholder farmers. It is based on the design and construction of an integrated information system at the field scale. This is useful to support strategic planning and decision-making driven by the data generated by the system components.

3.5.1 Technological Architecture

Figure 3.2 shows the technological and functional architecture of the information system for the collection, processing and exploitation of agricultural production system data through the deployment of technological components.

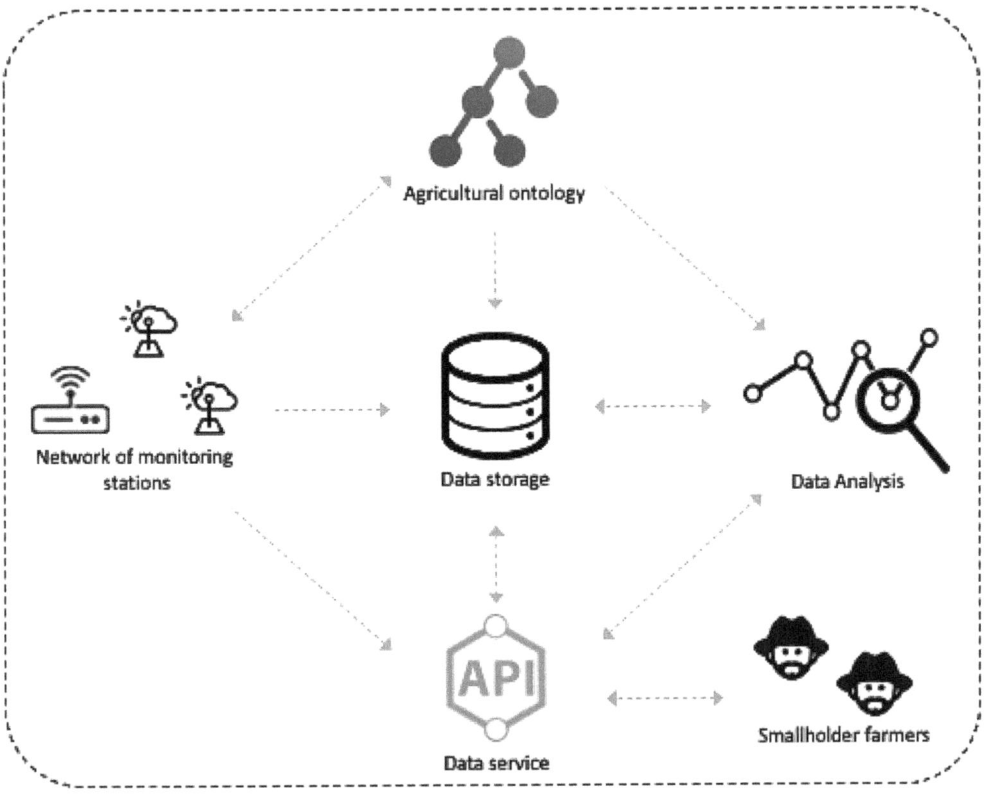

FIGURE 3.2 Technological architecture.

Proposed technological components are:

1. Agricultural ontology performs the semantic definition of the data within the agricultural context. It represents the static and dynamic properties of the components and actors involved in agricultural production. Agricultural ontology allows the acquisition and representation of knowledge to operate the system. It also provides semantic and operational consistency to the different processing phases of the data generated by the system components.
2. Network of monitoring stations is an active physical component composed of automatic microclimate monitoring stations. Which were enabled to make observations of climatological variables and soil properties. As well, they have the capacity to communicate the observed values to a component that receives and makes the data persistent in a storage repository. Communication of the observed data is done through a wireless network.
3. Data storage is a software component that is based on the implementation of a database in a private repository to receive, validate and store the data generated by the system components. Data integration is done efficiently and securely according to the semantic definitions of the agricultural ontology.
4. Data analysis, software services that take care of the integration, processing and validation of the data stored in the database. Analysis algorithms guarantee the integrity and consistency of the data to be used by end-user-oriented data services.
5. Data services, implement software services oriented to the end users of the system. That system provides tools for the visualization of information to be used in the tasks of strategic planning, operation and decision-making in the agricultural production system.

3.5.2 Technologies Deployed

Figure 3.3 describes the different technologies implemented for data processing and functional purpose in the proposed system. The technologies are classified into sensors, communication networks, data integration, intelligence and augmented behavior.

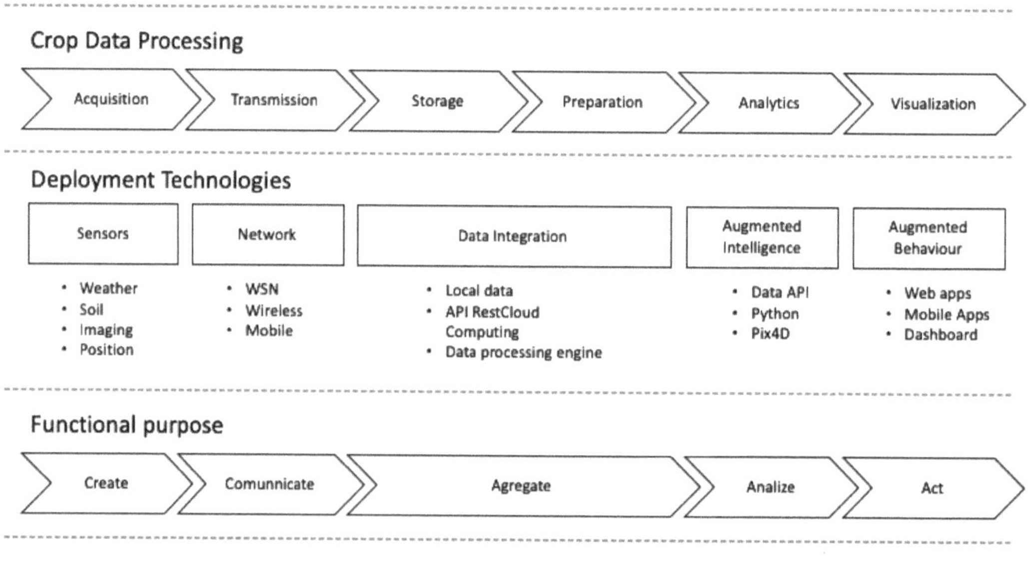

FIGURE 3.3 Technologies deployed. (Authors.)

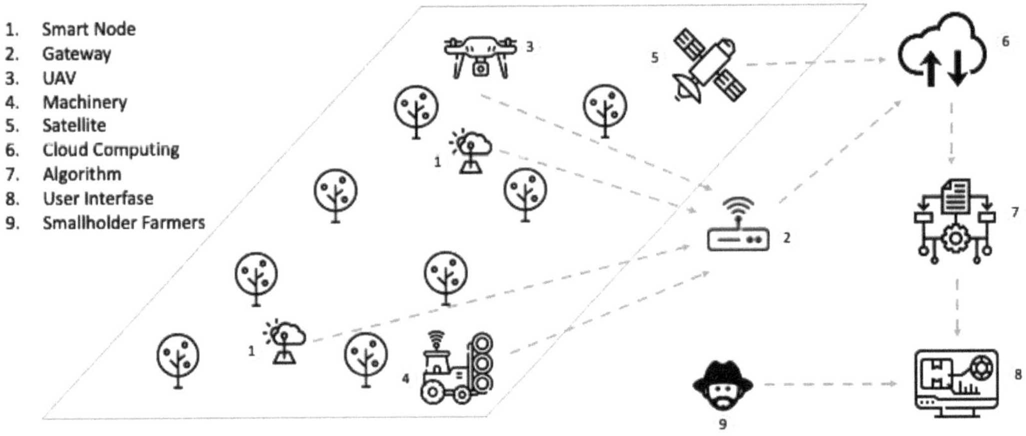

FIGURE 3.4 Agricultural technology framework.

3.5.3 Agricultural Technology Framework

Deployment of the integrated information system in the crop plot achieves the integration of technologies in an architecture to reconfigure the value chain of the agricultural production system. Figure 3.4 presents the integrated digital technologies: sensors, integrator nodes, private storage clouds, computational algorithms and data visualization tools.

From the functional perspective, the sensors are installed in Smart Nodes distributed over the surface of the crop plot. Observed values are communicated to the concentrator node (Gateway) to be sent to the storage space, private cloud. In addition, integration of the field logbook with the programmed and applied field practices related to the costs of the inputs used is performed. With the data integrated into the private cloud, data analysis algorithms and information visualization tools presented in interfaces designed for the end user of the system, the smallholder farmer, are employed. Agricultural technology framework can be extended by integrating aerial imagery using RGB and multispectral cameras on board drones that perform flight plans at different stages of crop development. Incorporation of agricultural machinery with geolocation and sensors can improve the differentiated application of activities and inputs. Finally, satellite images and macro-level weather forecasts can be incorporated to complement the results of the analysis algorithms.

3.6 Conclusions

Agricultural technology framework has been deployed in apple and chili bell pepper crops, both with high market value, using low-cost technologies to increase opportunities for smallholder farmers. Technological components have been implemented to generate, communicate and process data from the agricultural production system at the field level so that farmers can know what is happening with their crop at any stage of development of their plantation. With the information generated, it is possible to carry out strategic planning of field activities for crop maintenance and input optimization, directly impacting production costs and ensuring product quality to achieve differentiation, thus strengthening the competitive advantages of small farmers.

REFERENCES

Bayala, J., Zougmoré, R., Dayamba, S.D. and Olivier, A. 2017. Editorial for the Thematic Series in Agriculture & Food Security: Climate-Smart Agriculture Technologies in West Africa: learning from the ground AR4D experiences. *Agriculture and Food Security* 6(1): 40.

Behera, U. K. 2013. A text book on farming systems. Udaipur: Agrotech Publishing House.

Capolupo, A., Kooistra, L., Berendonk, C., Boccia, L. and Suomalainen, J. 2015. Estimating plant traits of grasslands from UAV-acquired hyperspectral images: a comparison of statistical approaches. *ISPRS International Journal of Geo-Information* 4(4): 2792–2820.

Costinot, A., Donaldson, D. and Smith, C. 2016. Evolving comparative advantage and the impact of climate change in agricultural markets: evidence from 1.7 million fields around the world. *Journal of Political Economy* 124(1).

Danes, M.H.G.I., Jellema, A., Janssen, S.J.C. and Janssen, H. 2014. Mobiles for agricultural development: exploring trends, Challenges and Policy Options for the Dutch Government (2501).

Domingues Franceschini, M., Bartholomeus, H., van Apeldoorn, D., Suomalainen, J. and Kooistra, L. 2017. Intercomparison of unmanned aerial vehicle and ground-based narrow band spectrometers applied to crop trait monitoring in organic potato production. *Sensors* 17(6):1428.

Donatelli, M., Bregaglio, S., Confalonieri, R., De Mascellis, R. and Acutis, M. 2014. A generic framework for evaluating hybrid models by reuse and composition – a case study on soil temperature simulation. *Environmental Modelling and Software* 62: 478–486.

Elliott, J., Kelly, D., Best, N., Wilde, M., Glotter, M. and Foster, I. 2013. The parallel system for integrating impact models and sectors (pSIMS). In Proceedings of the Conference on Extreme Science and Engineering Discovery Environment Gateway to Discovery – XSEDE'13 (p. 1). New York, NY: ACM Press.

FAO 2015a. Ground on understanding the challenges of climate change and food security. http://www.fao.org/3/a-i3817e.pdf

FAO 2015b. Climate-Smart Agriculture: A call for action. Synthesis of the Asia-Pacific Regional Workshop Bangkok, Thailand. http://www.fao.org/3/a-i4904e.pdf

FAO 2017. El estado mundial de la agricultura y la alimentación. Roma, Italia. http://www.fao.org/3/a-I7658s.pdf

Foster, I. 2015. Lessons from Industry for Science Cyberinfrastructure. In Proceedings of the 1st Workshop on The Science of Cyberinfrastructure Research, Experience, Applications and Models – SCREAM'15 (pp. 1–2). New York, NY: ACM Press.

Gartner 2016. Top 10 Technology Trends Signal the Digital Mesh - Smarter With Gartner. https://www.gartner.com/smarterwithgartner/top-ten-technology-trends-signal-the-digital-mesh/

Hall, M., Dixon, J., Gulliver, A. and Gibbon, D. 2001. Farming Systems and Poverty: Improving farmer's livelihoods in a changing world. Rome: FAO and World Bank.

Janssen, S.J.C., Porter, C.H., Moore, A.D., Athanasiadis, I.N., Foster, I., Jones, J.W. and Antle, J.M. 2017. Towards a new generation of agricultural system data, models and knowledge products: Information and communication technology. *Agricultural Systems* 155: 200–212.

Kuhl, L. 2020. Technology transfer and adoption for smallholder climate change adaptation: opportunities and challenges. *Climate and Development* 12(4): 353–368.

Lokers, R., Knapen, R., Janssen, S., van Randen, Y. and Jansen, J. 2016. Analysis of Big Data technologies for use in agro-environmental science. *Environmental Modelling and Software* 84: 494–504.

Mahapatra, I.C. 1994. Farming systems research: A key to sustainable agriculture. *Indian Journal of Fertilizer Research* 39(11): 13–25.

Porter, M.E. 1990. The competitive advantage of nations Harvard business review. *Competitive Intelligence Review* 1(1): 72–91.

Rehman, A. and Hussain, M.I. 2016. Modern agricultural technology adoption its importance, role and usage for the improvement of agriculture. *American-Eurasian Journal of Agricultural and Environmental Sciences* 16(2): 284–288.

Singh, R. and Singh, G. 2017. Traditional agriculture: a climate-smart approach for sustainable food production. *Energy, Ecology and Environment* 2(5): 296–316.

Sridhar, J., Kiran Kumar, K., Murali-Baskaran, R.K., Senthil-Nathan, S., Sharma, S., Nagesh, M., Kaushal, P. and Kumar, J. 2020. Impact of climate change on communities, response and migration of insects, nematodes, vectors and natural enemies in diverse ecosystems. In: Venkatramanan V., Shah S., Prasad R. (eds) Global Climate Change: Resilient and Smart Agriculture. Springer, Singapore.

Struik, P.C. and Kuyper, T.W. 2017. Sustainable intensification in agriculture: the richer shade of green. A review. *Agronomy for Sustainable Development* 37(5): 39.

Valentini, R. 2017. The global agricultural System and climate change: challenges and opportunities for the Russian federation. *Finance: Theory and Practice* 21(6): 5671–2017.

Zilberman, D., Goetz, R. and Garrido, A. 2018. Climate smart agriculture building resilience to climate change. Roma: Springer.

4

Analyses of Rural Infrastructure for Agriculture Development

Firuza Begham Mustafa[1] and Benjamin Ezekiel Bwadi[2]
[1]*Department of Geography, University Malaya, Kuala Lumpur, Malaysia*
[2]*Department of Geography, Taraba State University, Jalingo, Nigeria*

CONTENTS

4.1 Introduction

Infrastructure has a significant role in the development of any agricultural development. The natural environment serves as a valuable requirement for agricultural activities but without infrastructure, productivity will plummet. In Malaysia, agriculture has developed and served as a means of employment to the marginalized community, food security, and economic development of the nation (Ghosh et al., 2017; Adnan et al., 2017). The growth in agriculture was as a result of favorable natural environment, application of modern technology, and other farm inputs such as fertilizers and high yielding seeds. However, this development in agriculture was common and associated with areas that were already developed in term of good infrastructural facilities such as roads, markets, electricity, and financial institutions. Despite government effort to improve the rural infrastructure facilities, the disparity continues to exist between these developed areas and some of the rural undeveloped areas (Kamal and Flanagan, 2012). The rural areas where infrastructure facilities are inadequate or lacking are being affected by low agricultural productivity. The provision of good infrastructure such as good rood network will reduce the cost of transportation (Manjunath and Kannan, 2017). Road networks enhance accessibility and transportation of farm input from the market to the farm and also transportation of farm produce to the market.

Other infrastructures, such as market and credit faculties, play significant roles in agricultural development (Bashir et al., 2010). Access to credit facilities institution reduces the farmer the burden of borrowing from people for farm inputs and enhances investment in the agricultural sector (Akram and Hussain, 2008). Having access to market boosts the farmer's productivity and profitability as his inputs can be purchased and his farm product can be sold within a short time (Bwadi and Mustafa, 2019; Pinstrup-Andersen and Shimokawa, 2006). Rural electric supply fulfills the power and energy requirements for irrigation and other farm machines (Pinstrup-Andersen and Shimokawa, 2006).

DOI: 10.1201/9781003164968-5

The area-specific characteristic should be put into consideration while making decision on rural infrastructure investment in order to reduce the imbalance and disparities in rural development to accelerate agricultural development. Satisfying the infrastructural needs of the rural agricultural sector in Malaysia without widening the existing regional disparity is needful. However, there is a need for a comprehensive analysis of the rural infrastructure and the evaluation of its influence on the agricultural productivity of the area. In Malaysia, there are limited empirical studies that focus on rural infrastructure and agricultural development. Some of these studies include Ibrahim and Razzaq (2010) on rural community in Malaysia, Islam and Siwar (2012) on urban agriculture in Malaysia, Alam et al. (2010) on socioeconomic profile of farmer, Ngah (2010) on regional development, and Baydar et al. (1990) on agriculture policies on migrant farmers. None of these studies investigated the infrastructure facilities and rural agricultural development in Malaysia, and they only focus on the provision of infrastructure but could not examine the importance and how the infrastructure can be used for the rural agricultural development. The current study collects and analyzes infrastructure data and generates a map indicating the level of its suitability for a sustainable rural agriculture development of Negeri Sembilan State of Malaysia.

The study uses the area data on infrastructure of Negeri Sembilan State of Malaysia for the analysis. Some of the important rural infrastructural parameters that have great influence on agricultural development are used for this study as suggested by expert's opinion. It is expected that areas with favorable natural environmental condition and good rural infrastructural facilities will have greater agricultural productivity (Manjunath and Kannan, 2017).

In the current study, analytic hierarchy process (AHP) and geographic information system (GIS) are applied to evaluate the significance of infrastructure to rural agricultural development. A suitability map was generated of the area to show the level of importance of each of the parameters of infrastructure to rural agricultural development.

4.2 Material and Method

4.2.1 The Study Area

Negeri Sembilan lies between latitude 2°43″ N and longitude 102°25″ E covering a land area of about 6646 km^2 in Malaysia. At the northeast, it borders the state of Selangor; at the north, it borders Pahang; at the east, the State of Johor; while at the south, it boarders Melaka with Seremban as the capital city. Negeri Sembilan is an agricultural area in Malaysia; its farmers are engaged in oil palm plantation, rubber trees, paddy, and aquaculture. The area enjoyed some infrastructure facilities such as road, electricity, markets, and some financial institutions, which facilitated economic activities of the region.

4.2.2 Determination of the Parameters and the Class Weights

The parameter for the evaluation of the infrastructure facilities for the development of rural agriculture was performed using the AHP was applied where expert's opinion was sort to determine the level of significance of each parameter by assigning weight to them. Thirty experts from the study area who are knowledgeable in rural agriculture were contacted and interviewed to find out which parameters or factors favor agriculture development. However, only 20 experts out of these 30 were found to be consistent. The four parameters were chosen including road network, electricity power supply, availability of markets, and credit institutions. Saaty (1980) suggested the pairwise comparison matrix to assign weight to the parameters based on their priority and preference and the consistency ratio (CR) to measure the consistency of the experts' judgments. The pairwise comparative matrix was used to compare a pair of the parameters in relation to their suitability to agricultural purposes. Weights were assigned to the parameters by experts using the Saaty's nine points scale for measurement (Table 4.1).

The AHP pairwise comparison matrix calculates the weights of the parameters (*wi*) and to determine the consistency of the judgment by normalizing the weight of the parameters.

$$\mathrm{CI} = (\lambda - n)/(n-1) \tag{4.1}$$

$$\mathrm{CR} = \mathrm{CI}/\mathrm{RI} \tag{4.2}$$

TABLE 4.1
AHP Nine-Points Scales

Level of Significance (Ranks)	Explanation
1	Two parameters with equal importance
3	Moderate importance over the other
5	Strongly important
7	Very strongly important
9	Extremely strongly important
2,4,6,8	The intermediary values
1/9,1/8,.... 1/2	Reciprocal values (inverse)

TABLE 4.2
Table of Random Index

n	1	2	3	4	5	6	7	8	9	10
RI	0	0	0.58	0.9	1.12	1.34	1.32	1.41	1.45	1.49

where CI is the consistency index, CR is the consistency ratio, RI is the random index (Table 4.2) λ is the average consistency vector, and n is the number of parameters in each pairwise matrix

In order to maintain the consistency of the expert judgments and check mistakes that could be made by the expert, Saaty (1994) introduced the CR. Saaty places the CR at 0.1. In a situation where the calculated CR is 0.10 or below, the judgment is said to be consistent, but in a situation where the CR is 1.0 and above, the judgment is said to be inconsistent, the evaluation process has to be repeated.

4.2.3 Integration with the GIS

Once the normalized layers and the weight of each parameter were calculated, the weighted sum was overlay in the ArcGIS 9.3 version to produce the suitability map of infrastructure for agricultural development of the study area. The integration of the AHP and GIS are functional tools for decision-making in a situation where you have multiple criteria or parameters to choose from. On the map that was generated, various levels of suitability were indicated as most suitable, moderately suitable, and not suitable.

4.3 Results and Discussion

4.3.1 Spatial Variability of the Infrastructural Parameters

The pairwise comparison matrix of infrastructure for rural agriculture shows in Table 4.3 that the CR for the parameter was 0.4, which fall within an acceptable level of the evaluation. The value of the evaluation indicated that the comparison of the parameters is consistent. The result shows that market

TABLE 4.3
The Pairwise Comparison Matrix

Parameters	Market	Roads 2	Credit Inst.	Electricity	Priority (%)	Rank
Market	1	2.00	3.00	3.00	45.5	1
Roads	0.50	1	2.00	2.00	26.3	2
Credit inst.	0.33	0.50	1	1.00	14.1	3
Electricity	0.33	0.50	1.00	1	14.1	3
		Consistency ratio CR = 0.4%				

ranked first with 45.5%, followed by road network with 26.3%, while credit institutions, and electricity power supply both ranked third having equal importance in the area with 14.1% each. The result revealed that infrastructural facilities have significant influence on agricultural productivity agreeing with Sangwan (2010).

4.3.2 Infrastructure Parameters Maps for Rural Agricultural Development

The infrastructure parameter layers used were the markets, the road network, electric power supply, and credit institutions. These layers were overlay into the Arc GIS 10.3 software to generate the suitability map for sustainable rural agricultural development of the area (Figure 4.1). The integration of the AHP and the GIS provides farmers with a good decision-making tool to decide valuable sites for maximum production (Ayodele et al., 2018). From the analysis, the land area was classified as 496,198.73 ha having sufficient sustainable infrastructure for rural agricultural development, 105,414.83 ha having moderate infrastructure, and 63,733.74 ha was regarded not adequately covered by sustainable rural infrastructure for agriculture development (Table 4.4). The areas closer to towns and cities on Negeri Sembilan were the areas that were favored with adequate and moderate infrastructures, while the remote hilly areas were less and inadequately covered. Agricultural activities in these marginalized areas in terms of infrastructures may not do well, although they may be naturally endowed with good physical environment.

Rural agriculture production was largely influenced by infrastructure facilities (Fakayode et al., 2008). From the results, market ranked first as an infrastructural parameter for rural agriculture. Without market the output from the farm may be low. Most of the agricultural products demand ready market to be sold. These markets also provide avenues where the farm inputs such as fertilizers, herbicide, and farm

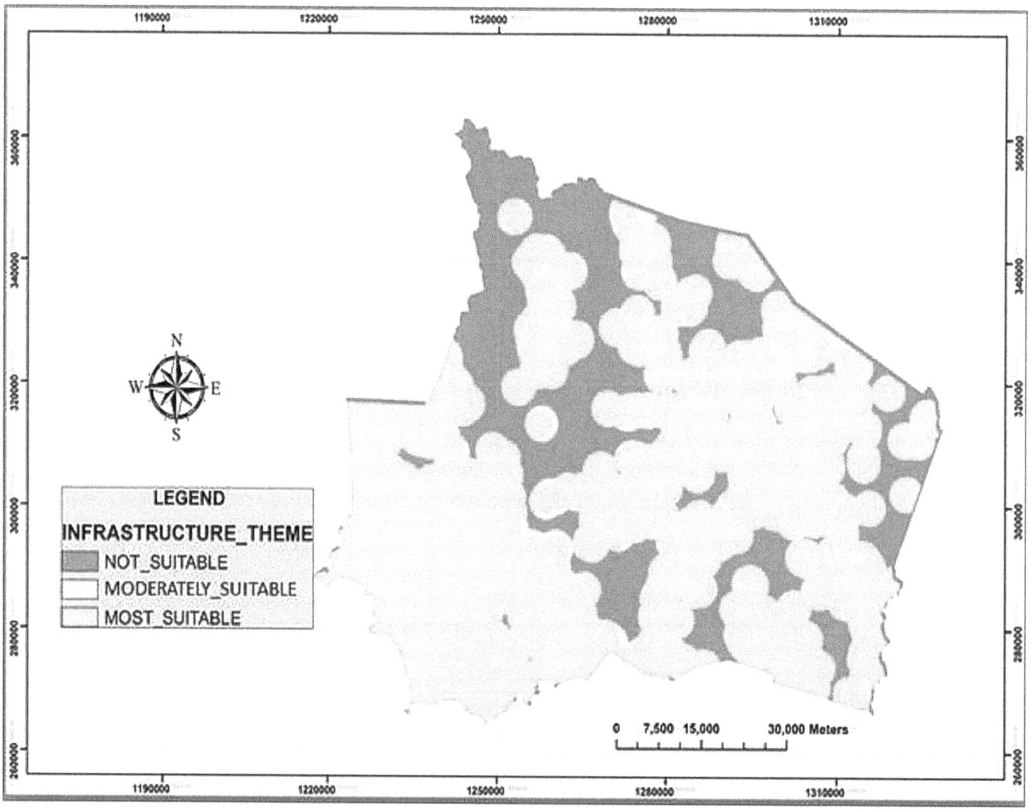

FIGURE 4.1 The infrastructure parameter suitability map for rural agriculture development in Negeri Sembilan, Malaysia.

TABLE 4.4
Area of Land with the Level of Infrastructures for Rural Agriculture

Area of Land (ha)	Level of Infrastructure Coverage	Level of Suitability for Rural Agriculture
496,198.73	Sufficient infrastructure	Most suitable
105,414.83	Moderate infrastructure	Moderately suitable
63,733.74	Not adequate infrastructures	Not suitable

machines can be obtained (Sheldon, 2017). Most of the rural goods are brought for sale by the farmers to the Seremban market.

One important infrastructure facility for rural agricultural productivity is the road network. From the AHP result, the road network parameter ranked second, which indicates the level of priority it has in the lives of the experts and farmers in Negeri Sembilan. Narayanamoorthy and Hanjra (2006) pointed out that government investments on roads and rural developments have by far great impact on agricultural productivity. Government expenditures on education have impact on poverty reduction and productivity, but expenditures that go to road construction and maintenance have direct impact on agriculture growth (Fan et al., 2000). Road network increases accessibility to markets and rural areas, reduction in the cost of transportation, therefore resulting to greater agricultural output (Negi et al., 2018). In the study area, there are major roads that connect the area; these roads boast transporting agricultural inputs and output between the rural and urban area. The major road connecting Seremban and Kuala Lumpur transports both farmers and buyers from the rural to the urban regions.

Electricity is an indispensable infrastructure for rural agriculture. From the AHP analysis, electricity ranked third in priority for agricultural development as suggested by experts in the study area. This result agreed with the study by Jain et al. (2009) that electricity and credit organizations have positive influence on the adoption of technologies for agriculture productivity. In Malaysia, electricity generation capacity connected to the Malaysian National Grid is 22,858 MW, with a maximum demand of 17,788 MW as of April 2016 according to Suruhanjaya Tenaga (Tenaga, 2008). Electricity powers most of the farm equipment, preserve perishable crops in a refrigerator, and aerate fish ponds. In Negeri Sembilan, many of the remote areas are neglected in terms of electricity and partly the unwillingness of many land owner to allowed their land for power installation (Sabo et al., 2017).

Credit institutions and organizations are another important infrastructure facility for rural agriculture activities. Credit institutions will reduce from the farmer the stress and cost of borrowing money for their farm inputs (Narayanamoorthy and Hanjra, 2006). Credit facilities, which have close links to the market, have a pivotal influence to rural agriculture; farmers need money to purchase most of the farming machines and other farm inputs, and access to credit institutions and market helps in doing their farm work (Linh et al., 2019).

4.4 Conclusion

It is not a debatable fact that infrastructure is still poor and inadequate specifically in many rural areas of Malaysia. The exact connection between infrastructural facilities and rural agricultural development was not known. Against this background, this chapter attempted to better understand the connection between infrastructure facilities and rural agricultural development. The study was to evaluate the importance of infrastructure facilities for rural agriculture development in Negeri Sembilan, Malaysia. An integration of AHP and GIS methods was applied to analyze infrastructural facilities and agricultural development. The methods have been effective in analyzing the infrastructural facilities for rural agricultural development. The expert's judgment in identifying the parameters was consistent with the CR = 0.4. The infrastructure facilities identified were market, roads, electricity, and credit facilities which have great influence on sustainable rural agriculture development. The AHP pairwise comparative matrix showed that market ranked first among the parameters, followed by road network, and electricity and credit institution took the third position in the order of their priority and weightage. Given a good natural environment and all the necessary technology without adequate infrastructural facility, agricultural productivity will plummet. The study has established that there are disparities in the distribution of infrastructural facilities in the region. From the map generated, this can be clearly seen. Government and stakeholder in rural development may use this in decision-making concerning infrastructure and rural agricultural development in the study area.

REFERENCES

Adnan, N., Nordin, S. M., Rahman, I., and Noor, A. (2017). Adoption of green fertilizer technology among paddy farmers: a possible solution for Malaysian food security. *Land Use Policy, 63*, 38–52.

Akram, W.,and Hussain, Z. (2008). Agricultural credit constraints and borrowing behavior of farmers in rural Punjab. *European Journal of Scientific Research*, 23(2), 294–304.

Alam, M., Siwar, C., Murad, M. W., Molla, R., and Toriman, M. (2010). Socioeconomic profile of farmer in Malaysia: study on integrated agricultural development area in North-West Selangor. *Agricultural Economics and Rural Development, 7*(2), 249–265.

Ayodele, T., Ogunjuyigbe, A., Odigie, O., and Munda, J. (2018). A multi-criteria GIS based model for wind farm site selection using interval type-2 fuzzy analytic hierarchy process: the case study of Nigeria. *Applied Energy, 228*, 1853–1869.

Bashir, M. K., Mehmood, Y., and Hassan, S. (2010). Impact of agricultural credit on productivity of wheat crop: evidence from Lahore, Punjab, Pakistan. *Pakistan Journal of Agricultural Science, 47*(4), 405–409.

Baydar, N., White, M. J., Simkins, C., and Babakol, O. (1990). Effects of agricultural development policies on migration in peninsular Malaysia. *Demography, 27*(1), 97–109.

Bwadi, B. E., and Mustafa, F. B. (2019). Site suitability analysis of infrastructure facilities for giant freshwater prawn farming emerging technologies. In *Environment and Research for Sustainable Aquaculture*. IntechOpen.

Fakayode, B. S., Omotesho, O., Tsoho, A. B., and Ajayi, P. D. (2008). An economic survey of rural infrastructures and agricultural productivity profiles in Nigeria. *European Journal of Social Sciences, 7*(2), 158–171.

Fan, S., Hazell, P., and Thorat, S. (2000). Government spending, growth and poverty in rural India. *American Journal of Agricultural Economics, 82*(4), 1038–1051.

Ghosh, S., Manna, S., Sahu, N., Dutta, A., and Goswami, R. (2017). Social, economic and production characteristics of freshwater prawn, *Macrobrachium rosenbergii* (De Man, 1879) culture in West Bengal, India. *Aquaculture International, 25*(5), 1935–1957.

Ibrahim, Y., and Razzaq, A. R. A. (2010). Homestay program and rural community development in Malaysia. *Journal of Ritsumeikan Social Sciences and Humanities, 2*(1), 7–24.

Islam, R., and Siwar, C. (2012). The analysis of urban agriculture development in Malaysia. *Advances in Environmental Biology, 6*(3), 1068–1078.

Jain, R., Arora, A., and Raju, S. (2009). A novel adoption index of selected agricultural technologies: linkages with infrastructure and productivity. *Agricultural Economics Research Review, 22*(1), 109–120.

Kamal, E. M., and Flanagan, R. (2012). Understanding absorptive capacity in Malaysian small and medium sized (SME) construction companies. *Journal of Engineering, Design and Technology, 10*(2), 180–198.

Linh, T. N., Long, H. T., Chi, L. V., Tam, L. T., and Lebailly, P. (2019). Access to rural credit markets in developing countries, the case of Vietnam: a literature review. *Sustainability, 11*(5), 1468.

Manjunath, S., and Kannan, E. (2017). Effects of rural infrastructure on agricultural development: a district level analysis in Karnataka, India. *Journal of Infrastructure Development, 9*(2), 113–126.

Narayanamoorthy, A., and Hanjra, M. A. (2006). Rural infrastructure and agricultural output linkages: a study of 256 Indian districts. *Indian Journal of Agricultural Economics, 61*(3), 444–459.

Negi, D. S., Birthal, P. S., Roy, D., and Khan, M. T. (2018). Farmers' choice of market channels and producer prices in India: role of transportation and communication networks. *Food Policy, 81*, 106–121.

Ngah, I. (2010). Overview of regional development in Malaysia. *Paper presented at the International Conference on Regional Development: Vulnerability, Resilience and Sustainability*, Universitas Diponegoro, Semarang.

Pinstrup-Andersen, P., and Shimokawa, S. (2006). Rural infrastructure and agricultural development. *World Bank.*
Saaty, T. L. (1980). The analytic hierarchy process: planning. *Priority Setting. Resource Allocation.* MacGraw-Hill, New York International Book Company, 287.
Saaty, T. L. (1994). Highlights and critical points in the theory and application of the analytic hierarchy process. *European Journal of Operational Research, 74*(3), 426–447.
Sabo, M. L., Mariun, N., Hizam, H., Radzi, M. A. M., and Zakaria, A. (2017). Spatial matching of large-scale grid-connected photovoltaic power generation with utility demand in Peninsular Malaysia. *Applied Energy, 191*, 663–688.
Sangwan, S. (2010). Occasional Paper-53. Infrastructure for Agriculture Development National Bank for Agriculture and Rural Development, Mumbai.
Sheldon, I. M. (2017). The competitiveness of agricultural product and input markets: a review and synthesis of recent research. *Journal of Agricultural and Applied Economics, 49*(1), 1–44.
Tenaga, S. (2008). Statistics of Interim on the Performance of the Electricity Supply in Malaysia for the First Half of Year of 2007. Viewed June, 10, 2009.

5

Gender Mainstreaming and Women Empowerment in Small Holdings

A Sustainable Way for Better Livelihood

Riti Chatterjee[1], Deepa Roy[2], Gunja Kumari[3], and Prithusayak Mondal[3]
[1]*Department of Agricultural Extension, Faculty of Agriculture, Bidhan Chandra Krishi Viswavidyalaya, Nadia, West Bengal, India*
[2]*Department of Agricultural Extension, Faculty of Agriculture, Uttar Banga Krishi Viswavidyalaya, Coochbehar, West Bengal, India*
[3]*Regional Research Station (Terai Zone), Uttar Banga Krishi Viswavidyalaya, Coochbehar, West Bengal, India*

CONTENTS

DOI: 10.1201/9781003164968-6

5.1 Introduction

The idea of gender orientation should be seen unmistakably as a crosscutting sociocultural variable. It is an all-encompassing variable as sexual orientation can likewise be applied to any remaining cross-cutting factors, for example, race, class, age, ethnic group, and so forth. Sexual orientation frameworks are set up in various sociocultural settings that figure out what is normal, permitted, and esteemed in a woman/man and young female/kid in these settings. Gender roles or responsibilities are found out through socialization measures; they are not fixed however are alterable over the long haul and between societies. Gendered frameworks are standardized through instruction frameworks, political and financial frameworks, enactment, culture, and conventions. In using a gender role approach, the attention is not on individual women and men, rather on the framework, which decides gendered duties, access to and control over assets, and decision-making authorities.

It is additionally essential to stress that the gender concept is not compatible with women. Gender alludes to the two, women and men, and the relations between them. Lately, there has been a lot more grounded direct spotlight on men in exploration on gender equality points of view. There are three fundamental methodologies taken in the expanded spotlight on men: firstly, right off the bat, the need to distinguish men as partners for gender equality and include them more effectively in this work. Secondly, the acknowledgment that gender equality is unimaginable except if men change their perspectives and conduct in numerous regions, for instance corresponding to conceptive rights and well-being. Furthermore, thirdly, the gender equality setup in numerous settings is impossible for men just as for women – making unreasonable requests on men and expecting men to carry on in barely characterized ways. A lot of fascinating examination is being attempted, by the women and men, on male characters and manliness. The expanded spotlight on men will have a huge effect on future systems for working with gender perspectives in development.

5.2 Why Should Gender Be Taken into Consideration in Agriculture?

Agriculture is considered as the backbone of the economy for most of the countries contributing a good portion of gross domestic product (GDP). In agriculture sector, women serve critical roles and are recognized as the strategic contributors as they make up almost 50% of the agricultural labor force in sub-Saharan Africa and had an increase from about 45% since 1980. They fulfill important roles in agriculture as farmers and farm workers, horticulturists, businesswomen, entrepreneurs, and community leaders that lead to the development of women economies.

Despite this significant contribution, farm women are often excluded from agricultural policies. At the same time, discriminatory laws and practices deprive them of their entitlement, land ownership, their various rights, and their livelihoods. The disparities covering women in agriculture also influence access to financing and agricultural resources. Sometimes, their return from their labor contribution becomes a misfit. Women do farming just as men when women farmers have the same degree of access to and control over productive resources, services, and economic facilities. These may create a significant increase in agricultural output and socioeconomic sustainability, mitigation of climate change effects contributing toward reduction in the number of poor people (Figure 5.1).

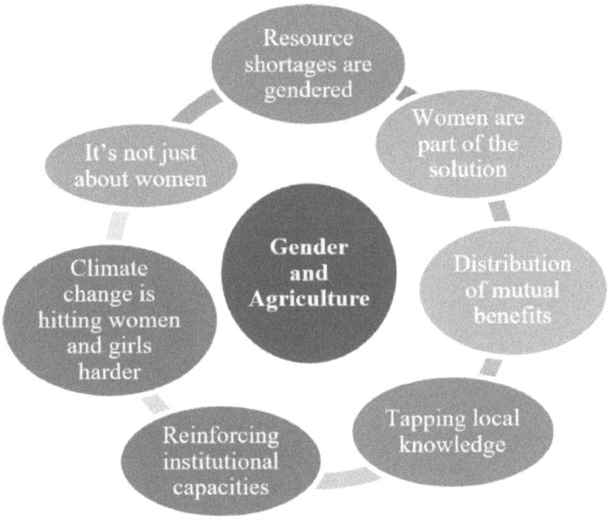

FIGURE 5.1 Some ways of gender mainstreaming in agriculture.

5.3 Why Is Gender Mainstreaming Important?

As per the agreed conclusions of UN ECOSOC (1997), gender mainstreaming is the process of assessing the implications for women and men of any planned action in legislation, policies or programs, and at all other levels. This entails strategic action for making women's as well as men's concerns and experiences as an integral dimension of the implementation, monitoring, and evaluation of development policies and programs in all societal, economic, and political spheres. The ultimate goal is to accomplish gender equality, i.e., women and men should benefit equally at all levels by identifying gaps in gender equality and also developing strategies to close those gaps.

5.3.1 Dimensions of Gender Mainstreaming

Gender mainstreaming integrates gender perspective to the context of the international and national policies ensuring the representation of women and men in the policy area. However, gender representation and gender response dimensions must be included in each and every phase of the policy-making process.

5.3.2 Gender Mainstreaming Tools and Examples (UNDP 2013)

There are a number of tools and dimensions of gender mainstreaming (Figure 5.2); however, six tools described below are generally used at different times throughout any gender issue-related project. The tools are as follows: (i) Harvard gender analysis framework, (ii) UNDP gender marker, (iii) UNDP checklist for gender mainstreaming in any project proposals, (iv) UNDP checklist for gender mainstreaming in work planning, (v) UNDP monitoring and evaluation, and (vi) reporting guidelines, reflection tool derived from UNDP's eight point agenda for women's empowerment and gender equality in crisis prevention and recovery.

5.3.3 Gender Equity or Equality?

Gender equity can be defined as an element of interpretation of social justice, based on tradition, custom, or culture, which is often found to be detrimental to women. UNFPA (2005) stated gender equity as the process of being fair to both women and men. This gender equity leads to gender equality. Gender

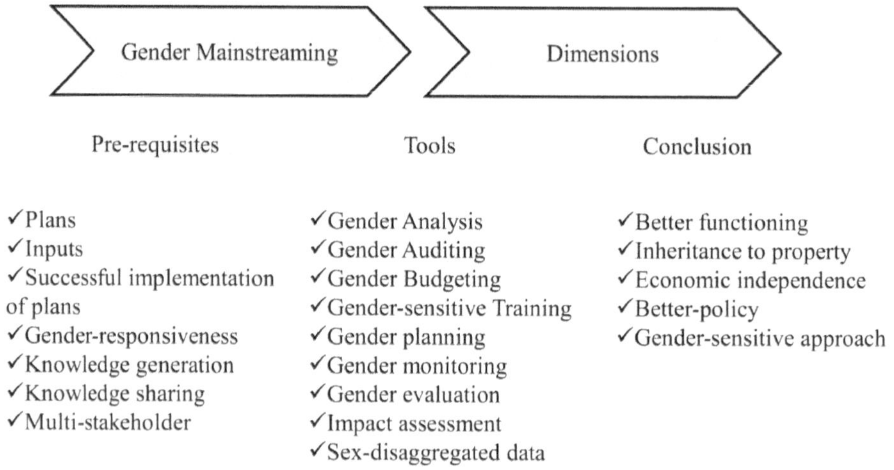

FIGURE 5.2 Gender mainstreaming: prerequisites and tools required.

equality must ensure the equal enjoyment of socially valued goods, opportunities, resources, and rewards by both women and men. It is mostly the women, who are excluded from any decision-making process and having no or little access to economic, social resources, and other opportunities, thus creating gender inequality.

5.4 Women in Agriculture: Global, National, and Local Scenario

Though women are involved in the whole crop production process, they never get the recognition as women farmers. Most of the time, they work on family farms or remain underpaid or unpaid. Despite varying patterns of participation, women's roles in agriculture remain critical both in terms of the quantity of their effort as well as the quality and nature of labor performed (Sharma 2012).

Women make indispensable contributions to the agricultural and women economies in each and every developing country (Ghosh and Ghosh 2014). Their roles vary significantly among and within regions and are changing rapidly in every part of the world, where economic and social factors are operating within the agricultural sector. Women manage everyday household chores and pursue multiple livelihood functions, i.e., crop production, rearing animals, caring family members and preparing food, work wage laborers in agriculture sector or any other women enterprises, and sometimes engage in trade and marketing side also (FAO 2011).

5.4.1 Do Women Make Up 60–80% of the Total Labor Force in Agriculture Sector?

The statement is often found that women contribute 60–80% of the agricultural labor force in developing countries. This seems to have originated in an earlier study carried out by the United Nations Economic Commission for Africa (UNECA), which stated: 'Few persons would argue against the estimate that women are responsible for 60–80 percent of the agricultural labour supplied on the continent of Africa ....' (UNECA 1972). In India, agriculture sector employs 80% of all economically active women, 33% of the agriculture labor force, and 48% of the self-employed farmers (Abbott et al. 2015).

5.4.2 Time Devoted by Women Farmers in Agricultural Activities

According to FAO (2011), time-use studies embrace a more complete record of time used by people than are accessible from the workforce measurements. Time-use studies that cover every agrarian action uncover extensive variety across nations, and some of the time, within the nations.

In India, women are extensively involved in both household and agricultural field activity. However, the nature and extent of their involvement differ with the need of the family and farm. Logeswari and Thiruchenduran (2016) in their study noted that the mode of women participation in agricultural activities varies with the landowning status of farm household. Their roles range from manager to landless laborers. In India, women's average contribution to overall farm production is estimated at 55–66% of the total labor (Vandana 1991).

5.4.3 Women Farmers' Contribution to Global Food Security

Doss (2002) gave a definite examination of the reasonable and observational difficulties engaged in assessing the portion of food produced by the women farmers. Difficulties incorporate, among others, (i) characterizing and estimating food creation, (ii) characterizing the assets to be remembered for the figuring, and (iii) assigning those assets as indicated by the gender of the individual who controls them. An outline of the accessible proof, utilizing an assortment of definitions and procedures, finds that the commitment of women farmers in farming is most likely considerable yet cannot be assessed with any level of insightful thoroughness.

5.4.4 Women in Livestock Rearing

Women are playing a pivotal role in domesticating the major livestock species. The term 'livestock' means the practice of keeping animals to kill as and when needed. In Asia, more than three-quarters of livestock-related tasks like feeding, taking care of young and sick animals, and milking are the sole responsibility of women (Tipilda and Kristjanson 2009). Here in India, the livestock industry is dominated by women who are providing nearly 55% of employed livestock-farming labor and more than 77% of the work involved in taking care of animals. Furthermore, 93% of people employed in dairying are women (RNCOS 2006). Women share their duty with men and youngsters for the consideration of creatures, and specific species and kinds of action are more connected with women than men. For instance, women frequently have a conspicuous function in overseeing poultry and dairy animals (Okali and Mims 1998) and think about different animals that are housed and taken care of inside the estate. At the point when assignments are isolated, men are bound to be engaged with lodging and grouping of grazing animals, and in advertising of items if women's mobility is obliged. Bravo-Baumann (2000) also found that the impact of women is solid in the utilization of eggs, milk, and poultry meat for home utilization and they regularly have command over showcasing and the pay from these items.

Given the more restricted capacity of women to begin their own organizations, they will in general become workers instead of independently employed. Women are obvious in any place where meticulous semi-talented work is required, especially, in exercises like the creation of day-old chicks, the arrangement of administrations, and butchering, preparing, and retail. However, almost no data is accessible about the degree of their association contrasted with that of men, or their control and access over assets.

5.4.5 Women in Fisheries and Aquaculture

In 2008, almost 45 million individuals worldwide were straightforwardly connected with, full-time or part-time, in the fishery primary sector (FAO 2014).

Findings of women engaged in aquaculture, particularly in Asia where aquaculture has a long convention, show that the commitment of women in labor is regularly more noteworthy than men's despite the fact that it is right around the total absence of full-scale level aquaculture related sex-disaggregated information. Women are accounted for to comprise 33% of the village level aquaculture labor force in China, 42% in Indonesia, and 80% in Vietnam (Kusabe and Kelker 2001). In certain nations, women have become significant business visionaries in fish handling; truth be told, most of the fish preparing is done by women, either in their own family level enterprises or as compensation workers in the large-scale processing industry.

5.4.6 Women Farmers in Modern Contract Farming

The rise of modern supply chains is significantly changing the pathway through which food and high-value farming items are delivered and exchanged in nonindustrial nations, with significant impacts for provincial women. There are many case studies showing that women are less likely than men to be included in contract farming schemes as signatories in the case of contract farming (Brewin and Murphy 2019). As a result, female farmers remain disempowered, with no control over the income earned from neither the contract nor how it is spent. This has knock-on effects for the household also. Some studies have also found that men are inclined to spend the income from contract farming on individual rather than family needs (Schneider and Gugerty 2010).

High-value contract farming may bring about diminished access to assets for female farmers worried about resource for food production and eventually lead to the weakening of the food security circumstance of women and the generations (Bravo-Baumann 2000).

5.5 Key Challenges in the Pathway

Women in the country world face numerous imbalances. Presently, about 33% of the area is involved by women; however, the information demonstrates that women farmers are more modest, less productive and have more challenges getting to credit and advancement. There is critical disparity regarding the work market as far as admittance to paid business, and consequently, numerous women need to pick independent work, with all the troubles it involves. Firing up a business in a rural area is substantially more troublesome than in a city, and there are numerous inadequacies out in the open administrations, correspondence, advances, and admittance to financing. In fact, there is a huge piece of the rural world in monetary avoidance, particularly women. As to administrations, there have been numerous cuts since the monetary emergency, which have emphasized the generally existing deficiencies in schooling, well-being, and transport. Moreover, there is another ramification for women. At the point when the government does not supply the essential administrations, the most important thing is that women are the ones responsible for supplanting the state in its obligation of care. At the point when a school cafeteria in a town is shut, it is women who arrange them to take care of the youngsters, and when a well-being place is shut, it is women who deal with the wiped-out one (Valdivia 2019).

5.5.1 Social Norms

According to Sharma et al. (2016), economic and social issues are firmly interwoven, one strengthening the other. In India and in the majority of South Asia, socioeconomic and strict elements keep on hindering the strengthening of women in fluctuating degrees. These limitations show themselves at different levels and structures for women. Man-centric structures and male predominance keep women at the outskirts of any dynamic programs. Additionally, the dynamic capacities of women are obliged by the low degree of instructions and total customarily ignorance. This speaks to a recurrent wonder of avoidance of women from dynamic cycles and delivers them undermined: socially and economically. In the greater part of the towns, notwithstanding a couple of special cases, ladies are unskilled and the current age of young women has begun going to class. This is hosted by the inaccessibility of schools in the region from the past days.

5.5.2 Patriarchal Institutions

Ladies' admittance to credit, which is basic for horticulture creation measure, is additionally restricted in light of the fact that both formal and casual credit foundations look for unmistakable insurance for advance. Admittance to institutional credit is a significant limitation as a larger part of women in India doesn't have property proprietorship or even rights. Low degrees of mindfulness, schooling, and restricted versatility outside the home further fuel the present circumstance. Notwithstanding the above-examined requirements looked at by women, the Thar district by uprightness of its agrobiological imperatives and restrictions presents further difficulties. The area gets negligible precipitation; water is scant in any event, for family use, drinking, and for domesticated animals, prompting an enormous portion of time spent by women in getting water from significant distances. Little youngsters and women have the essential duty of getting water, bringing about young ladies exiting from schools to satisfy these undertakings. The time spent on this movement and the drudgery engaged with bringing water restricts their experience of self-care affecting adversely their well-being and prosperity. Despite the fact that there have been enhancements in the accessibility of water assets, women actually invest critical measures of energy in overseeing water assets inside families, and to that degree, their occasion to take an interest in monetarily enabling exercises stays confined. The legislatures and other actualizing organizations consequently must be conscious of this and offer creative choices for far-off women's people groups to have opportune admittance to new information, mechanical enhancements, and institutional plans (Sharma et al. 2016).

5.6 Designing Gender-Friendly Strategies

The status of women taking an interest in decision-making just as advantages they in the long run harvest from agribusiness and associated exercises, as compared to difficulties seen by them, were found out at the beginning of the undertaking. Helpful bits of knowledge assembled through a thorough planning of status and limitations educated the venture plan and execution. A perplexing arrangement of vital choices is associated with the promoting of the rural produce that can ultimately decide the vocation status of the family. These decisions may go from the following.

In almost 50% of the example families, such choices were taken simply by male members without talking with their female individuals from the families, while in about 30% cases such choices were taken together, which by and by implies a choice taken by the male counterpart in meeting with. Just a little extent of women (~3%) really markets the produce. Farming, as a monetary and social action, includes numerous undertakings that involve an assortment of choices. These choices may go from crop inputs, timings of collect, get-togethers, and interest in well-being and training of relatives, recruiting of work and transportation of produce.

Notwithstanding the way domesticated animals are being raised and the board is generally done by women and youngsters in women cultivating networks, a totally unbalanced condition exists among guys and females concerning choices for this venture. Additionally, access, degree, and patterns in dynamic relating to regular property assets and utilization of field land are customarily taken simply by senior male individuals from the network, frequently without speaking with females. In situations where choices on the utilization of neglected land, cultivable no man's land, and town timberland wood are taken by the justly chosen town level specialists, the women's voice stays unheard and their advantages unattended.

5.6.1 Use of Gender Sensitive Tools and Interventions

There are scopes of mediations that can be presented in the dry season-stricken agribusiness in the Thar. A large number of them could be founded on automation or may require actual strength. In any case, major essential for sex mainstreaming is to apply devices and procedures that are considerably more sexual orientation touchy in application, reasonable for use by females and are practical. Presentation of preparing strategies, market access impetuses, and pay age activities should observe these elements.

Bundling of mediations should likewise be done that suits and supports women, for example, supports climate protection adaptation techniques offered through women's Self-Help Groups (SHGs) to increase yield of crop and domesticated animals.

5.6.2 Gender Inclusive Approach

Women face different imperatives in all circles of life and one limitation normally strengthens the other. Acquiring such obstructions and contriving methodologies to defeat at the phase of planning an undertaking or an activity is a significant piece of sexual orientation comprehensiveness. The upfront investment accomplished through including men from the beginning can loan genuinely necessary stimulus to sex explicit intercessions.

5.6.3 Filling the Capacity Deficit

Capacity building and awareness generation is the initial move toward strengthening and guaranteeing fair and just networks. Preparing in different strategies identifying with specific livelihoods, for example, food handling, different water the board methods, and monetary administration, should be bestowed consistently. Advancement of improved apparatuses and methods among women should be started for a huge scope. This would guarantee scattering of new advancements, improved creation, decreased remaining burden, and drudgery for women ranchers.

5.6.4 Mobilization and Awareness

The making of SHGs of women in distant towns has assisted with assembling women, which is the initial move toward mindfulness age and strengthening. It is imperative to extend the extent of this activity particularly given that there have been positively no such mediations in any of the distinguished towns before this venture. Notwithstanding sidelong development, a heightening of endeavors toward this path is required, which would involve giving more roads to cooperation among the individuals to empower conversations on different issues of concerns, handholding in setting up frameworks for exchanges, and building capacities with regards to dealing with the frameworks.

5.6.5 Facilitation through Addressing Socioeconomic and Political Context

The socioeconomic milieu in the women zones is not favorable for the mobility and interest of women in economic activities. There is a need to work around different settings, organizations, and gatherings to guarantee that women approach and occasion to assemble their ability and partake in an impact dynamic. Helpful admittance to water, kindling, and markets loosens up women's time requirements and can acquire wanted change.

5.6.6 Political Participation and Decision-Making

To guarantee that the choices taken at the community level in nearby self-government organizations advantage women, a powerful portrayal of women in neighborhood self-administration structures is required. It additionally adds to mainstreaming sexual orientation in any choice taken for the community via having their perspectives included during the time spent dynamic at network and family levels.

5.6.7 To Reach the Unreachable

To guarantee that the decisions are taken at the community level in local self-government institutions advantage women, a compelling portrayal of women in neighborhood self-administration structures is required. It likewise adds to mainstreaming sexual orientation in any choice taken for a network via having their perspectives included during the time spent dynamic at network and family levels. Neighborhood bodies could be instrumental in invigorating, utilizing, and operationalizing government plans to help women in different areas.

5.6.8 Larger Advocacy Agenda

Public and state-level agribusiness approaches perceive the requirement for sex mainstreaming in rural cycles and there are a few arrangements and projects intended to encourage them. It is, nonetheless, important to guarantee that such expectations are granted with a genuinely necessary operational and utilitarian nature. Promotion should be worked around all these arrangement components that should have been changed over into feasible yields.

5.7 Gendered Approach in Ensuring Sustainable Livelihood

5.7.1 Development through Group Approach

Women make important commitments to agrarian cycle, family pay, food security, and guaranteeing reasonable vocation for the families. The activity of formation of self-improvement gatherings of women in couple of towns has given stage to women to communicate, talk about regular issues, unite their sentiments, and settle on it at network level.

The essential reason for the SHGs was the part of month-to-month reserve funds by the individuals for their monetary strengthening. This will possibly prompt other aggregate chances and exercises, making of resources and control of assets by them. Different women likewise were energized and propelled by SHGs and have utilized this occasion to learn and assemble their abilities. There have been various activities presented and executed through SHGs. Advancement of women-driven exercises, for example, goat raising, kitchen gardens, plantation the board, and formation of gathering resources, including capacity drum for seed banks, have energized the exercises of women's gatherings. During the gatherings, women examined different issues concerning their jobs, including cultivating and unified exercises (e.g., creature farming). The difficulties looked by women and men in women regions of parched Rajasthan were examined (e.g., giving instruction to their kids) alongside circumstances present and finding conceivable arrangements or roads for job upgrade.

5.7.2 Capacity Building and Income Generating Activities

Gender-inclusive capacity building of individual ranchers and the farming community as a whole that will fill the information holes and does arrangement on promising advancements has been a critical component of the task. While limited advancement was done on improved farming practices at the village level with a specific spotlight on women, explicit aptitudes-based training was likewise coordinated for women's gatherings. Limit reinforcing of women on elective types of pay age is commenced on the essential principles of upgrading accessible assets, conventional exercises, and monetary feasibility of the undertaking. Advancement of development of therapeutic plants, for example, Sankhpuspi, Arna, and Jivanti inside the conventional cultivating framework is an illustration of utilizing the dry and bone-dry conditions. Building limit of women in developing such plants adds to monetary flexibility at the family unit level. The homestead women were likewise important for limit reinforcing exercises: on soil well-being and ripeness the board, utilization of bio composts and pesticides, bone-dry cultivation, creature farming, water gathering, setting up and sustaining agriculture units, the executives of network field lands, seed creation and capacity, improved assortments of seed and different strategies for editing for wise and ideal utilization of water assets in the field, for example, line trimming. Other than ICRISAT and GRAVIS, different accomplices, for example, ICAR-Central Arid Research Institute (ICAR-CAZRI) in Jodhpur, KVK Barmer, National Seed Corporation, Dabur India Ltd., and state line offices assumed key function in fortifying farmer's ability. On an average, women contribute 30–40% of the total members in these trainings.

5.8 Mainstreaming Gender and Fulfillment of Nutritional Security in Farm Families

In Asia, rural women play a critical role in supporting the main three pillars of food security like food production, economic access to available food, and nutritional security (FAO 2005). However, women's productive roles have historically undervalued their contribution. Bridge (2014) stated that food and nutrition insecurity is a gender justice issue. Women and girls are the most disadvantaged by the inequitable global economic processes because of their low status and lack of access to resources.

5.9 Global Innovations in the Light of Gender Mainstreaming and Women Empowerment in Small Holdings

5.9.1 'Farming for Food' to 'Farming for Money' Approach: Story of Annapurna Self-Help Group in West Bengal

Bimala Barman from Paschim Borochowki town is an energetic individual from the Annapurna women's self-improvement gathering. She takes care of the gathering's rice bank that is set up in her home. They gather overflow rice from every one of their individuals' rice produce and save it in their rice bank. They offer this to individuals in the town, including their individuals, during the lean season when individuals need more to eat. The equivalent is returned after the gathering. The Annapurna self-improvement gathering grows a scope of vegetables together utilizing self-arranged vermicompost. The individuals, being very much aware of its advantages, show solid abhorrence toward synthetic substances. They are certain that their confidence in the regular methods of keeping the dirt sound and rich will pay off. Bimala's neighbor Kalpana Sutradhar has a farm, which reflects a natural/forest ecosystem – where there is enough food for both men and animals. She grows multiple crops that complement, so she has food all year. On a little plot of land, she has set a few pots of water covered under the ground, which supply steady controlled dampness to the soil. The remainder of the homestead is watered with water, which she gets in a lake that is likewise home to numerous sorts of fish. She likewise gets ready fertilizer utilizing special earthworms (vermi), called vermicompost. This – and more has added to her certainty – she has figured out how to take care of her family with healthy sound food without relying much upon the market (Muinga 2020).

5.9.2 Potato Farming: A New Way to Be Independent as a Woman Farmer

In the wake of expending her whole time on earth as somebody's girl and afterward a spouse, Shamima Begum from West Bengal's Molaypur town is presently cutting her own way of life as a farmer. She lately offered 12 tons of potatoes to PepsiCo India, and with the income created, she took care of her significant other's obligation. She likewise bought family unit fundamentals like the ice chest and cooking gas and saved aside some cash for her girl's clinical profession (Karelia 2020).

5.10 Conclusions

Agriculture unarguably establishes an integral piece of improvement talk in India. It is certain that agricultural development in India is significantly dependent upon aptitudes and the limit of farmers and agrarian workforce. There is notwithstanding, an innate gender imbalance noted in this labor force. Farmers' support in agribusiness is a basic component of agrarian economies of the developing world. The commitments of women, the non-acknowledgment of their job, and restricted decision-making power keep on being an issue of concern. Social and economic costs of overlooking this imperative part of economy may imbalance improvement gains for the country.

REFERENCES

Abbott, P., Mutesi L. and Norris E. 2015. Gender Analysis for Sustainable Livelihoods and Participatory Governance. Oxfam International, Kigali. Retrieved from: https://www.google.com/url?sa=t&rct=j&q=&esrc=s&source=web&cd=&ved=2ahUKEwiShMHK69fvAhVUzDgGHT7wDPAQFjAAegQIAxAD&url=https%3A%2F%2Fcore.ac.uk%2Fdownload%2Fpdf%2F43540241.pdf&usg=AOvVaw0eQ3x_b7Y33Fjkifp7CiCh

Bravo-Baumann, H. 2000. Gender and Livestock. Capitalisation of Experiences on Livestock Projects and Gender. Working Document. Swiss Development Cooperation, Bern. Retrieved from: http://www.bridge.ids.ac.uk/sites/bridge.ids.ac.uk/files/docs_genie/sdc/Gender_and_Livestock.doc

Brewin, S. and Murphy, S. 2019. The Farmer and Her Husband: Legal Innovations for Women in Contract Farming. Investment in Agriculture Policy Brief #8. International Institute for Sustainable Development. Retrieved from: https://www.iisd.org/publications/farmer-and-her-husband-legal-innovations-women-contract-farming

Bridge. 2014. Gender and Food Security. Towards Gender-Just Food and Nutrition Security Overview Report. ISBN: 978-1-78118-203-1.

Doss, C. 2002. Men's Crops? Women's Crops? The Gender Patterns of Cropping in Ghana. World Development, 30(11): 1987–2000.

FAO. 2005. Rural Women and Food Security in Asia and the Pacific: Prospects and Paradoxes. RAP Publication 2005/30. ISBN: 974-7946-80-7.

FAO. 2011. The Role of Women in Agriculture. ESA Working Paper No. 11-02. Retrieved from: https://www.empowerwomen.org/en/resources/documents/2013/9/the-role-of-women-in-agriculture-esa-working-paper-no-1102?lang=en

FAO. 2014. The Role of Women in Fisheries. Retrieved from: https://www.youtube.com/watch?v=CB1Vsw3bIwA

Ghosh, M. M. and Ghosh, A. 2014. Analysis of Women Participation in Indian Agriculture. Journal of Humanities and Social Science, 19(5): 01–06. DOI:10.9790/0837-19540106, Corpus ID: 54812850

Karelia, G. 2020. Women Used Potato Farming to Become Independent. The Better India. Retrieved from: https://www.thebetterindia.com/244135/west-bengal-potato-farming-women-farmers-empowerment-india-gop94/

Kusabe, K. and Kelker, G. Eds. 2001. Gender Concerns in Aquaculture in Southeast Asia. Gender Studies Monograph 12. Gender and Development, Studies, School of Environment Resources and Development. Asian Institute of Technology, Bangkok. Retrieved from: http://www.fao.org/3/ad070e/ad070e08.htm

Logeswari, S. and Thiruchenduran, S. 2016. Empowerment of Women Farmers for Agricultural Development. Imperial Journal of Interdisciplinary Research, 2(8): 991–992.

Muinga, G. 2020. Why Should Gender Matter in Agriculture? Leadership. Opinion. Retrieved from: https://leadership.ng/2020/09/29/why-should-gender-matter-in-agriculture/

Okali, C. and J. Mims. 1998. Gender and Smallholder Diary Production in Tanzania. Report to the Livestock Production Programme of the Department for International Development (DFID): Appendix 1 and 2, pp. 37–38.

RNCOS, 2006. Indian Livestock Industry – An Industry Analysis. RNCOS Research. Retrieved from: http://www.prweb.com/releases/2006/05/prweb380863.htm

Schneider, K. and Gugerty, M. 2010. Gender and Contract Farming in Sub-Saharan Africa Literature Review. University of Washington, March, pp. 1–2. Retrieved from: https://epar.evans.uw.edu/research/gender-contract-farming-sub-saharan-africa

Sharma, K. 2012. Role of Women in Informal Sector in India. Journal of Humanities and Social Science, 4(1): 29–36. DOI:10.9790/0837-0412936

Sharma, N., Kumar, S., Ravula, P. and Tyagi, P. 2016. Mainstreaming Gender and Empowering Women in Agriculture in the Thar Region of India. Research Report 69. Patancheru 502 324. International Crops Research Institute for the Semi-Arid Tropics, Telangana. 24 pp. ISBN: 978-92-9066-591-5.

Tipilda, A. and Kristjanson, P. 2009. Women and Livestock Development: A Review of the Literature. ILRI Innovation Works Discussion Paper 01-09. ILRI, Nairobi. Retrieved from: http://www.ilri.org/innovationworks

UNDP. 2013. United Nations Development Programme. Gender Mainstreaming Made Easy: Handbook for Programme Staff. Somalia. Retrieved from: http://library.unccd.int/Details/fullCatalogue/681

UNECA (United Nations Economic Commission for Africa). 1972. Women: The Neglected Human Resource for African Development. Canadian Journal of African Studies/Revue Canadienne des Études Africaines, Special Issue: The Roles of African Women: Past, Present and Future, 6(2): 359–370.

UN ECOSOC. 1997. UN Economic and Social Council Resolution 1997/2: Agreed Conclusions. Retrieved from: https://www.refworld.org/docid/4652c9fc2.html

UNFPA. 2005. The Promise of Equality: Gender Equity, Reproductive Health and the Millennium Development Goals. ISBN: 0-89714-750-2.

Valdivia, A. G. 2019. The Challenging Life Of Female Farmers: Why A Gender Mainstreaming Is Necessary in Agriculture. Retrieved from: https://www.forbes.com/sites/anagarciavaldivia/2019/03/18/the-challenging-life-of-female-farmers-why-a-gender-mainstreaming-is-necessary

Vandana, S. 1991. Most Farmers in India are Women. FAO, New Delhi.

6

A Pandemic Resilient Framework for Sustainable Soil Health and Food Security

Response beyond COVID-19

Sudip Sengupta[1,4], Shubhadip Dasgupta[1,2], Kallol Bhattacharyya[1], Somsubhra Chakraborty[2], and Pradip Dey[3]
[1]*Department of Agricultural Chemistry and Soil Science, Faculty of Agriculture, Bidhan Chandra Krishi Viswavidyalaya, Nadia, West Bengal, India*
[2]*Agricultural and Food Engineering Department, IIT Kharagpur, Kharagpur, West Bengal, India*
[3]*ICAR-Indian Institute of Soil Science, Bhopal, Madhya Pradesh, India*
[4]*School of Agriculture, Swami Vivekananda University, Barrackpore, Kolkata, West Bengal, India*

CONTENTS

6.1 Introduction

Since the inception of the outbreak of COVID-19 in December 2019 (WHO, 2020), the new coronavirus in the form of severe acute respiratory syndrome, coronavirus 2 (SARS-CoV-2), is wreaking havoc throughout the world. The sudden phenomenon has crippled the global development and emerged as a

DOI: 10.1201/9781003164968-7

pandemic, affecting and killing lakhs (Hui et al., 2020). The contagious and infectious nature of the virus has culminated in serious decision-making by the governments to prioritize among life and livelihood. In the majority of the countries, the emphasis was given to sustenance of life through lockdown and social distancing has been enforced (Deaton and Deaton, 2020). Earlier pandemic experiences reveal that such quarantines have an impact on societal and economic growth (Hanashima and Tomobe, 2012) with increased incidences of hunger leading to malnutrition. Such has been the havoc, there has been plunge in global growth that halved to 1.5% and further spiraling down (*http://www.oecd.org/coronavirus; retrieved 12.06.2020*). In India, an estimated 348-million-dollar trade impact has been assessed through a UN report, which would possibly affect all the sectors in a devastating way (Kumar et al., 2020). As the agriculture and food sectors are the mainstay of sustenance, any disturbance in these sectors may culminate into disastrous famine (World Bank, 2020). The progress of the disease and the associated movement restrictions have further created labor shortage for harvest of crops, sowing of succeeding crops, or even the marketability of the produce, thereby breaking the supply chain badly. Several cases of burning down of crops or leaving them in the field to rot have been reported throughout the country, which emerges as a serious threat to the future food availability, food security, and nutritional sustainability of the exorbitantly large population. The prognosis of the situation to end is uncertain; however, what can easily be ascribed is that post COVID-19 period will not be the same as earlier.

Research to find any suitable drug or vaccine for the disease is underway; however, till date, we have no specific medications in hand. This emphasizes the need for alternative strategies to control or combat the disease. Such interventions may also enable us to tide over such pandemics in future as well. One of the most common interventions yet of tremendous significance is having a well-functioning immune system – a person with an impaired immune system has always the greatest risk of such infectious pathogens. This emphasizes adequate intake of easily assimilating sources of proteins, minerals, vitamins, etc. (Taghdir et al., 2020). The easiest available source is through food where the agricultural sector stands flag-bearer.

This necessitates a highly focused and goal-oriented research toward rearing of such food crops though proper soil and crop management practices for tackling current as well as future pandemics.

6.2 Need of Nutritional Security: Diet-Based Minerals and Vitamins to Boost Immunity

To emphasize the vehement need of nutritional security, we should apprehend the postulate of more than 2,500 years ago by Hippocrates as: '*Let food be thy medicine and not medicine be thy food.*' The experiences of the previous pandemics have revealed that the nutrient availability and intake among populations and incidence of the disease are closely interrelated especially in the developing countries, where malnutrition incidences are aplenty. Although the increasing reports of asymptomatic coronavirus patients pose significant public health issues (Al-Tawfiq, 2020), if we try to unearth and isolate this population, we can find them to be mostly healthy people. To illustrate in other words, it includes the population that has a very well-developed immune system that enables the combating potentiality and can be considered as a measure of resilience (Aman and Masood, 2020). Ensuring optimal level of nutrient through dietary intake facilitates gene expression, the activation of cells for signaling molecule generation, and even the microbial constituents in the gut to augment immune responses (Aslam et al., 2017). On the contrary, improper diet-mediated malnutrition not only increases mortality, morbidity, and the high rate of infectious diseases but also bears significant economic impacts on health-care systems. The incidence of the COVID-19 pandemic along with the associated measures of self-isolation, lockdown, and social distancing often leads to severe repercussions toward a person's social life. The sedentary behavior affects both the mental and physical health (Naja and Hamadeh, 2020) to combat which balanced and healthy diet is the prima facie necessity.

In the aforesaid pretext, several vitamins and essential nutrient elements have been found to have potential benefits to have a well-immune system. Apart from the supplementation of vitamins and minerals through diet (Jayawardena et al., 2020), nutraceuticals and probiotics have also found to accrue tremendous significance in this regard (McCarty and DiNicolantonio, 2020). The common sources of the vitamins, viz., Vit-A (pumpkin, carrots, broccoli, spinach, squash, sweet potato, mangoes, and papaya), Vit-C (amla, lemon, guava, oranges, fruit juices, and tomatoes), Vit-E (almonds, peanuts, spinach, asparagus, pumpkin, and broccoli), and Vit-B9 (whole grains and green leafy vegetables) are needed to be incorporated in diet. Further augmenting the availability of minerals like zinc, copper, iron, and selenium (Sharma, 2020) through food fortification and dietary diversification must also be ensured for increasing their bioavailability.

6.3 How Soil Research is Linked with Food and Nutritional Security

The present pandemic and the asymptomatic cases have quite evidently put forward the importance of food security to tide over the situation. The importance of agricultural research to augment the food production and thereby ensure two square meals for the population cannot be denied. In this purview, soil is the invaluable natural resource that serves as a medium for crop growth, provides nutrient and water to the growing plants in a judicious manner, and even enables its diversification through interaction with climatic counterparts. Soil can be of paramount importance to the human health by directly affecting ingestion, inhalation, and absorption of its constituents on one hand while indirectly through agriculture mediated food quality and quantity on the other (Oliver and Gregory, 2015). In association with this, soil is also the habitat of a wide range of organisms some of which impart negative consequences on human health through their pathogenic mode and toxic substances (Brevik and Burgess, 2013). Healthy soils are therefore a must for healthy life and survival. Any form of soil degradation, if not well addressed, will impact soil functions and severely hinder food security and human well-being (Rojas et al., 2016). Several research activities pertaining to food security bring forward the complexity of the process (Beddington et al., 2012). Taking due consideration of all the components, it can be well established that soil system is the core of food production, nutritional security, and plant-animal-human health (FAO, 2011) emphasizing its research potential for future sustainability.

6.4 Balanced Nutrition and Trade-Off between Productivity and Quality of Produce

With the ever-increasing trend of the global population surge as well as the squeezing of the available arable land for cultivation of crops, the post green revolution period has been marked by agricultural research for vertical expansion or rather increasing production per unit area of land. This production and productivity-oriented system through wide-scale fertilizer and other input application is unleashing its wrath in the long-term benefits of food security. The endeavor to increase the quantity of the produce has largely ignored the quality parameter of the food, thus having effect on nutritional security and human well-being. As the well-immune, nutritious human body successfully alleviated the COVID-19 pandemic, it is a timely reminder for us that it is not only the quantity, but also equilibrium between the quality and quantity of the produce, which is of paramount importance for combating pandemics. The conceptual 'nutrition-sensitive agriculture' is the key thespian to address the gap between available food and food needed for ensuring healthy and balanced diet among people.

The report of more than 870 million undernourished, 2 billion affected by nutrient deficiencies, and 1 billion obese people (Jaenicke and Virchow, 2013) suggests that somehow down the line we have missed the trick of nutritionally enriching the global food production system. Our failure further paved the way of such dreadful pandemics to occur over time to time. The vagaries of climate, social insecurity, and natural resource degradations have further intensified the problem. The emphasis should not be only limited to the production site, as the marketing channels also need to be thoroughly monitored to perform its dual role of cash provision to producers for nutritious food as well as for fresh or even the processed food to the consumers for sustaining balanced diet. The agricultural systems need to be made more biologically diverse by incorporating underutilized and even the minor grain crops, pulses, fruits, vegetables, root crops, tuber crops, etc. to the existing staple crops for balancing the diet and ensuring the environmental pliability. Rearing and incorporating alternative yet nutritious food crops can also broaden the horizon of the agroecosystems in even the climatic aberrations against biotic and abiotic stresses (Keatinge et al., 2010; Kahane et al., 2013).

Therefore, an imminent need of a paradigm shift toward a nutritionally enriched agricultural system by balancing or rather having a trade-off among quality and quantity of produce is perceived in the post COVID-19 period. Further supply channel modifications to reach the vulnerable groups of malnourished women, children, elderly, and sickly people and allowing their accessibility to nutritious food must be the cue in future.

6.5 Nutraceuticals and Functional Food Components – Role in Combating Pandemics

The significance of functional foods, nutraceuticals, and other natural health products has emerged greatly in relation to human health, disease reduction, and thereby a reduction in health costs. The natural tendency of consumption of whole grains, nuts, etc. often leaves the other parts like the skins and processing by-products of foods. These materials through research have been found to be a concentrated source of diverse nutritional components having health benefits and thus have been tried to be inculcated in human diets through the functional foods (Shahidi, 2009). A classic example is the presence of higher levels of nutritive like vanillic, *p*-coumaric, ferulic, and sinapic acid like phenolics in wheat bran than edible processed flour (Liyana-Pathirana and Shahidi, 2007). In general terms, functional food can be categorized as products resembling traditional food but having variation in physiological functions, while nutraceuticals are food derivatives often available in medicinal forms of pills, capsules, potions, or liquids. They include herbal products, dietary supplements, genetically engineered designer foods, specific diets, and processed foods (Jayawardena et al., 2020). These materials contain different phytochemicals and phenolic derivatives that through different mechanisms render disease resistance and combating ability against future pandemics. Several nutraceuticals like polyphenol-rich protein powder, prebiotics, plant stanol ester, arabinoxylan rice bran, broccoli sprout homogenates, aged garlic extract, and cranberry polyphenols upon testing have been effective against encapsulated RNA viruses and their infections and thereby usher a hope against the current COVID-19 and other future pandemics.

Another fascinating area of functional food components is the probiotics. They can be categorized as living microorganisms conferring health benefit to host when administered adequately and also stimulate immunity through antibody production. Some studies have found to reduce the disease severity of common cold through supplementation and can thus be attempted in other viral infections as well (Sengupta and Dey, 2020). The presence of *Bifidobacterium longum, Lactobacillus plantarum, Lactobacillus paracasei, Lactobacillus casei, Lactobacillus gasseri,* etc. in fermented food items like yoghurt, curd can serve the purpose of probiotics in diet quite vividly.

It is quite evident that the functional food research is highly correlated with agricultural research. Majority of the natural functional foods are the agricultural crops and their processed components; thereby judicious soil-based water and nutrient management can augment the nutritive capacity of the produce. As soil health is the fundamental factor for good quality nutritive diet, its maintenance and equilibrating crop quality and quantity must be further researched upon. Further as in the case of functional foods, the processed foods apart from whole grain are included in diet, soil pollution must also be critically looked upon. The presence of toxic heavy metals, pesticide residues or harmful pathogens in the milling products can increase their bioaccumulation and nullify the benefits accrued.

6.6 Artificial Intelligence-Mediated Remote Sensing Analytics Pertaining to Food Traceability

Foodborne illness can be attributed to significant societal costs. This emphasizes the need of holistic food safety measures (probability of a food to deter consumer health risks) and their traceability (ability to detect the origin and spread of the hindrance to safe food consumption from farm gate to the consumer's plate) (Souza-Monteiro and Hooker, 2013). However, the inability to link food chain records, inaccuracy and errors, and delays in obtaining essential data serve as major setbacks to authentic food safety and traceability (Badia-Melis et al., 2015). To ensure food safety a traceability system must encompass breadth (quantity of information), depth (tracking of information in both forward and backward directions), and precision (accuracy and assurance of food transshipment).

The current pandemic has quite aptly rendered an emphasis of the fact that food quality should not be synonymously coined with food safety. A product that may appear to have high quality (i.e. well colored, appetizing and flavorful, etc.) may be unsafe because it might be contaminated with undetected pathogenic organisms, toxic chemicals, or physical hazards. Thus, thorough evaluation approaches through hologram, barcode, radio-frequency tags, geographical identification tags, biotracing, nano sensor for precise Global Positioning System (GPS) identification through proper artificial intelligence-mediated remote analytics must be ensured (Dandage et al., 2017).

6.7 Post COVID Researchable Issues for Soil Management

The COVID-19 pandemic has raised an alarming situation with grave consequences on global economy, trade, agriculture, educational as well as industrial sectors. Among these entire sectors, still many believe that agriculture through ensuring nutritive production can revive the other sectors (Dev and Sengupta, 2020). This primarily puts forward the prioritization of multidisciplinary agricultural research with a central focus on soil as the medium for growth (Figure 6.1). Post COVID-19 soil research can thereby encompass the following.

6.7.1 Carbon Sequestration in the Soil

If one tries to unearth the direct relationships between the climate change and occurrence of pandemics, it turns intricate. However, reports of several scientists suggest that climate change amplifies the spread of infectious diseases beyond geographical limits, thereby aggravating the cataclysmic consequences. The compound risk exacerbated by both the factors becomes even more gruesome for policy making by governments (Phillips et al., 2020). The decadal rise of global temperature and elevated levels of CO_2

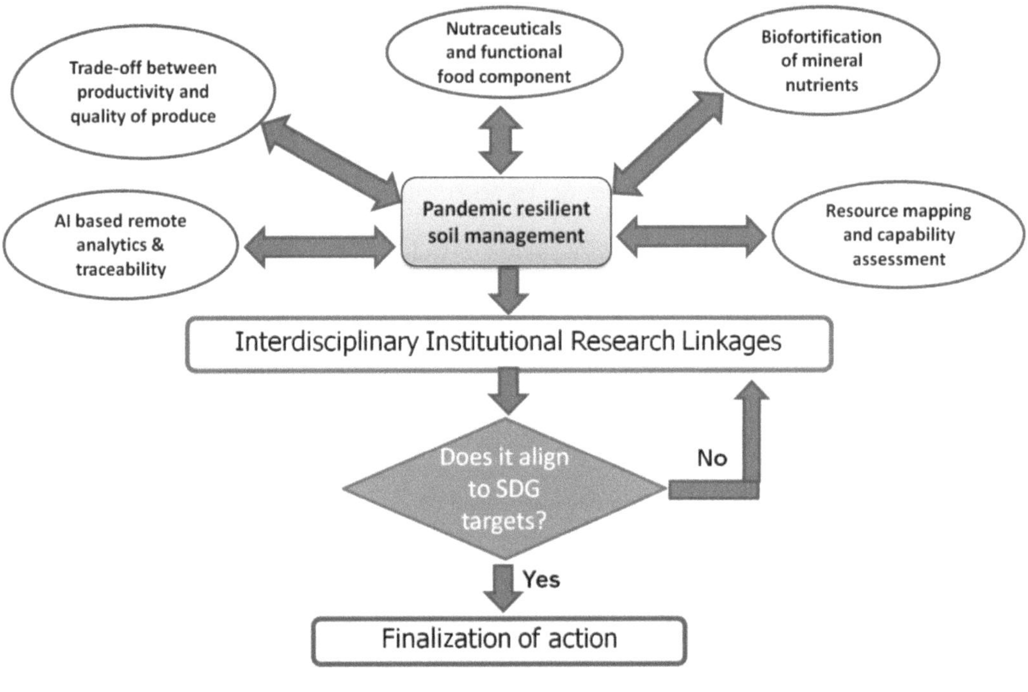

FIGURE 6.1 Framing a pandemic resilient soil-food research linkage pertaining to global food security.

concentration have severely affected the well-being of the planet. The pandemic puts forward the necessity of further promulgating the ensuing research of carbon sequestration in the soil and thereby curtails its atmospheric congregation. Thereby post pandemic soil research should encompass conservation agriculture practices of no or minimum tillage, agroforestry, incorporation of straw and other crop residues, crop rotation, ground cover crops, organic interventions, use of biochar, etc. for sequestering carbon (Lal, 2019) in conjunction with its critical estimation by wet digestion, dry combustion, spectroscopic technique, eddy covariance estimation, neutron scattering, etc. (Nayak et al., 2019).

6.7.2 Biofortification of Mineral Nutrients in Plant Edibles

Mineral nutrient deficiencies in human body are the primary reason for the wide-scale pandemic severity. A well-nourished immune body has the potency to alleviate the risks of mortality and morbidity. This necessitates a further outlook into the food security arena with the option of increasing the load of vital nutrients in the edible food grains through agricultural research. Agronomic and genetic biofortification is the researchable issue that can garner success (Zhang et al., 2018). Future researchable soil issue through soil application of mineral-nutrient sources of Zn, Fe, etc. at the right rate, time, and stage to improve phyto availability as well as interlinking with genetic research of high accumulating cultivars may be the pathway (Sengupta et al., 2019). The soil application of water-soluble fertilizers, seed treatment to boost initial growth, seedling root dipping as well as foliar application along with its evaluation through phytic acid and Zn/Fe molar ratio, and other methodologies should be taken into consideration. Further, the selenium fertilization and/or foliar application to improve its bio-accessibility through rice grain (Li et al., 2017) can also be researched in post pandemic period for rendering resilience.

6.7.3 Balanced and Need-Based Fertilization of Crops

The indiscriminate fertilizer application to augment the production hitherto has robbed off the inherent nutritive capacity of the soil. The soil health has been severely affected for which good quality and quantity of food production is impaired. The imminent need of balanced fertilization of crops to recreate equilibrium between quality and quantity of the produce over diversified ecosystems is thus felt. Apart from being a key to sustainable crop production by ensuring adequate availability of all the essential nutrients (primary, secondary, and micronutrients) in readily available form, balanced fertilization can safeguard environment, reduce C and N footprints, improves crop yield and quality for ensuring food and nutritional security (Bhattacharya et al., 2019). The strategic options for adopting balanced fertilization should take into account soil test-based fertilizer scheduling, integrative nutrient management, site-specific nutrient management, customized fertilizer, fortified fertilizers, nanocomposites of nutrients, precision farming, integrated plant-nutrient supply programme, etc. for better crop growth and nutrition assurance.

6.7.4 GPS-Based Resource Map Generation and Capability Assessment for Sustainable Agriculture

The nutrient deficiency and toxicity in the soils can never be uniform throughout the globe, rather it exists in pockets. The deficiency in soil often is the reason behind insurmountable malnutrition among the population. Further, it is also quite evident that the entire land resource can never be brought into cultivation. There exists a dire need to assess and classify the land based on their capability classification to register which land can be arable, which can be brought under afforestation or for pasture, etc. for ensuring sustainability in agricultural land use planning. The use of advanced geospatial tools like remote sensing, geographic information system (GIS), and GPS can be the torchbearer for effective land resource mapping and agricultural land use planning. Integrating the spatial data with real-time on-site data can effectively manage land resources, enable agroecological zoning, mapping land degradation assessing soil fertility and moisture status, crop coverage, yield forecasting, water resource delineation, precision agriculture, etc. (Reddy et al., 2013). A greater emphasis on this soil-based researchable issue can accentuate decision support systems to ensure agricultural sustainability and thereby food security to combat pandemics.

6.7.5 Composting Techniques and Use of Beneficial Microbes in Soil

With the emerging concern of unscrupulous inorganic fertilizer applications in the soil and deterioration of soil microbial ecology and diversity, a greater impetus is currently given on the organic sources of nutrient application. The application of compost, farmyard manure, vermicompost, phosphor compost, and other microbially enriched composts to augment the soil health (Maheshwari, 2014) has gained tremendous importance. Integrating organic and inorganic source to increase the availability of nutrients and ameliorate problematic soils has been widely practiced. Therefore, the post pandemic soil research should encompass wide-scale production of compost, their judicious application to soil and augment quality produce for sustainable agriculture. Another significant research area is the use of suppressive compost that creates an environment where the incidence of plant diseases reduces even when pathogen and host coexist through metabolic activities of beneficial microbial constituents (Hadar and Papadopoulou, 2012).

The existence of many beneficial microbes in soil is related to improving nutrient availability, acquisition of resources through beneficial plant-microbe interactions, alleviation of stress, phytohormones production, control of pathogens, volatile chemical productions to deter harmful microbes, bioremediation of polluted soils, synthesis of bioactive compounds, osmotic regulation, etc. (Singh et al., 2016). Thus, research initiatives to identify the soil beneficial microbes, culturing, and rearing them to obtain the benefits should be given emphasis to strive against pandemics by virtue of increased food production security.

6.7.6 Wastewater Treatment Prior to Use in Agricultural Soils

Assessment of the etiology of the causal agent of the COVID-19 pandemic revealed the presence of the viral pathogen even in the stool of human population that suggests the existence of the pathogen even in municipal sewage and wastewater (Ahmed et al., 2020). The wastewater is often used widely for irrigating of crops and the sewage material as the source of nutrient to the plants. The wastewater and biosolids containing different harmful chemical constituents, pesticide residues, and heavy metals in conjunction with harmful pathogens like bacteria, virus, protozoans, and helminths (Jaramillo and Restrepo, 2017) necessitates a wide-scale research on disinfection prior to agricultural application to reduce food-chain-mediated bioaccumulation.

6.7.7 Assessment of Soil Pollution and Their Remediation

Soil acts as the sink of a wide range of pesticide residues, toxic heavy metals, pathogens, and its pollution can disembark the equilibrium of natural resources. Crops grown under contaminated soils are always prone to the risk of their accumulation in above threshold levels; thus their consumption is often correlated with serious health ailments. The elevated levels of heavy meals, particularly in processing raw materials like bran (6–8 times higher arsenic load than grain) and husk (2–3 times) (Signes-Pastor et al., 2008), further accentuate the risk in functional food components. In the purview of nutrition-sensitive agriculture, this issue should be seriously looked upon and strategic remediation strategies through water management, organic and inorganic interventions, tolerant crop and variety selection, microbial remediation, and phytoremediation of each of the toxic substances must be encompassed. A healthy diet from a healthy soil is the epitome of nutritious food and healthy immune population.

6.8 Policy Issues in Post COVID-19 Period

The current pandemic has given us a timely reminder of how much research and policy initiatives are needed to sustain nutrition for tackling future pandemics. Ensuring nutrition to the population should be the prime focus of policies to fight the pandemics in future. Strengthening of public food distribution system as in India by different schemes of National Food Security Mission (NFSM), Bringing Green Revolution to Eastern India (BGREI), PM Garib Kalyan Anna Yojana (PMGKAY) through governmental interference and cross-linking with different nongovernmental organizations (NGOs) must be ensured for easy and free flow of nutritious food items to the weaker section of the society and thereby combat malnutrition. The supply-chain moderations for easy marketability of the perishable food items must also be categorized to allow easy selling as well as availing good quality produce. The emerging scenario from different parts of the globe suggests that neither the scientific technologies alone nor the traditional knowledge exclusively can completely solve the threats of food and nutritional security challenges emanating from climate change; however, a fusion of the two can. The current pandemic has provided the planners and policy makers yet another tool and dimension to initiate participatory action plan involving farmers and their rich reserve of traditional knowledge in order to develop adoptable technology that will enable the mitigation of water scarcity and problem of climate change for financial inclusion and mainstreaming of indigenous population. Moreover, region-specific amalgamated technological prescriptions refined with targeted policy analysis are required for effective implementation and obtaining positive outcomes within a finite time horizon (Dey and Sarkar, 2011). Financial policy enactment for agricultural research must also be guaranteed for enabling nutrition-mediated production system. Hitherto, the often forgotten critical elements in the euphoria of including technical content, viz., land governance, extension & advisory services, finance & markets, local governance, and cooperation models, together with monitoring and evaluation at all stages for developing a robust workable policy for sustainable soil management and climate resilient agriculture must be taken into consideration. Success stories showed addressing tenure insecurity at intra-household level, giving women access to land use rights, legal recognition and protection of land tenure, involving different stakeholders in policy-making processes, integration of sustainable land (SLM) management into community development plans, adoption of information and communications technology (ICT), incorporate externalities for making economic decisions, accounting for capital invested in the face of climate change, strengthening local and community governance structures, and development of voluntary guidelines on SLM in sync with local law are essential elements for development of such policy (Dey, 2020).

Owing to its importance, land degradation features prominently addressed in the Sustainable Development Goals (SDGs) under Goal 15 (protect, restore, and promote sustainable use of terrestrial ecosystems, sustainably manage forests, combat desertification, and halt and reverse land degradation, and halt biodiversity loss). The establishment of Land Degradation Neutrality (LDN) as Target 15.3 in the SDGs highlights the environmental importance and the land restoration dimension. Reversing land degradation is essential if we are to achieve the goals of the United Nations Convention to Combat Desertification (UNCCD), Convention on Biological Diversity (CBD), or the United Nations Framework Convention on Climate Change (UNFCCC) as well as to continue meeting related sustainable development targets. The translation of global targets into national ones, such as LDN, will help position the interconnected challenges of desertification, land degradation, and drought at the center of the conservation sector and will provide impetus toward more integrated responses to climate change and the other major environmental crises of our time. The interlinking of soil, crop production, crop protection, their genetic diversification research, extension and advisory services, and information technologies through institutional linkage can be the forerunner to sustainable natural resource management for climate resilient smart agriculture that may serve global food and nutritional security.

6.9 Conclusions

The current COVID-19 pandemic has quite evidently unearthed the reality that under circumstances when pandemics arise unavailing medications; proper nourished food and dietary supplementation can usher a hope. The developing world, crippled by balancing life and livelihood, becomes severely affected by such infections. To avail resilience to pandemics, a proper cognizance of viable options to ensure nutritive food availability must be made. Research and policy initiatives should all be directed toward the common target of fulfilling the United Nation SDGs to end hunger, ensure food, nutritional security and promote sustainability to agriculture. The rejuvenated impetus to fulfill such objective in the post COVID-19 world can only be the cue to strive in future.

Acknowledgment

The work was not funded. The work has no affiliation with any public or private entity. The author thanks the anonymous peer reviewers for their input.

Conflict of Interest

The authors declare that they have no known competing financial interests or personal relationships that could have appeared to influence the work.

REFERENCES

Ahmed, W., Angel, N., Edson, J., Bibby, K., Bivins, A., O'Brien, J. W., Choi, P. M., Kitajima, M., Simpson, S. L., Li, J. and Tscharke, B. (2020). First confirmed detection of SARS-CoV-2 in untreated wastewater in Australia: A proof of concept for the wastewater surveillance of COVID-19 in the community. *Science of the Total Environment*, **728**, 138764.

Al-Tawfiq, J. A. (2020). Asymptomatic coronavirus infection: MERS-CoV and SARS-CoV-2 (COVID-19). *Travel Medicine and Infectious Disease*, **27**(101608), 1–2.

Aman, F. and Masood, S. (2020). How nutrition can help to fight against COVID-19 pandemic. *Pakistan Journal of Medical Sciences*, **36**(COVID19-S4), S121.

Aslam, M. F., Majeed, S., Aslam, S. and Irfan, J. A. (2017). Vitamins: Key role players in boosting up immune response—A mini review. *Vitamin and Mineral*, **6**, 153.

Badia-Melis, R., Mishra, P. and Ruiz-García, L. (2015). Food traceability: New trends and recent advances. A review. *Food Control*, **57**, 393–401.

Beddington, J. R., Asaduzzaman, M., Bremauntz, F. A., Clark, M. E., Guillou, M., Jahn, M. M., Erda, L., Mamo, T., Van, B. N., Nobre, C. A. and Scholes, R. J. (2012). Achieving food security in the face of climate change: Final report from the Commission on Sustainable Agriculture and Climate Change.

Bhattacharya, P., Sengupta, S. and Halder, S. (2019). Customized, fortified and nano enabled fertilizers-prioritizing and profiteering sustainability in agriculture. In: Advances in Agriculture Sciences, Vol. 19 (R. K. Naresh, Eds.). Akinik Publications, New Delhi, 69–97 (ISBN 978-93-5335-728-3).

Brevik, E. C. and Burgess, L. C. (2013). Soils and Human Health. CRC Press, Boca Raton, FL.

Dandage, K., Badia-Melis, R. and Ruiz-García, L. (2017). Indian perspective in food traceability: A review. *Food Control*, **71**, 217–227.

Deaton, B. J. and Deaton, B. J. (2020). Food security and Canada's agricultural system challenged by COVID-19. *Canadian Journal of Agricultural Economics*, **68**(2), 143–149.

Dev, S. M. and Sengupta, R. (2020). *Covid-19: Impact on the Indian Economy*. Indira Gandhi Institute of Development Research, Mumbai, April.

Dey, P. (2020). Policy matrix for sustainable soil management and climate-resilient agriculture. *Current Science*, **118**(2), 199–201.

Dey, P. and Sarkar, A. K. (2011). Revisiting indigenous farming knowledge of Jharkhand (India) for conservation of natural resources and combating climate change. *Indian Journal of Traditional Knowledge*, **10**(1), 71–79.

FAO. (2011). *The State of the World's Land and Water Resources for Food and Agriculture*. Routledge, Abingdon.

Hadar, Y. and Papadopoulou, K. K. (2012). Suppressive composts: Microbial ecology links between abiotic environments and healthy plants. *Annual Review of Phytopathology*, **50**, 133–153.

Hanashima, M. and Tomobe, K. (2012). Urbanization, industrialization, and mortality in modern Japan: A spatiotemporal perspective. *Annals of GIS*, **18**(1), 57–70.

Hui, D. S., Azhar, E. I., Madani, T. A., Ntoumi, F., Kock, R., Dar, O., Ippolito, G., Mchugh, T. D., Memish, Z. A., Drosten, C. and Zumla, A. (2020). The continuing 2019-nCoV epidemic threat of novel coronaviruses to global health—The latest 2019 novel coronavirus outbreak in Wuhan, China. *International Journal of Infectious Diseases*, **91**, 264–266.

Jaenicke, H. and Virchow, D. (2013). Entry points into a nutrition-sensitive agriculture. *Food Security*, **5**(5), 679–692.

Jaramillo, M. F. and Restrepo, I. (2017). Wastewater reuse in agriculture: A review about its limitations and benefits. *Sustainability*, **9**(10), 1734.

Jayawardena, R., Sooriyaarachchi, P., Chourdakis, M., Jeewandara, C. and Ranasinghe, P. (2020). Enhancing immunity in viral infections, with special emphasis on COVID-19: A review. *Diabetes & Metabolic Syndrome: Clinical Research & Reviews*, **14**(4), 367–382.

Kahane, R., Hodgkin, T., Jaenicke, H., Hoogendoorn, C., Hermann, M., Hughes, J. D. A., Padulosi, S. and Looney, N. (2013). Agrobiodiversity for food security, health and income. *Agronomy for Sustainable Development*, **33**(4), 671–693.

Keatinge, J. D., Waliyar, F., Jamnadas, R. H., Moustafa, A., Andrade, M., Drechsel, P., Hughes, J. D. A., Kadirvel, P. and Luther, K. (2010). Relearning old lessons for the future of food—By bread alone no longer: Diversifying diets with fruit and vegetables. *Crop Science*, **50**, S51.

Kumar, S., Thombare, P. B. and Kale, P.A. (2020). Impact of coronavirus (COVID-19) on Indian economy. *Agriculture & Food: E-Newsletter*, **2**(4), 301–302.

Lal, R. (2019). *Management of Carbon Sequestration in Soil.* CRC Press, Boca Raton, FL.

Li, Z., Liang, D., Peng, Q., Cui, Z., Huang, J. and Lin, Z. (2017). Interaction between selenium and soil organic matter and its impact on soil selenium bioavailability: A review. *Geoderma*, **295**, 69–79.

Liyana-Pathirana, C. M. and Shahidi, F. (2007). Antioxidant and free radical scavenging activities of whole wheat and milling fractions. *Food Chemistry*, **101**, 1151–1157.

Maheshwari, D. K. (2014). *Composting for Sustainable Agriculture*, Vol. 3. Springer, Switzerland.

McCarty, M. F. and DiNicolantonio, J. J. (2020). Nutraceuticals have potential for boosting the type 1 interferon response to RNA viruses including influenza and coronavirus. *Progress in Cardiovascular Diseases*, **63**(3), 383.

Naja, F. and Hamadeh, R. (2020). Nutrition amid the COVID-19 pandemic: A multi-level framework for action. *European Journal of Clinical Nutrition*, **74**(8), 1117–1121.

Nayak, A. K., Rahman, M. M., Naidu, R., Dhal, B., Swain, C. K., Nayak, A. D., Tripathi, R., Shahid, M., Islam, M. R. and Pathak, H. (2019). Current and emerging methodologies for estimating carbon sequestration in agricultural soils: A review. *Science of the Total Environment*, **665**, 890–912.

Oliver, M. A. and Gregory, P. J. (2015). Soil, food security and human health: A review. *European Journal of Soil Science*, **66**(2), 257–276.

Phillips, C. A., Caldas, A., Cleetus, R., Dahl, K. A., Declet-Barreto, J., Licker, R., Merner, L. D., Ortiz-Partida, J. P., Phelan, A. L., Spanger-Siegfried, E. and Talati, S. (2020). Compound climate risks in the COVID-19 pandemic. *Nature Climate Change*, **10**(7), 586–588.

Reddy, G. O., Nagaraju, M. S. S., Ramteke, I. K. and Sarkar, D. (2013). Terrain characterization for soil resource mapping using IRS-P6 data and GIS – A case study from Basaltic terrain of Central India. *Journal of the Indian Society of Remote Sensing*, **41**(2), 331–343.

Rojas, R. V., Achouri, M., Maroulis, J. and Caon, L. (2016). Healthy soils: A prerequisite for sustainable food security. *Environmental Earth Sciences*, **75**(3), 180.

Sengupta, S., Bhattacharya, P. and Hazra, S. (2019). Ensuring nutritional security through zinc biofortification of rice grain in Indian scenario: A review. *International Journal of Chemical Studies*, **7**(6), 2129–2144.

Sengupta, S. and Dey, S. (2020). Unearthing asymptomatic COVID-19 cases: How nutrition and dietary management can render immunity against pandemics? *Agriculture & Food: E-Newsletter*, **2**(6), 552–554.

Shahidi, F. (2009). Nutraceuticals and functional foods: Whole versus processed foods. *Trends in Food Science & Technology*, **20**(9), 376–387.

Sharma, L. (2020). Dietary management to build adaptive immunity against COVID-19. *Journal of PeerScientist*, **2**(2), e1000016.

Signes-Pastor, A. J., Mitra, K., Sarkhel, S., Hobbes, M., Burló, F., De Groot, W. T. and Carbonell-Barrachina, A.A. (2008). Arsenic speciation in food and estimation of the dietary intake of inorganic arsenic in a rural village of West Bengal, India. *Journal of Agricultural and Food Chemistry*, **56**(20), 9469–9474.

Singh, J. S., Koushal, S., Kumar, A., Vimal, S. R. and Gupta, V. K. (2016). Book review: Microbial inoculants in sustainable agricultural productivity-Vol. II: Functional application. *Frontiers in Microbiology*, **7**, 2105.

Souza-Monteiro, D. M. and Hooker, N. H. (2013). Food safety and traceability. In: *US Programs Affecting Food and Agricultural Marketing* (pp. 249–271). Springer, New York, NY.

Taghdir, M., Sepandi, M., Abbaszadeh, S. and Parastouei, K. (2020). A review on some nutrition-based interventions in Covid-19. *Journal of Military Medicine*, **22(2)**, 169–176.

World Bank. (2020). India: Issues and Priorities for Agriculture. Retrieved from https://www.worldbank.org/en/news/feature/2012/05/17/india-agriculture-issues-priorities.

World Health Organization (WHO). (2020). Novel Coronavirus (2019-nCoV); Situation Report 3; https://www.who.int/emergencies/diseases/novel-coronavirus-2019/situation-reports.

Zhang, C. M., Zhao, W. Y., Gao, A. X., Su, T. T., Wang, Y. K., Zhang, Y. Q., Zhou, X. B. and He, X. H. (2018). How could agronomic biofortification of rice be an alternative strategy with higher cost-effectiveness for human iron and zinc deficiency in China?. *Food and Nutrition Bulletin*, **39**(2), 246–259.

Section II

Strategies and Platform for Small Farm Livelihoods: Research and Development

7

Technology and Policy Options: Opportunities for Smallholder Farmers to Achieve Sustainable Agriculture

Rakesh S[1], Ranjith Kumar G[2], Anil D[3], Ravinder Juttu[4], Kamalakar Jogula[5], Sharan Bhoopal Reddy[6], Bairi Raju[5], Jogarao Poiba[7], Umarajashekar A[8], and Ranvir Kumar[9]
[1]Department of Soil Science & Agricultural Chemistry, Uttar Banga Krishi Viswavidyalaya, Coochbehar, West Bengal, India
[2]Department of Soil Science and Agricultural Chemistry, Dr. Panjabrao Deshmukh Krishi Vidyapeeth, Akola, Maharashtra, India
[3]Department of Agronomy, Agricultural Research Station, Professor Jayashankar Telangana State Agricultural University, Peddapalli, Telangana, India
[4]Department of Soil Science & Agricultural Chemistry, Regional Sugarcane and Rice Research Station, Professor Jayashankar Telangana State Agricultural University, Nizamabad, Telangana, India
[5]Department of Soil Science & Agricultural Chemistry, Agricultural College, Professor Jayashankar Telangana State Agricultural University, Warangal, Telangana, India
[6]Department of Soil Science & Agricultural Chemistry, University of Agriculture Sciences, Raichur, Karnataka, India
[7]Regional Agricultural Research Station, Visakhapatnam, Andhra Pradesh, India
[8]Department of Agricultural Microbiology, Agricultural College, Professor Jayashankar Telangana State Agricultural University, Jagtial, Telangana, India
[9]Bhola Paswan Shastri Agricultural College, BAU, Purnea, Bihar, India

CONTENTS

DOI: 10.1201/9781003164968-9

7.1 Introduction

Global food production is more than tripled between 1960 and 2015 while enhancing the technologies since the Green Revolution period and a significant growth in utilization of available natural resources such as land and water for agricultural purposes. Agriculture sector comprises many other allied sectors, including forestry, dairy, horticulture, poultry, sericulture, beekeeping, and mushroom, generating employment opportunities to youths in the form of processing, marketing, and distribution while playing a critical role in the global economy. Indian agriculture engages almost 2/3rd of the population, ensures food and nutritional security, and accounts for 13% of our country's gross domestic products (GDPs). Agriculture is the mainstay of India's population as around 58% of people depended on it for income and livelihood security. Population of the country reached to 1.38 billion in 2020 which is 17.7% of the world's population as per the estimations of global population data report. The latest agricultural census shows that the average size of landholding per state in India is 1.08 ha. More than 80% of our country's farmers are small (1–2 ha area) and marginal (<1 ha) as they have very small landholdings. But, the share of small and marginal farmers for the food production is almost 60% in total food grain (Agricultural Census 2015–16). They have been facing several problems such as input supply, credit availability, proper transport, and market facility, since years. As per the estimates, India's food production for 2021–2022, government has set the target up to 307.33 Mt. However, several critical challenges are hindering the country's food security which include climate change and intensive farming practices that are degrading the natural resources and ecosystem. Rural sectors are highly vulnerable to vagaries of weather, since people who live in there entirely depend on natural resources and agriculture. In fact, about 70% of the population live in rural areas of developing countries adopted agriculture as their main source of livelihood. Agriculture production must be enhanced by 65% to meet the requirement of burgeoning population by 2050. The critical challenges for the country's food security such as climate change-induced droughts, floods, and irregular rainfall patters are degrading the soil productivity thereby affecting the overall crop yields and environmental quality and fluctuating the income of small and marginal farmers. Therefore, humankind should realize the mistakes done in the past and transform the current production system to sustainable production system to ensure the global food security and environmental safety.

7.2 Critical Need of Achieving Sustainability in Agriculture Production

Agriculture has changed dramatically since the Green Revolution period. Productivity of food and fiber has soared due to new and innovative mechanization in agriculture, increased use of chemical fertilizers and pesticides, specialization, and government policy programs that favored the greater productions and minimized food prizes. Although these developments turned into many positive effects in agriculture system, they depleted the topsoil fertility, contaminated the groundwater, air pollution, decline of family farms, greenhouse gas (GHG) emissions, and new threats to human health and safety, and ultimately disintegrated the rural communities. The overreliance on chemicals, fertilizers, and mono-cropping is the primary cause of environmental degradation and low-quality agricultural products. Such practices are the key contributor to degradation of around 20% of global land, and 30% of global GHG emissions (Brodt et al. 2011). It is also important to note that both natural and anthropological factors drive the land degradation and improvement process. Both agriculture and urban cities are contaminating the soil in the form of complex mixture of organic chemicals, metals, and municipal and domestic septic waste, farm animal waste, insecticides, herbicides, and other biowastes. Such wastes filled with chemicals and heavy metals contaminating the ecosystem both directly and indirectly. Soil with good amount of organic matter, well-developed structure, and appropriate physicochemical properties creates conducive environment to the plant growth that ultimately benefits with higher yields and are thus important for food security. Soil degradation through erosion, loss of soil structure and nutrient content minimizes the crop productivity and threatens food security (Srinivasarao et al. 2021a). Ceaseless addition of chemical fertilizers may improve the food security but not the quality of agricultural produce and healthy living life of humankind in a clean environment. Therefore, this is high time to realize the dreadful impacts of such intensive crop production on environment and ecosystem. There is a great need of taking account of the relative economic, environmental, and social costs and benefits of these systems in terms of yields, animal and human health, soil and water depletion, pollution, landscape, and wildlife habitats. The adverse effect of chemicals on ecosystem can only be eliminated by adopting new and sustainable agricultural technologies that yield multiple goods and services.

7.3 Technological Options Suitable for Smallholder Farmers to Achieve Sustainability in Agriculture System

7.3.1 Integrated Nutrient and Pest Management

Integrated Nutrient Management (INM) provides several benefits to soil, plant, and environment. The three key components of INM practice are soil reserves, mineral fertilizers, and organic nutrient sources. It aids in provision of maintaining physicochemical and biological properties of soil, enhances water holding capacity and amount of nutrients in soil, increases the withstand capacity of soil to drought situations, and improves the overall productivity of the crop (Srinivasarao et al. 2021b). Different options for INM practice for different cropping systems under various soil orders are presented in Table 7.1.

Integrated Pest Management (IPM) is an effective and alternative option to synthetic pesticides. It is nothing but producing more output from the same piece of land by minimizing the negative impacts on environment and increasing the contributions to natural capital (Pretty and Bharucha 2014). Various IPM options to avoid/reduce usage of synthetic pesticides for sustainable management is presented in Table 7.2.

TABLE 7.1

INM Options for Different Cropping Systems

Crop Production Systems	Soil Order	INM Option
Pearl millet	Aridisols	50% N (inorganic fertilizer) + 50% N (FYM)
Soybean	Vertisols	6 t FYM ha^{-1} + 20 kg N + 13 kg P
Sorghum	Vertisols	25 kg N ha^{-1} (through FYM) 25 kg N ha^{-1} (through urea)
Groundnut	Alfisols	50% RDF + 4 t groundnut shell ha^{-1}
Finger millet	Alfisols	FYM 10 t + 100% RDF(NPK)

Source: Adapted from Srinivasarao et al. (2009).

TABLE 7.2

IPM Options for Sustainable Management

Type	Application
Substitution for pesticidal products	Agrobiologicals or biopesticides derived from neem extracts
Management options for pesticides	Targeted spraying and threshold spraying by observing the incidence level
Pheromone compounds	Sticky and pheromone traps to capture the pests
Agroecological habitat design	Seed and seed bed preparation Use of wild crops/plants to push/pull pests, predators and parasites Crop rotations and multiple cropping to limit the pests Adding host-free periods into rotations

Source: Pretty and Bharucha (2015).

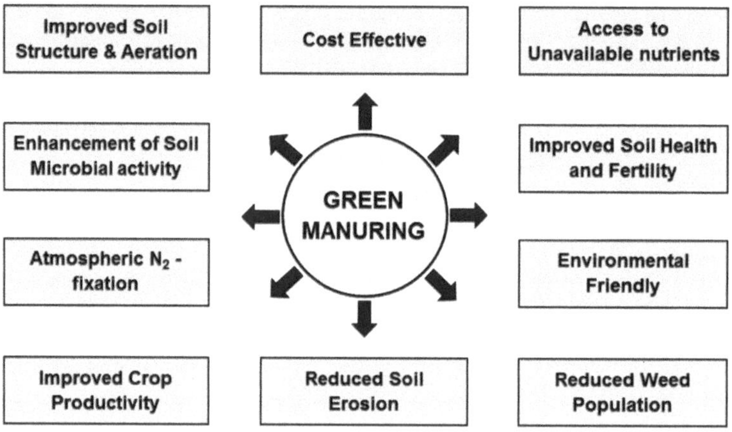

FIGURE 7.1 Advantages of green manuring.

7.3.2 Green Manuring and Organic Mulching

Green manuring has several benefits on improving soil health, crop yields, and income (Figure 7.1). There are two methods of green manuring: (a) *in situ* green manuring – growing short-duration crops and mulched on-site within 60–80 days of growth (b) *ex situ* green leaf manuring – leaves or young shoots of shrubs/trees from the outside of land are brought and applied on the soil and mulched before crop cultivation.

Incorporation of leaves of *Dalbergia sissoo* and *Cassia fistula* significantly improved the soil fertility and rice productivity in tropical dryland conditions. Apart from improved yields, organic farming through green manuring reduces the cost of expenditure with more profits. *In situ* green manuring species such as *Vigna unguiculata, Cyamopsis tetragonoloba, and Sesbania aculeata; ex situ* green leaf manuring such as *Millettia pinnata, Azadirachta indica, and Gliricidia sepium* are efficient crop species for green manuring.

Mulching covers the soil surface which improves soil health by facilitating moisture retention, preventing soil erosion, improving fertility, suppressing weeds, regulating temperature, plant nutrition, and preventing pests and diseases. Common organic mulching materials are compost, groundcovers, plant and animal residues. Organic mulching practice increases soil health, fertility, and moisture; maintains soil temperature; and reduces nutrient loss.

7.3.3 Pulses in Intercropping, Cover Cropping, and Crop Rotation

Legume crops play a significant role in sustainability principles by contributing to GHG mitigation, sequestration of carbon in soils, inducing cost saving on fertilizer through nitrogen fixation with their environmental and socioeconomic benefits (Srinivasarao et al. 2020). Intercropping is practiced obtaining sustainable production under adverse climatic conditions. Based on the legume species and amounts of N fixed, the yield advantage of subsequent cereal crops will be derived. Field pea (*Pisum sativum*) and faba bean (*Vicia faba*) fix atmospheric nitrogen in their aboveground biomass at the rate of 130 and 153 kg N ha^{-1}, respectively, and significant quantities may also be stored in belowground biomass. An increase in yield of 65 and 71% noticed in a system of 1 and 2 rows of maize alternated with 3 rows of soybean as compared with both crops grown as monoculture (Gao et al. 2010).

Cultivation of short-duration varieties of mung bean, urdbean, pigeon pea, etc. has paved way for crop diversification and intensification in North India. In hilly regions, cowpea, rice bean, French bean, urdbean, and mung bean act as an excellent cover crop. Pulses have replaced the traditional crop rotations like rice-wheat-mung bean, rice-urdbean/mung bean, rice-chickpea/lentil, pigeon pea-wheat, potato + rajmash, groundnut + pigeon pea, and soybean + pigeon pea, etc. Pulses like lentil, urdbean, mung bean, and lathyrus are grown as relay crop in humid regions of North-East India and drier regions of central and coastal regions of South India, which facilitates the sustainable production of the systems.

7.3.4 Integrated Farming System and Agroforestry

An integrated farming system (IFS) plays an important role in improving soil health and fertility levels. The IFS controls pest and disease infestation in effective manner. In IFS practice, amount of usage of herbicides, insecticides, fertilizers, and fungicides were decreased by 10.1, 28.3, 41.3, and 89.8%, respectively, as compared with conventional system (Walia and Kaur 2013).

Agroforestry system is a successful tool specifically for acid- and salt-affected soils, because perennial trees recycle nutrients, maintain soil organic matter (SOM), and reduce surface soil erosion and runoff. The major environmental benefits of agroforestry are water-quality enhancement, C sequestration, and soil improvement. Soil erosion can be minimized in alkaline soils with Babul (*Prosopis juliflora*) and Coastal sheoak (*Casuarina equisetifolia*). Cultivation rice-berseem (*Trifolium alexandrinum*) in an agroforestry system containing tree species (Acacia [*Vachellia nilotica*], Eucalyptus [*Eucalyptus globulus*], and Populus [*Populus deltoides*]) improved the soil microbial biomass C and N by 42 and 13%, respectively, than mono-cropping (Kaur et al. 2000).

7.3.5 Organic Farming

Organic farming relies more on sustainable method of cultivating crops to restrict chemical inputs that harms natural ecosystem. It contributes to socioeconomic and ecologically sustainable development, due to the application of organic principles, of using local resources *viz.*, indigenous seed varieties, manure, etc. and therefore cost-effective. India's North-Eastern state, Sikkim, covered 75,000 ha of land under organic cultivation and declared as an *Organic State* in India. By 2030, Meghalaya also eyes to transfer 2 lakh ha of land into organic cultivation. Around >1 lakh farmers of Kerala approved the organic farming practice (NPOF 2015-16).

7.3.6 Improved Water Using Techniques

Efficient use of water enhances the crop productivity. Adoption of drip irrigation and micro-sprinklers system is being considered as a potential alternative to achieve the yield target. There was around 28% increase in wheat yield in drip irrigation compared to surface irrigation. Drip fertigation with 100% water-soluble fertilizers recorded a net profit of potato i.e., 38,742 ha^{-1}. Seed cotton yield was significantly higher under drip irrigation than check basin (Thind et al. 2008). There was about 65% water saving in sugarcane crop with the adoption of drip irrigation noticed by Narayanamoorthy (2004).

7.3.7 Azolla and Biofertilizers

Azolla is a water fern which has maximum atmospheric-N fixing ability and is often used as a biofertilizer in rice cultivation especially in the parts of Southeast Asia. It is associated with blue-green algae "Anabaena" that helps in fixing atmospheric nitrogen (N_2) as ammonia (NH_3), can be utilized by rice plant. Azolla contains 2–5% N and 0.3–6.0% potassium (K). Incorporation of Azolla aids in provision of plant nitrogen and improves the overall crop productivity. An increase in growth attributes such as plant height, number of effective tillers, dry mass, and N content of rice plants was observed by the alone usage of Azolla or in combination with N fertilizers (Bhuvaneshwari and Singh 2015).

Biofertilizers are the products containing live cells of efficient strains of nitrogen-fixing phosphate solubilizing or cellulolytic microorganisms. Biofertilizers fix N_2, in association with plant roots, solubilize insoluble soil phosphates (P), and produce plant growth substances in the soil. Its major role and importance in sustainable agricultural productivity has been reviewed by several authors. Bacterial inoculation in combination with 1.0-kg molybdenum ha^{-1} in the cultivation of soybean significantly increased the weight of nodules, plant dry weight, total N content, and test weight and pod weight and also produced significantly higher yield (27.3 g $plant^{-1}$) as compared to control (8.1 g $plant^{-1}$) (Jabbar et al. 2014). Application of *Rhizobium* + phosphorus solubilizing bacteria (PSB) in chickpea significantly enhanced the nodule number, nodule fresh weight, nodule dry weight, shoot dry weight, and leghemoglobin content and also the overall grain and straw yields (Tagore et al. 2014).

7.3.8 Nutrient Cycling through Crop Residues under No-Till Farming

Crop residues which are the parts of the harvested plants left in the field have been regarded as waste materials, but it has become increasingly realized that they are important natural nutrient resources. Crop residues are good sources of plant nutrients, majorly nitrogen (N), phosphorus (P), and potassium (K). At the maturity stage, rice plant residues absorb about 40% of the N, 30–35% of the P, 80–85% of the K, and 40–50% of the S (Dobermann and Fairhurst 2000). Similarly, wheat residue accumulates 25–30% of N and P, 35–40% of S, and 70–75% of K. Residues maintain the soil physicochemical and biological properties in balanced condition and improve the overall ecological balance of the crop production system. Soil under zero tillage (ZT) tended to have higher soil organic carbon (SOC) than conventional tillage (CT) (Sinha et al. 2019).

7.3.9 Stress-Tolerant Varieties to Salt, Drought, and Waterlogging

Salt, drought, and moisture are the foremost abiotic stresses that decline crop productivity and weaken the global food security. Million hectares of crop lands are affected by these stresses. However, plants have their own nature of developing resistivity at various levels allowing them to escape and/or adapt to unfavorable environmental situations. Stress-resistant crops are most essential to achieve sustainability in crop production by mitigating the environmental impacts. Some of different stress-tolerant crop varieties are presented in Table 7.3.

7.3.10 Slow-Release Fertilizers

Slow-release fertilizers release plant nutrients as and when required by the crop that allows optimum uptake and nutrient use efficiency (NUE) thereby minimizing losses through leaching and volatilization. If the nutrient release is slower, plants utilize it at different critical crop growth stages and get its maximum benefits that help in achieving higher NUE. Slow-release fertilizers save the time and energy for application, indirectly reduce the fertilizer input cost which is beneficial for most of the small farmers of our country as it improves the crop yields. Globally, most of the farmers apply N fertilizer in the form of urea fertilizer which is highly soluble and volatile in nature. Neem-coated urea (NCU) significantly reduces the gaseous loss of NH_3 compared to prilled urea (Khandey 2017). NCU showed a yield improvement in Punjab (14%), Haryana (10.4%), and UP (9.6%) over ordinary urea. Main advantages of slow-release fertilizers such as long-term persistence of nutrients in the field that are utilized for later

TABLE 7.3

Crop Varieties Tolerant to Various Stresses

Stress	Crop	Varieties	Source
Salt	Rice	CSR-10, 13, 23, 27, 30 and 36	CSSRI, Karnal
	Wheat	KRL-1–4, 19, 35, 99, 210, 213	
	Mustard	CS-52, 54, 56	
Drought	Chickpea	JG 11, JAKI 9218, KAK 2, and Vihar	ICRISAT, Telangana State
	Ground nut	ICGV-91114	
Water	Rice	FR-13A and FR-43B	Ahmed et al. (2013)
		Savitri and IR-42 developed by CRRI	

TABLE 7.4
Some Important Slow-Release Fertilizers

Fertilizer	N Content (%)	Description	Source
Urea formaldehyde (UF)	30	Product of urea and formaldehyde in water-soluble form.	Varadachari and Goertz (2010)
Sulfur-coated urea (SCU)	31–38	Prepared by spraying molten sulfur over granular urea to yield.	
Neem-coated urea (NCU)	46	Urea is coated with neem tree seed oil.	Book: Fundamentals of Soil Science, ISSS
Polymer-coated urea (PCU)	43	Organic compound which is slowly soluble.	Ransom et al. (2020)
Isobutylidene-diurea (IBDU)	32	N content is water insoluble.	Saitoh and Watanuki (1997)
Natural organics	–	Activated sewage, sludge, and manure.	–

stages of crop growth, eco-friendly, easy to handle, reduce the loss of fertilizer nutrients, and remain inactive in soil for longer periods, hence do not pollute the groundwater and economical. Various slow-release fertilizers that are available in market have been presented in Table 7.4.

7.4 Government Policies and Programs

Neem-coated urea (NCU) scheme was initiated by the government of India (GOI) to boost the productivity of wheat and paddy crops especially. It releases plant-required nitrogen slowly and supplies at the critical growth stages of the crop growth that ultimately benefit the crop with maximum yields. *Soil Health Card (SHC)* scheme was launched in February 19, 2015 by the GOI to issue SHCs to the farmers to raise awareness and to disseminate SHC-based recommendations. *Mahatma Gandhi National Rural Employment Guarantee Act (MGNREGA)* was established to eliminate unemployment problem and provide a legal guarantee of minimum 100 days of wage employment to the families in a financial year. GOI had launched *Agroforestry Policies* to deal with sustainable agriculture (SA). All the programs give equal importance to the significant role of agroforestry in maintenance of vegetation cover, efficient nutrient cycling, and organic matter addition for SA. The Ministry of Water Resources adopted the first Indian *National Water Policy* (NWP) in 1987. Since then, the Indian NWP has been revised twice by Ministry of Water Resources in 2002 and 2012. *National Programme for Organic Production* (NPOP) was initiated by the GOI in 2001 involved with several objectives such as accreditation program for certification agencies, encouragement of organic farmers, and promotion of organic production. Recently, GOI announced the state "Sikkim" as an "*Organic State*" of India. The *Paris Agreement* a "4 per 1000/4 per mille" initiative launched by France in 2015 and adopted by the United Nations Framework Convention on Climate Change (UNFCCC) at a Conference of the Parties (COP) 22 intends to build the SOC through agroforestry, conservation agriculture for sustainable intensification (CASI), and landscape management. India is also committed to UNFCCC and the Paris Climate Change Agreement.

7.5 Conclusion and Way Forward

Promoting SA in India plays an imperative role in safeguarding the food security and ensuring environmental safety as it brings sustainability in food production system, restores the degraded natural resources, and improves the overall income of small and marginal farmers. Thus, a long-run investments are utmost important for the sustainable ecosystem and national food security. The SA interventions involved efficient utilization of natural resources, minimal external chemical inputs, and soil health maintenance are critical to meet the food demand of burgeoning population while balancing the agricultural production and livelihood security. Yet, collaboration of stakeholders such as governments, private industries, research centers, funding agencies, and the farmer organizations is important to understand the challenges in agriculture sector and decide the site-specific strategies to promote SA in developing country like India.

- Creating awareness among farmers regarding soil health deterioration and environmental pollution and its impacts on production is crucial.
- Implementation of sustainable technologies based on locally available resources such as cropping systems, water management practices, tillage operations, and biotic and abiotic stresses.
- Identifying policy interventions needed to promote SA that are effective at regional and national level.
- Encouragement of scientific community for conducting research and develop the suitable technologies related to natural resource management in agriculture system is required.
- Implementing a protocol for payment to farmers to adopt SA in order to strengthen the ecosystem services.

REFERENCES

Agricultural Census 2015-16. All India Report on Number and Area of Operational Holdings. Ministry of Agriculture and Farmer Welfare, Government of India.

Ahmed, F., Rai, M., Ismail, M. R., Juraimi, A. S., Rahim, H. A., Asfaliza, R. and Latif, M. A. 2013. Review article waterlogging tolerance of crops: Breeding, mechanism of tolerance, molecular approaches, and future prospects. *BioMed Research International*, Article ID 963525, 10 pages, Hindawi Publishing Corporation. http://dx.doi.org/10.1155/2013/963525.

Bhuvaneshwari, K. and Singh, P. K. 2015. Response of nitrogen-fixing water fern Azolla biofertilization to rice crop. *3 Biotech* 5: 523–529. DOI: 10.1007/s13205-014-0251-8.

Brodt, S., Six, J., Feenstra, G., Ingels, C. and Campbell, D. 2011. Sustainable agriculture. *Nature Education Knowledge* 3(10): 1.

Dobermann, A. and Fairhurst, T. H. 2000. Rice: Nutrient disorders and nutrient management. Singapore and Los Baños: Potash & Phosphate Institute (PPI), Potash & Phosphate Institute of Canada (PPIC), and International Rice Research Institute (IRRI), pp. 1–191.

Gao, Y., Duan, A., Qiu, X., Liu, Z., Sun, J., Zhang, J. and Wang, H. 2010. Distribution of roots and root length density in a maize/soybean strip intercropping system. *Agricultural Water Management* 98: 199–212.

Jabbar, K. A., Halimi, M. S., Radziah, O., Sheikh, H. H., Hossain, K. and Saikat, H. B. 2014. Effect of Azospirillum in association with molybdenum on enhanced biological nitrogen fixation, growth, yield and yield contributing characters of soybean Bassam. *Journal of Food, Agriculture and Environment* 12(2): 302–306.

Kaur, B., Gupta, S. R. and Singh, G. 2000. Soil carbon, microbial activity and nitrogen availability in agroforestry systems on moderately alkaline soils in Northern India. *Applied Soil Ecology* 15: 283–294.

Khandey, N. 2017. Evaluation of neem coated urea and prilled urea on rice yield and nutrient use efficiency (NUE) in vertisol of CG. Doctoral Dissertation, Indira Gandhi Krishi Vishwavidyalaya, Raipur.

Narayanamoorthy, A. 2004. Impact assessment of drip irrigation in India: The case of sugarcane. *Development Policy Review* 22(07): 443–462. DOI: 10.1111/j.1467-7679.2004.00259.x.

Pretty, J. and Bharucha, Z. 2014. The sustainable intensification of agriculture. *Annals of Botany*. DOI: 10.1093/aob/mcu205.

Pretty, J., and Z. Bharucha. 2015. Integrated Pest Management for Sustainable Intensification of Agriculture in Asia and Africa. *Insects*. 6, 152–182; doi:10.3390/insects6010152.

Ransom, C. J., Jolley, V. D., Blair, T. A., Sutton, L. E. and Hopkins, B. G. 2020. Nitrogen release rates from slow- and controlled-release fertilizers influenced by placement and temperature. *PLoS One* 15(6): 0234544.

Saitoh, H. and Watanuki, T. 1997. Properties of a new slow-release nitrogen fertilizer "SUPER-IB." *Plant Nutrition for Sustainable Food Production and Environment*: 639–640. DOI: 10.1007/978-94-009-0047-9_203.

Sinha, A. K., Ghosh, A., Dhar, T., Bhattacharya, P. M., Mitra, B., Rakesh, S., Paneru, P., Shrestha, S. R., Manandhar, S., Beura, K. and Dutta, S. 2019. Trends in key soil parameters under conservation agriculture-based sustainable intensification farming practices in the Eastern Ganga Alluvial Plains. *Soil Research* 57(8): 883–893. https://doi.org/10.1071/SR19162.

Srinivasarao, C., Rakesh, S., Ranjith Kumar, G., Manasa, R., Somashekar, G., Subha Lakshmi, C. and Kundu, S. 2021a. Soil degradation challenges for sustainable agriculture in tropical India. *Current Science* 120(3): 10.

Srinivasarao, C., Ramesh Naik, M., Subha Lakshmi, C., Ranjith Kumar, G., Manasa, R., Rakesh, S., Kundu, S. and Prasad, J. V. N. S. 2021b. Economic and environmental benefits ofIntegrated nutrient management in Indian agriculture. *Indian Journal of Fertilisers* 16(11): 1124–1137.

Srinivasarao, C., Ravindra Chary, G., Venkateswarlu, B., Vittal, K. P. R., Prasad, J. V. N. S., Singh, S. R. S. K., Gajanan, G. N., Sharma, R. A., Deshpande, A. N. and Patel, J. J. 2009. Carbon Sequestration Strategies Under Rainfed Production Systems of India, Central Research Institute for Dryland Agriculture, Hyderabad (ICAR), Hyderabad, p. 102.

Srinivasarao, C., Subha Lakshmi, C., Kundu, S., Ranjith Kumar, G., Manasa R. and Rakesh, S. 2020. Integrated nutrient management strategies for rainfed agro-ecosystems of India. *Indian Journal of Fertilisers* 16(4): 344–361.

Tagore, G. S., Namdeo, S. L., Sharma, S. K. and Kumar, N. 2014. Effect of Rhizobium and phosphate solubilizing bacterial inoculants on symbiotic traits, nodule leghemoglobin, and yield of chickpea genotypes. *International Journal of Agronomy* 581627: 1–8.

Thind, H. S., Aujla, M. S. and Buttar, G. 2008. Response of cotton to various levels of nitrogen and water applied to normal and paired sown cotton under drip irrigation in relation to check-basin. *Agricultural Water Management* 95(1): 25–34. DOI: 10.1016/j.agwat.2007.08.008.

Varadachari, C. and Goertz, H. M. 2010. Slow-release and controlled-release nitrogen fertilizers. In *ING Bulletins on Regional Assessment of Reactive Nitrogen*, Bulletin No. 11, (Ed. Bijay Singh), SCON-ING, New Delhi, pp. i–iv & 1–42.

Walia, S. S. and Kaur, N. G. 2013. Integrated farming system – An ecofriendly approach for sustainable agricultural environment: A review. *Journal of Agronomy, Forestry and Horticulture* 1: 1–11.

8

The Role of Agricultural R&D within the Agricultural Innovation Systems Framework

P. Anandajayasekeram
Independent International Service Provider, Agricultural and Rural Innovation Support. Melbourne, Australia

CONTENTS

8.1 Introduction

Since the introduction of formal scientific theory of inquiry into the agricultural sector in the 19th century, agricultural research and development (R&D) processes have gone through a number of paradigm shifts. Currently, the most widely used framework is the Agricultural Innovation Systems (AIS) framework. Through application, R&D practitioners have managed to successfully integrate value chain (VC) analysis into this framework. This chapter will first briefly trace the evolution of the AIS framework and then discuss key components and concepts. The third section deals with the agricultural innovation process and the role of agricultural research. Then, the factors influencing the innovation process are identified. Finally, key challenges in integrating AIS framework within agricultural R&D processes are also outlined.

DOI: 10.1201/9781003164968-10

8.2 Changing Paradigms within Agricultural Research and Development (R&D)

Until the 1960s, most agricultural research efforts tended to be commodity and factor oriented. Agricultural science found great success in this so called reductionist approach (Winkelmann 1987). The underlying assumption was that an understanding of the whole comes from an understanding of the parts, and solutions appropriate for the parts will lead to suitable solutions for the whole.

Early success encouraged emphasis on commodity and disciplinary research to improve understanding of the parts, which, in turn, fostered increased understanding of the ever more refined parts. Therefore, most agricultural research rested on individual discipline within a commodity focus. The emphasis was on Technology Development and Transfer (TDT) model for enhanced farm-level production and productivity. This reductionist approach produced dramatic success in the developed and developing world, particularly in favorable environments. The so-called Green Revolution or the seed fertilizer revolution had very little success in sub-Saharan Africa (SSA) and parts of Latin America and Asia.

Work done globally revealed that smallholder farmers (SHF) are rational decision makers. Very often they operate complex systems, follow risk-mitigating strategies in farming, demonstrate step-wise adoption behaviors, and possess a wealth of indigenous technical knowledge. Many problems are location-specific and farmer-specific. The search for alternate approaches to address the problems, constraints, and opportunities of the majority of SH resource-poor farmers resulted in a research approach which came to be known as Farming Systems Research (FSR) (Tripp et al. 1990; Anandajayasekeram 1997; Collinson 2000).

The key differences between the FSR approach and traditional reductionist approach are as follows: (i) it uses systems framework; (ii) research thrusts are derived from users through diagnostic activity; (iii) technology is first tested in farmers' environments; (iv) system interactions are given explicit consideration in identifying problems, technical interventions as well as evaluation of technologies; and (v) the evaluation criteria match those used by farmers.

Initially, FSR focused on technology generation and, subsequently, expanded to address policy and institutional support systems. Despite the various terminologies used in the literature (Anandajayasekeram 2011), in general, it has been widely accepted that the FSR approach has a set of research procedures that emphasize the following activities: diagnosis, planning (both biophysical and socioeconomic research, including experimentation), evaluation (replanning if needed), and recommendation and wider dissemination (scaling-up/out).

In the 1990s, the next phase of participatory approaches, tools and methods were added to the existing system perspective and on-farm orientation. This development shift was based on the premise that the non-adoption of technologies was not due to farmer ignorance, but deficiency in the technology and process that generated it, specifically inadequate participation in all stages by those intended to benefit. Proponents of this method argued that earlier FSR work could be seen as an extension of the TDT model, where outside professionals obtained information from farmers, analyzed it, and decided experiments and solutions for farmers.

By contrast, in the participatory approach, analysis, choice, and experimentation are conducted with and by farmers, with outside professionals providing analytical and facilitation support (Chambers 1993; Anandajayasekeram 1997). The salient feature of the participatory approach is reverse learning, where researchers and extension workers learn from farmers. The new paradigm put emphasis on people rather than things, decentralization, empowering participants, and learning from beneficiaries rather than teaching them. Locations and roles are also reversed, with farms and farmers central instead of research stations, labs, and scientists.

The continuous evolution and application of the participatory approaches and systems thinking in agricultural R&D eventually resulted in the IS framework. The VC concept is explicitly integrated into this IS framework.

While researchers were incorporating the systems concept in processes, development practitioners, and policy makers began to use it in organizational analysis. After the Second World War, the National Agricultural Research Institute (NARI) framework was the first to facilitate major investment in agricultural technology to increase food production. In many developing countries, NARIs were set up as organizational structures for agricultural research by colonial powers to protect and promote export cash crop production. Realizing that effective R&D requires more than research, the National Systems Framework (NSF) was introduced. The NSF included the National Agricultural Research Systems (NARS), the National Agricultural Extension System (NAES), and the National Agricultural Education and Training System. This trend of thinking continued to include other organizations involved in agricultural R&D and resulted in concepts such as Agricultural Knowledge and Information System (AKIS; Rolling 1988) and the AIS (Hall et al. 2005; World Bank 2006). Thus, the application of system's concept within agricultural R&D evolved in two directions – as a framework for organizational analysis, and for technology development, dissemination, and utilization – both leading to the development of the AIS framework as shown in Figure 8.1.

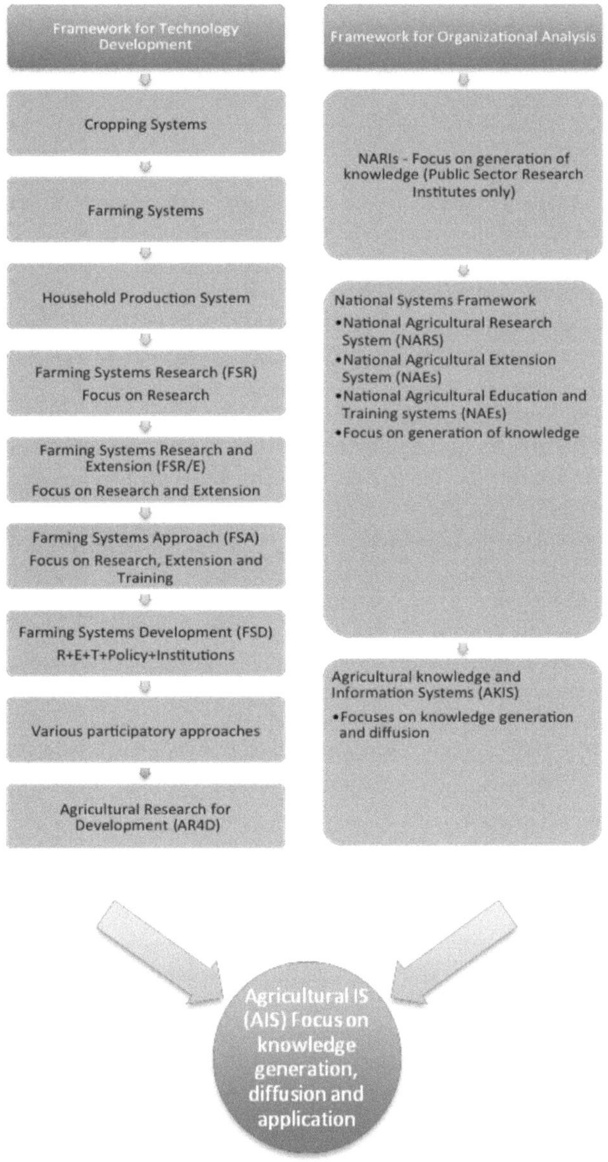

FIGURE 8.1 The evolution of systems thinking and its application in agriculture.

Source: Anandajayasekeram et al. (2005).

A number of sources contributed to the adoption of the IS framework in agricultural R&D. These include the successful application of the National Innovation Systems (NIS) perspective in the industrial sector; the inadequacy of traditional linear models to explain the innovation process; the inadequacy of existing frameworks for organizational analysis; and, increased demand for developmental impacts on agricultural R&D investment. Beyond empirical demonstrations of nonlinearity in innovation, an interactive model is considered to be attractive because of the interdependence and potential complementarities that arise in an environment in which diverse actors invest in knowledge production and utilization at comparable levels.

8.3 The Agricultural Innovation Systems Framework: Key Components and Concepts

The key components of the AIS framework are outlined in this section.

8.3.1 Innovation

In the literature, there are many definitions for "innovation" (Anandajayasekeram 2011). The simplest is "anything new introduced into an economic or social process" (OECD 1997). The most relevant definition for agricultural R&D is "the economically successful use of invention" (Bacon and Butler 1998). This distinguishes between knowledge and innovation. Knowledge generated through research is necessary for innovation to begin, but not sufficient. Responsibility of the R&D community does not end with production of new technology or knowledge only. Success is when inventions are disseminated, adapted, and widely adopted.

The most useful definition of innovation in the VC context is "a continuous learning process in which individuals/groups of individuals/organizations and firms, master and implement the design, production and marketing of goods and services that are new to them, although not necessarily new to their competitors – domestic or foreign" (Metcalfe and Ramlogan 2008). This definition highlights that innovation can rely on both new technologies/knowledge and novel combination/adaptation of existing technology/knowledge.

In the past, science and technology generations were equated with innovation. But innovation is a new creation of economic significance. Moreover, such innovations are not limited to technology but also include institutional, organizational, managerial as well as service delivery. Thus, the term innovation in its broadest sense covers activities and processes associated with the generation, production, distribution, adaptation and use of new technical, institutional, organizational, managerial knowledge and service delivery (Hall et al. 2005). IS should not be confused with invention system. IS incorporates the invention system as well as complementary economic processes required to turn invention into innovation into utilization.

8.3.2 Innovation System

In its simplest form, an IS has five elements: (i) the organizations and individuals involved in generating, diffusing, adapting, and using new knowledge; (ii) their actions and interactions; (iii) the interactive learning that occurs when organizations and individuals engage in these processes; (iv) the way this leads to new products and processes (innovations); and (v) the institutions (rules, norms, and conventions, both formal and informal) that govern how these interactions and processes take place (Horton 1990). An IS can be defined at different levels: national, sectoral, commodity, and intervention based.

8.3.2.1 National Innovation System (NIS)

NIS is defined as a set of functional institutions, organizations, and policies that interact constructively in pursuit of common social and economic goals and objectives, and that introduce innovation as the promotor of change (World Bank 2007). NIS is a generic concept with three components: knowledge domain, business domain, and environment. NIS started with relatively simple descriptive analysis to explain the difference in innovative activity and performance between countries. Since then, the theoretical underpinning of NIS has been improved by insights from various streams, including evolutionary economics, theories of learning, institutional thesis, and systems theory. NIS is an analytical tool for planning and policy making to enhance innovations at the national level, covering all sectors of the economy.

8.3.2.2 Agricultural Innovation System (AIS)

A collaboration of organizations working toward technological, managerial, organizational, and institutional change in agricultural R&D can be called AIS. A typical national AIS includes traditional innovation sources (indigenous technical knowledge); modern actors (research institutes and universities); private sector, including agro-industrial firms and entrepreneurs (local, national, and multinationals); civil society organizations (NGOs, farmer and customer organizations, pressure groups); and institutions (laws, regulations, beliefs, customs, and norms) that affect how innovations are developed and delivered (see Figure 8.2). AIS focuses on all actors needed to stimulate innovation and growth and emphasizes knowledge generation and adoption in the sector. This is also called "innovation ecology."

8.3.2.3 Commodity-Based Innovation System

A commodity-based IS incorporates the various actors, their actions, and interactions, as well as the enabling environment, facilitating institutions and services that condition various forms of innovation along the VC of a particular commodity/enterprise. Innovation can occur anywhere along the VC, not just at farm level, thus broadening the agenda to incorporate biophysical and socioeconomic research. It is a subset of AIS.

8.3.2.4 Intervention-Based Innovation System

ISs do not occur automatically. The problem defines a particular innovation opportunity. Problem-focused ISs are created for a purpose, they change in context and pattern of interactions as the problem solution evolves, and they are constructed both at the macro and micro levels. Intervention-based IS solves "local" innovation problems anywhere along the VC. Intervention-based ISs are temporary and dissolve once the problem is solved. A problem-focused IS cuts across national boundaries, may be spatially unconstrained, and is most relevant and used by the agricultural R&D community.

Concepts such as NIS, AIS, and commodity-focused IS are generic. In order to explain the difference between generic systems and intervention-based ISs, Metcalfe and Ramlogan (2008) coined the phrase "Innovation Ecology." This refers to a set of individuals usually working within organizations who are the repositories and generators of existing and new knowledge. Included in this ecology are organizations that generate, store, and retrieve information as well as those who manage the general flow of information. These actors collectively exhibit a division of labor that is characteristic of knowledge production. These ecologies are typically national, often generic, necessarily reflecting rules of law, social and business practice, and political regulation in the economies in which they are located (Metcalfe and Ramlogan 2008). The ecologies are more permanent in nature.

On the other hand, intervention-based ISs address a specific problem. Based on the problem(s), there can be multiple intervention-based ISs supported by the same innovation ecology. Since the solution of one problem typically leads to different and new problems, one expects that as the problem evolves, actors in the system as well as their interconnectedness also vary, making the intervention-based IS very dynamic.

8.3.3 Innovation Systems Perspectives (ISP)

An IS perspective (ISP) uses an innovation lens in the design, implementation and evaluation of agricultural R&D activities. A major feature of ISP is that innovation is the organizing principle. The analytical implications of ISP are that a range of activities and organizations related to agricultural R&D need to

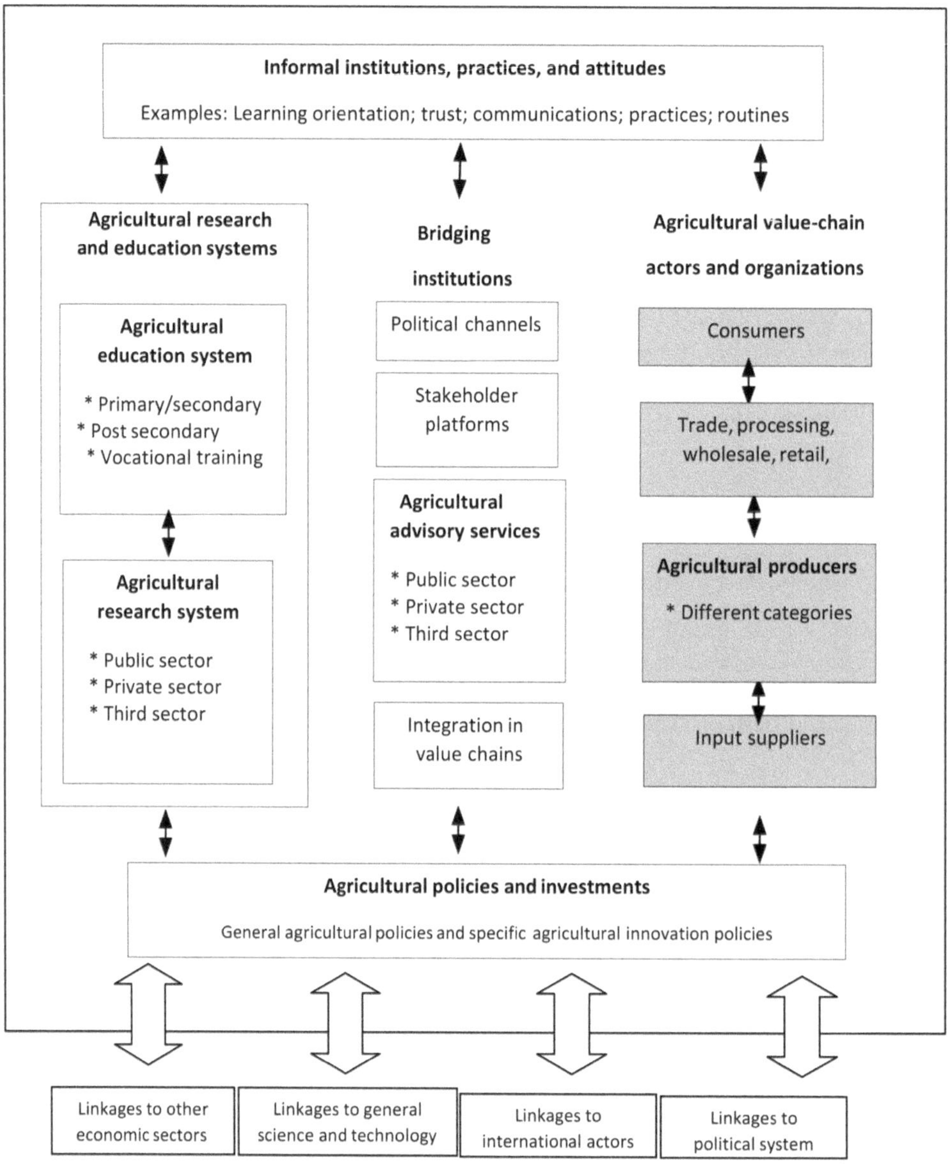

FIGURE 8.2 A typical national agricultural innovation system framework.

Source: World Bank (2007).

be considered, how these might function collectively, and to locate R&D planning and implementation in the context of local cultural and political norms (i.e., wider organizational and institutional context). Partnerships and linkages are integral to the IS.

8.4 Innovation Process and the Role of R&D

Taking a brilliant idea through an often-painful journey to become something that is widely used by a community involves many steps, resources, and problem solving. Transformation does not follow a linear path but rather is characterized by complicated feedback mechanisms and interactive relations involving science, technology, learning, production, partnerships policy, and demand. Thus, research and invention are necessary, but not sufficient, conditions for innovation. Success is when the output(s) of the research systems are disseminated, adopted, and used to generate much-needed socioeconomic outcomes.

8.4.1 Innovation Process

As discussed earlier, "innovation" in its broadest sense covers the activities and processes associated with generation, production, distribution, adaptation and use of new technical, institutional, organizational, or managerial knowledge (Hall et al. 2005).

The four stages involved in the innovation process are (i) invention: creation of knowledge and development of appropriate solutions to the priority problems; (ii) translation/realization: transforming knowledge into usable technologies, products, and processes through strategic, applied, and adaptive research; (iii) commercialization: dissemination, scaling-up, and provision of support services needed; and (iv) adoption: utilization in a meaningful way by end users or customers to create socioeconomic benefits. This process of course must be replicable at an economic cost, satisfy social and economic needs, and be supported by government, society, institutes, and institutions. If any of these elements change, the overall system becomes unstable. Therefore, in innovation planning, all these aspects should be considered.

The most important enabling factor for innovation is understanding end user needs and translating knowledge into actions that meet those needs. Within the innovation framework, the process and activities involved in translating knowledge into innovation are outlined in Figure 8.3. This figure highlights the approach to planning and implementing R&D activities.

8.4.2 Role of Research in the Innovation Process

The term "Research" refers to many things and can be grouped under different categories. Based on the intent/focus, research can be basic or strategic and applied. Basic research is curiosity driven and motivated by a desire to expand knowledge. It does not have immediate commercial objectives and, although it certainly could, it may not necessarily result in an invention or a solution to a practical problem. Strategic and applied research, on the other hand, is designed to answer specific practical problems – problem driven – and has specific commercial objectives in the form of products, procedures, and services. The term "adaptive research" is used to fine tune solutions to specific situations. For continuous and sustainable innovation, both types of research are important. Basic research is the foundation of knowledge and strategic/applied research is its practical application. This practical application can provide an information base for theory building and knowledge generation. If basic research were long neglected, then development will stagnate. Thus "knowledge frontier" research remains important and includes global technological fixes.

It is also worth noting that innovations may be developed through less formal mechanisms. This may involve modifying practices on the job, through the exchange and combination of professional experiences. Most organizational, managerial, institutional, and service delivery innovations emanate from these informal processes. Meta evaluation/analysis is another form of secondary source knowledge that can lead to innovation.

At the heart of innovation is the process of creating, sharing, and putting knowledge into productive use, which recognizes research as a fundamental part of a wider knowledge, and essential ingredient in innovation. All research is relevant, but development-oriented applied and adaptive research really contributes to innovation and socioeconomic impact.

8.4.3 Importance of "Scalability" in Agricultural Innovations

Another concept that is gaining importance as a result of ISP application is scaling up of R&D processes. Effective scaling-up of pilot-tested invention is vital for innovation to occur. Scaling-up is defined as the "deliberate efforts to increase the impact of innovations, that is successfully tested in pilot or experimental projects, in different places (expanding, replicating, adapting and sustaining) over time so as to benefit more

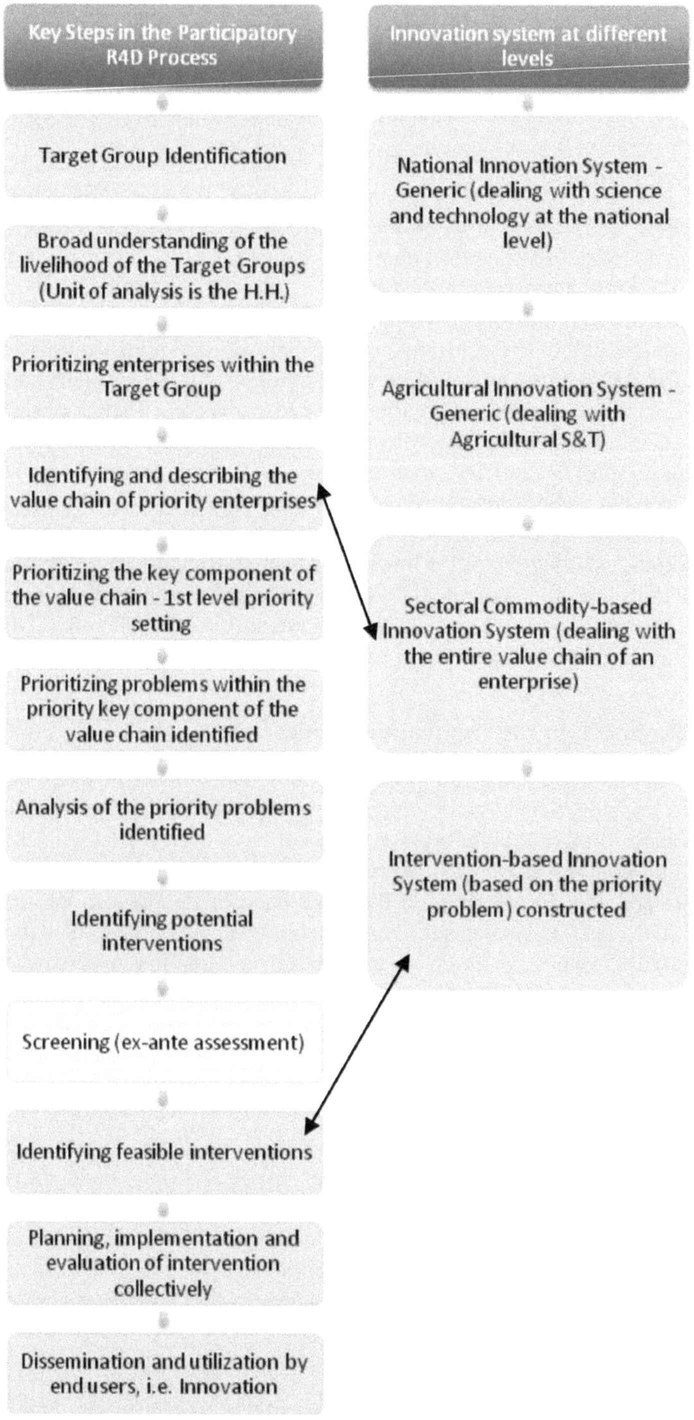

FIGURE 8.3 Integration of an innovation systems perspective and value chains in the R&D process.

Source: Anandajayasekeram et al. (2009).

Note: This is not a linear process; it is often characterized by constant iterative interactions and feedback loops, based on the "action–reflection–action" principle.

people and to foster policy and programme development on a lasting basis" (Hartmann and Linn 2008). Scalability is the potential of a particular innovation or change to be expanded, adapted, and replicated.

As shown in Figure 8.3, scaling-up is a continuum of agricultural R&D process aimed at realizing the full potential of an innovation. A scalability index (Anandajayasekeram 2018) can be used to assess the scalability of any pilot-tested technology/invention, which not only allows decision makers to assess scalability but also serves as the diagnostic tool to identify what can enhance the scalability of a pilot-tested intervention/invention in promoting innovations.

8.5 Factors Affecting Innovation Processes

Based on experience from the industrial sector, Rothwell (1992) identified several characteristics of successful innovation. These are emphasis on satisfying users' needs; establishing good internal and external communications; treating innovation as a corporate task; committing resources early; screening new projects openly; appraising projects regularly; and ensuring high quality management. In addition, the following aspects are also relevant for various entities involved in agricultural R&D activities.

8.5.1 Factors Contributing to Accelerated Innovation Processes

Over the years, practitioners have used a number of mechanisms to acquire and share knowledge and accelerate innovation within the agricultural sector. The key practices are use of open innovation (OI) concept; innovation intermediaries; innovation platforms (IPs); and "best-fit" as opposed to "best practices" for scaling up.

8.5.1.1 Open Innovation (OI)

OI is closely linked to the spillover effects of technological solutions. The applicability of research results over a range of agricultural production conditions or environments, often cutting across geographical and national boundaries, generally referred to as "spill-over effects" (Evenson 1989). "Spill-inns" refer to a situation where a country adopts a technology developed elsewhere – intelligent borrowing. This practice reduces national research costs and time required to generate innovation. Gains from spill-inns are important to all research organizations. But experience reveals that smaller NARS gains more than the larger NARS.

The concept of OI assumes that firms/organizations/groups can and should use external as well as internal ideas as they advance their technology in creating innovations (Chesbrough 2003). It is based on the idea that someone, somewhere has already solved the problem currently being faced.

A systematic OI protocol includes the following steps: (i) formulating the right problems; (ii) calling for OI solutions; (iii) ranking the selection; (iv) identifying and resolving the "yes, buts"; and (v) transferring tacit knowledge.

Common problems encountered in implementing OI initiatives are the initial problem posed maybe wrong; lack of objective means to determine whether the new solution is better than the existing one (solution maybe theoretically/technically possible but not practical given the circumstances); failure to adequately solve the "yes, but" problems; and failure to adequately transfer tacit knowledge. Tacit knowledge transfer is the most difficult problem.

OI as a concept makes considerable sense if properly used, and its use is increasing because of ICT. This has great potential for accelerating the creation of appropriate solutions and innovations through careful selection and intelligent borrowing of technologies. OI is necessary but not sufficient to drive continuous innovation.

8.5.1.2 Innovation Intermediaries

Empirical work demonstrates that innovation does not just happen within the supply side (based on new technological possibilities), or because of articulating user demands (based on social needs and market requirements). It occurs through a complex set of processes that link many different actors – not only developers and users – but also a variety of intermediary organizations, platforms, individuals, and institutional arrangements. Innovation intermediaries are the individuals, firms, and agencies that facilitate innovation by bridging, brokering, and transferring knowledge necessary to bring together a range of different organizations and knowledge needed to create successful innovation (Klerkx and Leeuwis 2009). Intermediate organizations often prove crucial to successful innovations, particularly when their task is to find out what producers (and their end users) want and to search through options within the stock of existing and new knowledge to find what best meets the need. Intermediaries are needed to bring organizations and knowledge together to build supply networks and markets.

Intermediaries or innovation brokers perform a range of management tasks related to innovation, including articulating demand for research; assisting in providing access to technical expertise, markets, and credit; facilitating the formation and strengthening of networks; and training and advocating for policy and regulatory changes (Ugbe 2010).

8.5.1.3 Innovation Platforms (IP)

Another institutional innovation that promotes innovations within the agricultural R&D arena is the IP. An IP is a physical or virtual forum that creates an environment within which to share and discuss ideas, listen and learn, think and talk, and collaborate with the view to innovate. IPs are used to facilitate interactions and learning among stakeholders selected from a commodity chain leading to participatory diagnosis of problems, and for joint exploration of opportunities and investigation of solutions leading to the promotion of agricultural innovation along the VC of the targeted commodity/enterprise (Adekunle et al. 2010; Ugbe 2010).

IPs are established at different levels (Adekunle et al. 2010). At the strategic level, IPs are established at the higher level of governance and management hierarchies, normally at the national or subregional levels (within the country or region). Strategic IPs target chief executives/senior managers of key stakeholder organizations to discuss and facilitate strategies to promote innovation along the targeted commodity or system. They also facilitate the functioning of IPs at the lower levels. IPs established at the lower levels have a different focus. They target frontline staff in organizations that facilitate operations, and who participate in the activities of the platform because of their current roles or skills. Grassroot IPs are also called innovation clusters. The composition of clusters depends on the nature of activities at hand and can change overtime. Cluster members do hands-on work in diagnosing, exploring, and investigating solutions and in facilitating adoption. Strategic IPs and innovation clusters are complementary and reinforce each other in promoting innovation. These platforms also provide an ideal opportunity for monitoring and evaluating impacts and sharing successes and lessons learned.

The key challenges in implementation of IPs are the low capacity of partner organizations (especially the skills required by farmers to understand and articulate key issues); dealing with persistent "hand out syndromes" (because many platforms are supported by external agencies); building new relationships between the private and public sectors and farmers for mutual and sustainable benefits; and ensuring inclusiveness and eliminating marginalization within the platforms (Tenywa et al. 2011). In many instances, these platforms are largely driven by external agencies and therefore sustainability is a key concern.

8.5.1.4 Moving from "Best Practices" to "Best-Fit"

Historically, recommendations for SHF have been in the form of packages called "best practices." However, it has been well documented that SHF demonstrate a stepwise adoption behavior. That means, some components of the package are easily adaptable, and some are not. To be effective and efficient, it is important to distil those components that are easily adaptable by the target group and focus on these in the scaling-up process. Proponents of this strategy (Birner et al. 2009) argue that even if "best practices" have worked elsewhere, to be successful, it is imperative to identify components that "best-fit" the specific conditions and priorities of the target group/development priorities of the country. This is the practical application of the "theory of the second best" in the agricultural R&D arena. This strategy will not only reduce the overall cost of research but also promote successful adoption of recommendations leading to accelerated innovation processes.

8.5.2 Factors Impeding the Innovation Processes

Based on the experiences in the industrial sector, Wycoff (2004) identified the top ten killers of innovation. These are: not creating a culture that supports innovation; not getting buy-in and ownership from business unit managers; not having a widely understood system-wide process; not allocating resources to the process; not tying projects to company strategy; not spending enough time and energy on the fussy front-end (planning); not building sufficient diversity into the process; not developing criteria and metrics in advance; not training and coaching innovation teams and not having an idea management team. These lessons from the industrial sector are highly relevant in creating the necessary preconditions for successful innovation within the agricultural sector.

8.6 Challenges Facing the R&D Systems in Integrating IS Framework

In many countries, the R&D systems are still focusing on research output(s)/inventions, hoping that other actors will play their part in converting them into innovations. An IS paradigm basically needs a mindset shift from "knowledge/technology" to "innovation" as the ultimate goal of R&D investments. Cultural change and high-quality management of R&D organizations are vital to promote continuous innovation.

A successful innovation capacity requires organizations to build and coordinate capabilities across all functions – a support structure. Fostering innovations requires a multidisciplinary and multi-organizational approach to R&D. Most learning institutions currently produce "domain experts" and not innovators. Networks, partnerships, and teams are critical elements in the innovation process. Additional skills in negotiation, facilitation, and conflict management are needed for all actors to foster innovations.

There should be good internal and external communication mechanisms and planning, monitoring and evaluation systems (PM&E) in place. Appropriate incentive structures and dynamic and open-minded leaders to attract and retain talented managers and researchers with the commitment to develop the necessary human capital on a sustainable basis are vital. In addition, innovation needs to be managed in a more structured manner and removed from "domain experts" control if it is going to produce the required results.

The key challenges that need to be addressed in fully integrating IS perspectives are (i) changing the organizational culture to incorporate innovation as the core value proposition of the organization and developing the necessary policy and governance mechanisms; (ii) creating the necessary capacity to generate continuous innovation – building the collective capability of the networks or systems of actors to interactively link with the view to innovate. The national agricultural higher learning institutes should play a significant role in this process; (iii) scaling-up innovations. Unless interventions are pilot-tested or scaled up sufficiently, it is not possible to create innovation. Attention should be paid to the political and economic context favoring the agricultural innovation processes; (iv) creating the necessary environment and incentive system as well as the investments needed to foster the partnership of the innovation actors and reducing the transaction costs or partnerships and collaboration; (v) collecting and disseminating empirical evidence of successful innovation stories to highlight the utility and value-add of using an IS perspective. This involves conducting credible empirical analysis and documenting and communicating results and experiences; (vi) an approach to addressing factors such as socioeconomic equity and environmental sustainability while generating new wealth and opportunities through the application of IS perspective; (vii) a coherent set of policies that foster innovation. This clearly demonstrates that action is required on many fronts to successfully integrate the IS perspective within the agricultural R&D process.

8.7 Conclusion

ISF is an analytical construct and its use in agricultural R&D is on the increase because of the demand for commercially relevant innovations from agricultural R&D investments. ISs can be defined at different levels – national, sectoral, commodity, and intervention based. However, the most widely used frameworks in agricultural R&D are the commodity-based and intervention-based ISs. Research is front and central in the innovation process. All types of research are still relevant, but development-oriented, applied, and adaptive research is well placed to contribute to innovation and socioeconomic impact. Not all innovations have their origin in a formal R&D system nor are they exclusively technical. Innovations can be technological, organizational, managerial, and institutional as well as service delivery mechanisms. While recognizing the centrality of the research process, innovation explicitly recognizes the importance and contribution of different actors, especially intermediaries, institutes, institutions, and the political economy as well as incentives.

The ISP and VC orientation underscores the need to invest not only in research that generates knowledge but also in high-quality and effective delivery channels, process mechanisms and the stakeholders who will use this knowledge once it emerges along the VC. Support structures, innovation intermediaries, IPs, OI and scaling-up processes are critical to accelerate as well as to mainstream ISP within the agricultural R&D systems. Effective integration of ISP within the R&D systems requires a number of preconditions to be in place. It is an evolutionary process that requires long-term commitment by the national governments, policy makers, and development partners.

REFERENCES

Adekunle, A. A., A. O. Fatunbi, M. P. Jones. 2010. How to Set Up Innovation Platforms. A Concept Guide for the Sub-Saharan African Challenge Program (SSACP). Forum for Agricultural Research in Africa, Accra.

Anandajayasekeram, P. 1997. Farming Systems Research-Extension: Concepts, Procedures and Challenges. *Journal of Farming Systems Research-Extension* 7(1): 1–28.

Anandajayasekeram, P. 2011. The Role of Agricultural R&D within the Agricultural Innovation Systems Framework. Conference Working Paper 6, Prepared for the ASTI/IFPRI-FARA Conference on Agricultural R&D: Investing in Africa's Future-Analysing Trends, Challenges and Opportunities, AccraGhana.

Anandajayasekeram, P. 2018. Assessing the Scalability of a Research and Development Project: Concepts, Framework and Assessment. A Contributed Paper Presented at the 30th International Conference of Agricultural Economists. Westin Bayshore Hotel, Vancouver.

Anandajayasekeram, P., Puskur, R. and Zerfu, E. 2009. Applying Innovation System Concept in Agricultural Research for Development. ILRI (aka ILCA and ILRAD), Addis Ababa, Ethiopia.

Bacon F. and Butler, T. 1998. Achieving Planned Innovation: A Proven System for Creating Successful New Products and Services. Free Press, New York.

Birner, R., Davis, K., Pender, J., Nkonya, E., Anandajayasekeram, P., Ekboir, J., Mbabu, A., Spielman, D. J., Horna, D., Benin, S. and Cohen, M. 2009. From Best Practices to Best Fit: A Framework for Designing and Analysing Pluralistic Agricultural Advisory Services Worldwide. *Journal of Agricultural Education and Extension* 15(4): 341–355.

Chambers, R. 1993. Challenging Professions: Frontiers for Rural Development. Intermediate Technology, London.

Chesbrough, H. W. 2003. Open Innovation: The New Imperative for Creating and Profiting from Technology. Harvard Business Press, Boston, MA.

Collinson, M., ed. 2000. A History of Farming System Research. CABI Publishing and Food and Agricultural Organization of the United Nations, Wallingford and Rome.

Evenson, R. 1989. Spill-over Benefits of Agricultural Research: Evidence from US Experience. *American Journal of Agricultural Economics* 71(2): 447–52.

Hall, A., Mytelka, L. and Oyeyinka, B. 2005. ISs: Implications for Agricultural Policy and Practice. ILCA Brief 2. International Livestock Centre for Africa, Addis Ababa.

Hartmann, A. and Linn, J. 2008. Scaling Up: A Framework and Lessons for Development Effectiveness from Literature and Practice. Wolfensohn Center for Development. Working Paper 5. Brooking Institute.

Klerkx, L. and Leeuwis, C. 2009. Establishment and Embedding of Innovation Brokers at Different Innovation System Levels: Insights from the Dutch Agricultural Sector. *Technological Forecasting and Social Change* 76(6): 849–860.

Metcalfe, S. and Ramlogan, R. 2008. ISs and the Competitive Process in Developing Economies. *Quarterly Review of Economics and Finance* 48: 433–446.

OECD (Organisation for Economic Co-operation and Development). 1997. National Innovation Systems. OECD, Paris.

Rothwell, R. 1992. Successful Industrial Innovation: Critical Factors for the 1990s. *R&D Management* 22(3): 221–240.

Tenywa, M. M., Tukarhirwa, K. P. C., Bruchara, R., Adekunle, A. A., Mugabe, J., Wanjiku, C., Mutabazi, S., Fungo, B., Kashja, N. I. M., Pali, P. and Mapatano, S. 2011. Agricultural Innovation Platform as a Tool for Development Oriented Research: Lessons and Challenges in the Formation and Operationalization. *Learning Publics Journal of Agriculture and Environmental Studies* 2(1): 117–146.

Tripp, R., Anandajayasekeram, P., Byerlee, D. and Harrington, L. 1990. Farming Systems Research Revisited. In "Agricultural Development in the Third World", Edited by C. K. Eicher and J. W. Staatz. The John Hopkins Press, Baltimore, MD.

Ugbe, U. 2010. What Does Innovation Smell like? A Conceptual Framework for Analysing and Evaluating DFID-RIU Experiments in Brokering Agricultural Innovation and Development. Research In to Use Discussion Paper 10. Department for International Development, London.

Winkelmann, D. 1987. Recent Views on Farming Systems. A paper Presented at the ISNAR Workshop, The Hague.

World Bank. 2006. Enhancing Agricultural Innovations: How to Go Beyond Strengthening Research Systems. World Bank, Washington, DC.

World Bank. 2007. Cultivating Knowledge and Skills to Grow African Agriculture. World Bank, Washington, DC.

Wycoff, J. 2004. The Big Ten Innovation Killers and How to Keep Your IS Alive and Well. http://www.thinkmart.com

9

Viable Nutrient Management Options for Sustaining Small Farm Agriculture

S. K. Pedda Ghouse Peera, Satya Prakash Barik, Himansu Sekhar Gouda, Sweta Shikta Mohapatra, Debashis Dash, and Santanu Kumar Patra
Department of Agriculture and Allied Sciences, C. V. Raman Global University, Bhubaneswar, Odisha, India

CONTENTS

9.1 Introduction

More than 80% of the farming community is made up of smallholder farmers. Their operational holdings make about 36% of their total. Despite having a smaller land area, marginal farmers outnumber small farmers by a rate of 3.3–1. (1:0.9). Tribal farmers make up as much as 8.6% of the total. In any case, the weaker parts of the smallholders prevail (GoI, 2019). When all important agricultural inputs are taken into consideration, the percentage of small farmers' expenditure on fertilizers is 41.7 (GoI, 2019).

Despite decades of research, the average N, P, and K recovery efficiency in agriculture has been quite low due to a variety of factors, including improper and imbalanced fertilizer use. This has resulted in a number of social and environmental implications, including a rise in the cost of cultivation, the combustion of fossil fuels, greenhouse gas emissions, contamination of water bodies, and so on.

In the context of small farms, plant nutrition has received little attention. It will be necessary to apply nutrient smart technology in order to maintain production and increase nutrient usage efficiency. These technical choices can assist increase the production system's resilience by decreasing the negative impact of extreme weather events like floods and droughts. Nutrient smart methods include piloting nutrient-efficient cultivars, soil test-based fertilizer delivery, customized fertilizers, improved efficient N fertilizers, site-specific nutrient management, real-time N application, and integrated nutrient management (INM), to name a few, discussed in the following sections.

DOI: 10.1201/9781003164968-11

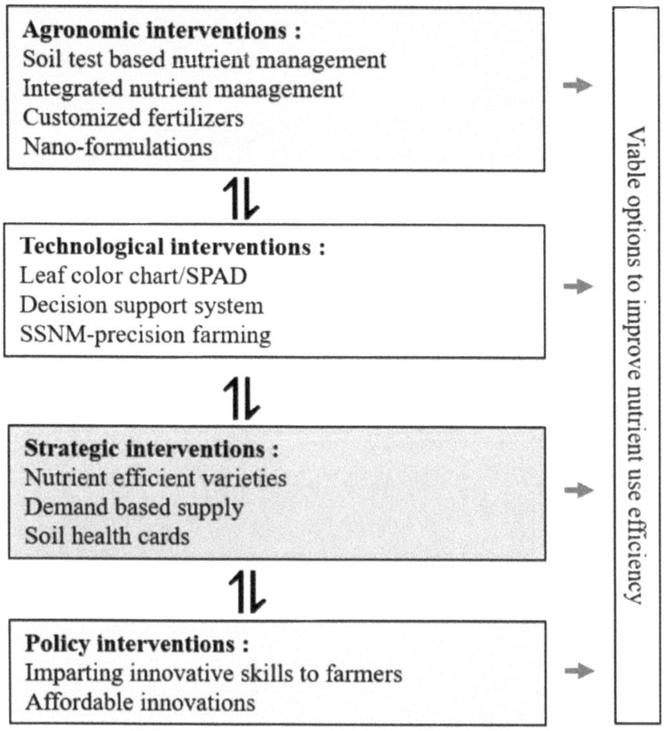

FIGURE 9.1 Interventions for high nutrient use efficiency.

9.2 Principles of Nutrient Smart Practices

The availability of nutrients in the soil and their use efficiency has a direct impact on crop yield. To ensure increased crop yield while being environmentally friendly, efforts should be made to improve fertilizer usage efficiency by applying the 4R principles (right source, right rate, right time, right place) and a variety of management alternatives (Figure 9.1).

The principles that govern effective nutrient management in the context of small farms are unlikely to alter much, but the tactics for improving nutrient usage efficiency may differ based on past management practices: (i) assess the farm's nutrient status using standard soil testing methods/sensing instruments and GIS platform, and generate the fertility rating chart and categorize the insufficient nutrient, if possible, utilizing soil variability map and variable applicator; (ii) evaluate the various nutrient resource possibilities available on the farmer's land; (iii) use a decision support system to bridge the gap between nutrient demand and supply; (iv) a short-term forecasting technique for agricultural fertilizer management that aids in estimating the crop's fertilizer need under varying weather conditions; (v) use decision support systems to recommend fertilizers based on soil test results, crop requirements, and available nutrient resource alternatives; (vi) use nutrient-efficient crop varieties (Shahid et al., 2021).

9.3 Agronomic Interventions

These interventions are viable crop management options i.e., soil test-based fertilizer recommendations, integration of available organic materials with chemical fertilizers, selection of efficient customized fertilizers and nano-fertilizers to reduce carrying and application cost.

9.3.1 Customized Fertilizers

According to the Fertilizer (Control) Order 1985, customized fertilizers are "multi nutrient carriers designed to contain macro and/or micro nutrient forms, both from inorganic and/or organic sources, manufactured through a systematic process of granulation, satisfying the crop's nutritional needs, specific to its site, soil, and stage, validated by a scientific crop model capability developed by an accrediting body, validated by a scientific crop model capability developed by an accrediting body. Customized fertilizers efficiency is influenced by edaphic and plant factors." These fertilizers are made up of urea, DAP, MOP, ZnS, bentonite S, and boron granules that are mixed and crushed to provide the correct proportions of N, P, K, S, and micronutrients. Using customized fertilizers, such as 16:22:14:4:1:0 and 8:15:15:0.5:0.15:0, has resulted in a 22.5% fertilizer decrease in paddy (N:P:K:S:Zn:B) (Gautam et al., 2019). Customized fertilizers are the greatest available alternative for correcting site-specific multi-nutrient deficits in soils and are thus beneficial in achieving target crop yields at a low cost. These fertilizers meet the nutritional needs of the crop in relation to the specific place, soil, and plant growth stage. The promotion of these fertilizers will increase fertilizer efficiency, and because these fertilizers also include micronutrients, the farmer will not have to purchase micronutrients separately, saving money.

9.3.2 Slow-Release Fertilizer

Slow-release fertilizers include less permeable coatings and inhibitors (either nitrification or urease inhibitors or both) inside the formulation or coating to control nitrification or urea hydrolysis or both, e.g. neem-coated urea, S-coated urea, and so on. After being applied to the soil, urea breaks down into ammonium ions (NH_4^+), which are subsequently oxidized to nitrite (NO_2^-) and finally nitrate (NO_3^-). Due to the quick conversion of urea to nitrate, nitrogen losses in the form of leaching loss as nitrate or gaseous emission as nitrous oxide are significant.

Several chemical and natural inhibitors for urea hydrolysis (urease inhibitors: NBPT – *N*-(*n*-butyl) thiophosphoric triamide [trade name – Agrotain], *N*-phenylphosphoric triamide [2-NPT], hydroquinone [HQ], phenyl phosphorodiamidate [PPD/PPDA]) and nitrification (nitrification inhibitors: nitrapyrin, DCD, N coating urea prills with less soluble compounds like S, polymers, and other materials like plaster of Paris, resins, and waxes results in slow- or controlled-release urea fertilizers. The use of neem oil to urea delays urea hydrolysis and nitrification, allowing for a more gradual release of nitrogen. The availability of nitrogen to the plant is increased as a result of the gradual release of nitrogen, as does the nitrogen use efficiency. When urea briquettes were used instead of conventional urea granules, nitrous oxide emissions were reduced by 17–25%. The use of urea coated with neem extends the time it takes for nitrogen to be released, resulting in extended periods of greenness in the plants. In July 2004, the Ministry of Agriculture added neem-coated urea to the FCO. When compared to prilled urea (PU), application of NCU resulted in 18% and 21% reductions in NO_3-N leaching and N_2O emission, respectively, and a 6% increase in yield (Mohanty et al., 2018).

9.3.3 Integrated Nutrient Management

INM is a practice that aims to achieve harmony by judicious use of chemical fertilizers in conjunction with organic manures, well-decomposed crop residues, green manures, recyclable waste, compost (including vermicompost), legumes in cropping systems, biofertilizers, and other locally available nutrient sources for sustaining soil fertility (Mahajan and Gupta, 2009). According to the findings, replacing half of the prescribed nitrogen with organic sources boosts crop yields and soil carbon in semiarid rainfed systems of India (Prasad et al., 2016), details of which are given in Table 9.1.

9.4 Technological Interventions

Recommendations made using site-specific nutrient management (SSNM) ensure that the proper quantity of fertilizer is applied at the correct time to the plant, minimizing losses. SSNM uses principles and tools for supplying nutrients to plants in order to achieve higher yields by synchronizing demand and supply

TABLE 9.1

Proven INM Recommendations for Higher Crop Yields and Soil Health

Crop	Proven INM Recommendations for Higher Crop Yields and Soil Health
Rice	• INM practice 50% recommended doses through chemical fertilizer + 50% through dhaincha green leaf manuring during wet season. • 75% inorganic N + 12.5% N through FYM + 12.5% N through well-decomposed poultry manure. • FYM @ 5 t ha^{-1} yearly once during kharif season + fertilizer NPK @80:40:40 kg ha^{-1} during both kharif and rabi season for rice-rice system of eastern Odisha. • Green manuring with *dhaincha* and application of urea at the rate of 15 kg N ha^{-1} three weeks after rice transplanting and at the rate of 15 kg N ha^{-1} at panicle initiation. • Dhaincha @ 8.25 t + 100 kg N ha^{-1} (urea) in four equal splits (basal 21 DAT, panicle initiation, and first flowering to rice). • BGA biofertilizer @ 10 kg ha^{-1} + 90 kg urea.
Maize	• Full dose of RDF (120-60-40 kg N-P-K ha^{-1}) + 10 t ha^{-1} FYM. • 25% RDF + *Azotobacter chroococcum* + phosphate solubilizing bacteria + green manuring with sunhemp + compost @10 t ha^{-1}.
Legumes	• Crop rotation with legume crops does not need a major increase in cultivation costs, but it does fix a considerable quantity of atmospheric nitrogen in the soil, which is accessible to the following crop in the cropping cycle. • Incorporating mung beans into a rice-wheat cycle boosted rice grain output by 10–14% and soil organic carbon (SOC) by 35%. • Adding mung beans to a maize-wheat rotation boosted wheat grain production by 5–11% and SOC by 24%. • In a rice-wheat cycle, replacing wheat with chickpea boosted rice grain production by 5–8% and SOC by 6%

after taking into account climatic yield potential, yield target, and nutrient availability from all possible indigenous sources, which vary from site to site. Following SSNM-based N application, field trials in different locations of South Asia revealed a 30–40% improvement in N-use efficiency of irrigated rice. The use of decision support tools for nutrient recommendations as well as additional tools such as the leaf color chart (LCC), Soil Plant Analysis Development (SPAD) meter, and green seeker for real-time nitrogen management to synchronize demand and supply is part of site-specific nutrient management.

9.4.1 Leaf Color Chart

Within a growth season, changes in rice leaf color are a sensitive indication of leaf nitrogen status. LCC, a low-cost and user-friendly real-time N management tool, can readily identify the change in leaf color. This tool can help farmers choose when and how much nitrogen fertilizer to use. It has four or more color panels ranging in tint from yellowish green to dark green. These color bands correspond to the color spectrum of rice leaves, which spans a range of leaf N deficiency to excessive leaf N content. CLCC-based N recommendations might save 25% of fertilizer while reducing N_2O emissions by 13–21% in rice (Mohanty et al., 2018; Nayak et al., 2017; Shahid et al., 2021).

9.4.2 RiceNxpert App

Based on leaf color analysis, this app recommends applying N fertilizer at the appropriate timing and dose to meet the crop's needs. This app is simple and easy to use. In terms of urea, it makes a direct recommendation. The following is the procedure for using this app: (i) capturing photographs of ten fully expanded healthy rice leaves against a white backdrop, (ii) cropping the acquired leaf image, (iii) extracting R G B values from image, (iv) pantone extraction for matching R G B values, (v) comparison with four standard rice leaf color categories, and (vi) if the green color is less than the required standard leaf color based on pantone, N fertilizer is recommended for distinct rice ecologies. Following RiceNxpert's advice over RDF, farmers' field validation demonstrated a 7–27% increase in partial factor production from applied nitrogen.

9.5 Strategic Interventions

9.5.1 Nutrient-Efficient Varieties

Nitrogen-efficient varieties are those that respond well to low fertilizer dosages and have greater internal and use efficiency even when soil nutrient levels are low. Reduced nutrient availability in soil arises owing to either an inherent nutrient shortage in the soil or because the nutrient present is not accessible to the plant owing to unfavorable soil conditions. In both cases, the agricultural plant is deprived of nutrients, resulting in lower growth and yield. Nutrient-efficient crop varieties can withstand nutrient deficiency through a variety of mechanisms, including well-developed nutrient transporters, a deeper and wider root system, rhizosphere engineering through root exudation, and a symbiotic relationship with microbes that promote nutrient availability, e.g. Swarna and Sarjoo-52 under low N; MTU 2400, Rasi under low P; CSR 10, IR-30864 under low Zn; Mahsuri, Phalguna under Fe toxicity.

9.5.2 Soil Health Card

The Soil Health Card (SHC) initiative was established by the Indian government in February 2015. A farmer's SHC is a printed report that is given for each of his land holdings. It provides data on 12 factors, including N, P, K, S, Zn, Fe, Cu, Mn, Bo, pH, EC, and OC.

This information is crucial for fertilizer recommendations and farm-specific soil amendments. Soil testing is done once every two years under this system so that required management activities may be taken to enhance the state of soil fertility. Phase I (2015–2017) of the Central Government's SHC plan delivered 10.74 crore cards, while Phase II (2017–2019) provided 11.69 crore cards to farmers. SHC offers farmers with a comprehensive fertility status, allowing them to administer major and micronutrients according to their desired yield and crop requirements. Farmers have increased their output by 5–6% as a result of the widespread use of SHC.

9.5.3 Nano-Fertilizers

Nanoparticles' unique properties, such as their high sorption capacity, higher surface-to-volume ratio, and controlled-release kinetics to targeted sites, make them a potential plant growth enhancer. Nano-fertilizers can be used as a smart delivery mechanism for nutrients to plants because of their features. When compared to traditional fertilizers, the release of nutrients from nano-fertilizers is highly gradual and regulated, resulting in little nutrient loss, increased nutrient usage efficiency, and less nutrient leaching into groundwater. Slaton et al. (2001) found that the smaller the particle size of ZnO, the higher the Zn solubility in soil, resulting in increased Zn absorption. Several researchers stressed the use of ZnO NPs to improve the effectiveness of traditional Zn fertilizer for improved absorption and crop yield (Tanha et al., 2020; Tirani et al., 2019, Shang et al., 2019). Small amounts of nano-fertilizers offered a benefit that was comparable to or larger than large volumes of conventional fertilizers, according to Kheyri et al. (2019). The studies focused on the use of nano-fertilizers in agriculture, particularly rice, on which more than half of the world's population relies directly for their food needs.

9.6 Conclusions

Only 30% of India's entire cultivable area is covered with fertilizers where irrigation facilities are available, while the other 70% of arable land is primarily rainfed. Organic manures are frequently used by farmers in these places as a source of nutrients that are easily available on their farm or in their community. In small farms, arresting the loss of soil organic matter is the most effective weapon in the fight against unabated soil degradation and threatened agricultural sustainability in India's tropical regions, particularly those under the influence of arid, semiarid, and subhumid climates, for the preservation of soil quality and future agricultural productivity. Economically feasible residue recycling technology, inexpensive and environmentally benign nano-fertilizers, and precise handheld yield prediction tools are all relevant policy interventions.

REFERENCES

Gautam, P., Lal, B., Nayak, A. K., Tripathi, R., Shahid, M., Meena, B. P., Singh, S., Srivastava, A. K., 2019. Nutrient management and submergence-tolerant varieties antecedently enhances the productivity and profitability of rice in flood-prone regions. J. Plant Nutr. 42, 1913–1927.

GOI (Government of India), 2019. Agricultural statistics at a glance – 2019. Ministry of Agriculture and Farmers Welfare, Government of India: New Delhi. http://agricoop.nic.in/agristatistics.htm

Kheyri, N., Norouzi, H., Mobasser, H. R., Torabi, B., 2019. Effects of silicon and zinc, nanoparticles on growth, yield, and biochemical characteristics of rice. Agron. J. 111(6), 3084–3090.

Mahajan, A., Gupta, R. D., 2009. Integrated nutrient management (INM) in a sustainable rice-wheatcropping system. Springer Science & Business Media.

Mohanty, S., Swain, C. K., Tripathi, R., Sethi, S. K., Bhattacharyya, P., Kumar, A., et al., 2018. Nitrate leaching, nitrous oxide emission and N use efficiency of aerobic rice under different N application strategy. Arch. Agron. Soil Sci. 64(4), 465–479.

Nayak, A. K., Mohanty, S., Raja, R., Shahid, M., Lal, B., Tripathi, R., Bhattacharyya, P., Panda, B. B., Gautam, P., Kasthuri-Thilagam, V., Kumar, A., Meher, J., Rao, K. S., 2017. Customized leaf colour chart (CLCC): A paradigm shift in real time nitrogen (N) management in lowland rice. ICAR-National RiceResearch Institute: Odisha.

Prasad, J. V. N. S., Rao, C. S., Srinivas, K., Jyothi, C. N., Venkateswarlu, B., Ramachandrappa, Dhanapal G.N., Ravichandra, K., Mishra, P.K., 2016. Effect of ten years of reduced tillage and recycling of organic matter on cropyields, soil organic carbon and its fractions in Alfisols of semi arid tropics of southern India. Soil Till. Res. 156, 131–139.

Shahid, M., Munda, S., Khanam, R., Chatterjee, D., Kumar, U., Satapathy, B. S., Mohanty, S., Bhaduri, D., Tripathi, R., Nayak, P. K., Nayak, A. K. 2021. Climate resilient rice production system: Natural resources management approach. Oryza. 58 (Special Issue), 143–167.

Shang, Y., Hasa, M., Ahammed, G. J., Li, M., Yin, H., Zhou, J., 2019. Applications of nanotechnology in plant growth and crop protection: A review. Molecules 24(14), 2558.

Slaton, N. A., C. E. Wilson, S. Ntamatungiro, R. J. Norman, and D. L. Boothe. 2001. Evaluation of zinc seed treatments for rice. Agronomy Journal 93: 152–157.

Tanha, E. Y., Fallah, S., Ali, R., Pokhrel, L., 2020. Zinc oxide nanoparticles (ZnONPs) as nanofertilizer: Improvement on seed yield and antioxidant defense system in soil grown soybean (*Glycine max* cv.Kowsar). Sci. Total Environ. 738, 140240.

Tirani, M. M., Haghjou, M. M., Ismaili, M., 2019. Hydroponic grown tobacco plants respond to zinc oxide nanoparticles and bulk exposures by morphological, physiological and anatomical adjustments. Funct. Plant Biol. 46(4), 360–375. doi:10.1071/FP18076.

10

Improving Livelihood and Farm Income of Small-Scale Farmers through Nutrition Sensitive Agriculture

Girijesh Singh Mahra[1], V. Sangeetha[1], Pratibha Joshi[2], Sujit Sarkar[3], and Renu Jethi[4]
[1]*Division of Agricultural Extension, ICAR-Indian Agricultural Research Institute, New Delhi, India*
[2]*CATAT, ICAR-Indian Agricultural Research Institute, New Delhi, India*
[3]*ICAR-Indian Agricultural Research Institute Regional Station, Kalimpong, India*
[4]*ICAR-Vivekananda Parvatiya Krishi Anusandhan Sansthan, Almora, Uttarahand, India*

CONTENTS

10.1 Introduction

Agriculture is the backbone of Indian economy and employs nearly half (54.6%) of the workforce in the country (Census 2011). Out of 320 million work force, 170 million are employed in agriculture. India is the largest producer of milk, spices, millets, jute, ginger, bananas, mangoes, papayas, coconut, pea and okra. India ranks second in several important crops including rice, wheat, groundnut and tobacco. At present India is the second-largest producer of fruits and vegetables with a 10.9% and 8.6% share of the global production respectively (FAO 2019). India has witnessed unprecedented growth in the livestock sector also. Since 1990s, India's milk production has more than tripled growing from 53.9 to 176.3 Mt (Ministry of Agriculture and Farmers Welfare, Government of India 2018). This growth has been primarily driven by large scale investments in agriculture which led to increases in per capita calorie

DOI: 10.1201/9781003164968-12

and protein availability (World Bank 2007). Overall, in terms of production India is among the top five countries having more than 80% of agricultural commodities. In terms of geographical features also, it shows great diversity being a homeland to most rainy places, arid regions, and snowy mountains. On the economic front, India's growth performance in the last three decades has been very special as there has been a notable shift in the course of economic development (Kotwal et al. 2011). The average annual real per capita GDP grew at 5% from 1990–1991 to 2017–2018 (Economic Survey 2020–2021). India made rapid progress in the agricultural sector also (Joshi et al. 2006). In the past three decades, food grain production increased from 176.39 to 168.38 Mt (Ministry of Agriculture and Farmers Welfare, Government of India 2018). During the same period, the horticulture sector also witnessed remarkable increases in area and production. The annual production grew rapidly, expanding from 127.7 to 311.7 Mt.

On one hand, India has achieved food security by the help of Green Revolution technologies and the massive adoption of high-yielding varieties in staple foods by farmers but on other hand the bigger goal of nutritional security still remains a daunting challenge. Despite significant economic growth, the performance of India in reducing malnutrition and undernutrition has not been noteworthy (Gillespie et al. 2012; Dixit 2011; Haddad and Zeitlyn 2009; von Braun et al. 2005; Subramanyam et al. 2011). International Food Policy Research Institute (2016) clearly indicates how India still lags behind many poorer countries but Africa is tackling malnutrition effectively. Though the country almost doubled the rate of stunting reduction in the past 10 years compared with the previous decade, it is still home to almost a third (31%) of the global burden of stunting (International Food Policy Research Institute 2018). Moreover, against the global nutrition targets of 2019, no progress has been achieved in reducing under-five wasting, adult female obesity, adult male obesity, adult female diabetes, adult male diabetes and WRA anemia. India has shown a dismal performance in reducing child overweight also which has slightly increased from 1.9% in 2006 to 2.4% in 2015 (Global Nutrition Report 2018). Another disturbing situation is the emergence of triple burden of malnutrition. In addition to hunger and micronutrient deficiencies, over-nutrition which is excess intake of calories leading to obesity has emerged in part from a lack of harmony between food systems and the promotion of human health where some are getting too little food while others are getting too much of the wrong food. According to the census 2011, total children in India are 1.587 million out of which male constitute 0.829 million and female constitutes 0.758 million. With nearly 20% of the 0–4 years' child population of the world, India is home to the largest number of children in the world where nearly every fifth child in the world lives in India. Around 40% of children remain undernourished with their growth and development impeded irrevocably, over the lifetime. According to UNICEF-WHO-The World Bank Joint Child Malnutrition Estimates 2014, child mortality rate in India is 56 per 1000 live births, 20% of total children (under 5 years) are wasting and 28% of total children (under 5 years) are born underweight. Table 10.1 summarizes the scenario of malnourished children in India.

India is home to the largest number of children in the world where nearly every fifth child in the world lives in India but around 40% of children remain undernourished with their growth and development

TABLE 10.1

Scenario of Malnourished Children in India (Census 2011, SRS 2010 and NFHS-3)

Total Children	1.587 Million (0.829 Million Male and 0.758 Million Female)
Childbirth	0.25 million annually
Child survival	0.175 million annually
Mortality	0.08 million annually
Child sex ratio	914/1000 (declined from 927/1000 – Census 2001)
Infant mortality rate	47 (male: 46, female: 49) per 1000 live births
Under five mortality rate	59 (male: 55; female: 64) per 1000 live births
Low weight babies	22% of total child population
Underweight	42.5% of children 0–5 years
Anemic children	79% children (6–35 months)
Immunization level of infants	44% of total child population

impeded irrevocably, over the lifetime. India has met only four out of ten Millennium Development Goals by 2015, lacking far behind in overcoming extent of child malnutrition, Infant mortality rate, maternal mortality rate and sanitation. Despite of many National level nutritional programs like Integrated Child Development Services (ICDS), Mid-Day Meals Program, Public Distribution System (PDS), community public works programs, and Annapurna Program, India is home to 40% of the world's malnourished children and 35% of the developing world's low-birthweight infants. The reason behind these gaps lies in inability of India to understand the location specific and family specific nutritional needs of more than six lakh villages having diverse climate, cropping pattern, eating habits, dietary pattern, socio-economic conditions, and communication needs. There is a gap between nutrition and agriculture which needs to be bridged.

10.2 Agriculture and Nutrition – The Connect and the Disconnect

A number of researchers in recent years exploring the linkages between agriculture and nutrition have reported that agriculture interventions can play a promising role in enhancing nutritional status (Gulati et al. 2012; Kadiyala et al. 2014; Kadiyala et al. 2012; Das et al. 2014; Bhaskar et al. 2017). Another study by Headey (2011) came out with similar findings suggesting that India needs to exploit the links between agriculture and nutrition. The recent establishment of the Sustainable Development Goals (at least 12 of the 17 SDGs are related to nutrition) and the UN's labeling of the decades as the "The Decade of action on Nutrition" shows that there is a global commitment to tackling the varied challenges of malnutrition. Furthermore, the need for agriculture to support health and nutrition has been reflected in United Nations 2030 agenda for sustainable development. It reflects nutrition's central role in achieving sustainable development, as well as its interrelationship with the majority of development sectors. The potential of agriculture for producing nutritious food has not been appropriately tapped for reducing malnourishment, especially in India. Production of quality food in adequate quantity alone may not improve nutritional outcomes, unless malnutrition is addressed by adopting a multisectoral approach (FAO 2013; Das et al. 2014). Making agriculture more nutrition sensitive requires a new way of thinking, planning, implementing, and partnering as well as the active engagement of a variety of stakeholders from multiple sectors. There is a dire need to identify critical entry points where nutrition goals can be incorporated into agro-food systems. Therefore, agricultural policies and programs need to be more nutrition-sensitive for improving the nutritional outcomes.

10.3 Nutrition Sensitive Agriculture for Improving Livelihood

10.3.1 Agri-Nutri Pathways

Agriculture sector cannot be overlooked in combating malnutrition because it plays a critical role in provision of food, livelihoods, and income. However, a major concern that prompts urgent attentions in which ways agriculture can be leveraged for improving nutrition and health. Agriculture and nutrition are next door neighbors as one complements the other and share a common entry point "food," but there has been a lack of synchronization in both aspects. In principle, reshaping agriculture to improve nutrition will require steps in four main areas: bridging knowledge gaps, ensuring that the agriculture, nutrition, and health sectors do not work in isolation or at cross-purposes, seeking out and scaling up innovations and successes, and creating an enabling environment (Fan and Pandya-Lorch 2012). Ruel et al. (2013) have also identified six pathways through which agricultural interventions can impact nutrition: (i) food availability and access from own-production; (ii) income from sale of commodities produced; (iii) food prices from changes in supply and demand; (iv) social and economic empowerment of women through increased access to and control over resources; (v) women's time through contribution in agriculture, which can be either positive or negative for their own nutrition and that of their children; (vi) women's health and nutrition status through engagement in agriculture, which also can have either positive or negative impacts, depending on exposure to toxic agents and the balance between energy intake and expenditure.

10.4 Nutrition Sensitive Agriculture and Food Systems

By now it is well recognized that the economic prosperity and agricultural growth and production in India has not significantly translated into nutritional status of the population (Fan and Pandya-Lorch 2012; Gillespie et al. 2012). Past strategy of agricultural development has been focused only on improving production of staples and providing remunerative prices to farmers. Though this strategy has helped in achieving self-sufficiency at the national level, it has not been able to curb poor nutritional outcomes which results primarily due to inadequate intake of micro and macro nutrients. Against this background, nutrition-sensitive agriculture (NSA), which aims to achieve nutritional and health objectives through sustainable agricultural development, has picked up momentum. It has emerged as a new agricultural development agenda that changes the way we think about food systems and looks at food security through a nutrition lens. As self-sufficiency in food production has not resulted in the nutritional transformation needed for reducing high levels of malnutrition, it aims to narrow the gap between available and accessible food and addresses the utilization dimension of food and nutrition security. In recent years, researchers from different organizations and disciplines have devoted considerable effort to the topic of NSA which has evolved our understanding of agri-nutri (A2N) linkages and has further changed nutrition and health scenarios in India.

In simple terms, NSA is an approach that explicitly incorporates nutrition goals in agricultural interventions and at all stages of the food chain from production, processing, retailing, and consumption. It seeks to ensure the production and consumption of a variety of affordable, nutritious, culturally acceptable, and environmentally sustainable foods in adequate quantity and quality to meet the daily dietary requirements of the population.

10.5 Interventions to Make Agriculture and Food Systems Nutrition Sensitive

10.5.1 Biofortification

As monotonous, staple-based diets are still the norm for poor households, making staple food itself more nutritious has been recognized as one of the strategies in tackling malnutrition. Biofortification is a way to improve the nutrient quality of food supplies with increased bioavailability by using modern biotechnology techniques, conventional plant breeding, and agronomic practices. It can be very effective in poverty driven areas where population rely on cheaper and more widely available staple foods such as wheat and rice for sustenance. The research focus of ICAR-Indian Agricultural Research Institute (ICAR-IARI) (New Delhi) has been continuously reoriented to address contemporary development challenges. With malnutrition emerging as one of the alarming problems in the country the institute has recognized the pressing need for the nutritional biofortification of the staples and thus have initiated many programs in different crops. In the recent years, many biofortified varieties of cereals, pulses, oilseeds, vegetables and fruits have been developed, for example, country's first provitamin-A, rich maize variety Pusa Vivek QPM9 Improved and Pusa HM4, HM8, and HM9. Improved mustard varieties like Pusa Mustard 30 and Pusa Double Zero Mustard 31 with low erucic acid content have been adopted by many villages. β-Carotene rich cauliflower (Pusa Beta Kesari 1) and Fe-rich pearl millet (Pusa Composite 701 and 443) were also released by the institute with much success.

Biofortified crops offer a rural-based intervention that, by design, initially targets the undernourished in rural and remote areas and poor households who eat large amounts of food staples daily. Gradually it penetrates to urban populations as production surpluses are marketed. Biofortification as a food-based intervention can be very effective in reducing micronutrient deficiencies along with other existing mix of micronutrient interventions such as fortification and supplementation programs (Bouis and Welch 2010). In the present context of agriculture, the challenge of global climate and rising food prices will negatively affect the nutrition of the poor forcing them to rely on staple foods like cereals to keep them from going hungry. Therefore, biofortification is a way forward to supplement the diets of the poor as this strategy relies on foods people habitually eat.

10.5.2 Agriculture Diversification

Diversification of agriculture has the potential to improve access to more diverse and nutritious foods, a key component of meeting the "Minimum Acceptable Diet" for nutrition of children. This necessitates synchronizing transitioning of food consumption with food production at farmer's field level. One of the farmers' most important tasks is to produce food of sufficient quantity (enough calories) and quality (with the vitamins and minerals needed by the human body) to feed the world population sustainably for healthy, productive lives. Evidence shows that diversification of agriculture has the potential to improve the quantity and quality of diets in households. It can also reduce income poverty through produce sales and agricultural labor. Furthermore, agriculture improvement decreases food price volatility by diversification and increases government revenues that can be used to finance health care, education, and nutrition interventions.

10.5.3 Post-Harvest Value Addition in Community Agri-Nutri Security Centres (CANSCs)

One of the primary objectives of Community Agri-Nutri Security Centres (CANSCs) is to bring value-added nutri products like soy nuts, pearl millet pop, soymilk, multigrain cookies, beetroot mango bar, etc., to the village and to integrate these nutrition rich snacks into the diets of the individuals, especially infants, young children and women of reproductive age. It adopts a community-based food system approach in which these CANSCs have been set up in villages at public locations with small-scale processing facilities comprising of spice coating pan, solar dryer, and microwave oven to impart skill in minimal processing techniques to rural women. In these centers, rural women are trained in processing and making value-added A2N products using machineries which makes clear the role of value-added A2N products in nutrition security. The centers are set up in remote rural areas, where many people are not able to access fresh, locally grown food. In the face of such state of rural areas where a majority of individuals are obese and children malnourished a set of complementary strategies are required to eradicate these maladies. One way out is to boost the development of small-scale enterprises on agriculture related to nutrition to increase the income as well as to ensure nutritional security. From this perspective, CANSCs give rural women the opportunity to get together to share the cost, planning and preparation of healthy, and nutri rich value-added foods. Moreover, it highlights how important it is for women to develop entrepreneurial skills to be financially independent and specialize in processing of locally available crops into nutri rich foods which can become a strategic lever for dynamic growth in developing countries where rural entrepreneurship plays an important role. Besides focusing on capacity building in rural women these centers act as a forum to impart A2N education, to conduct awareness programs, to stream videos on agri-nutrition, to organize rural women-scientists interface etc. It focuses on how social and behavioral change principles can guide these interventions and affect diets of rural people (women). Furthermore, it facilitates social learning, inter exchange of ideas and will help rural people to discover common interests that can lead to the formation of new groups focusing on a variety of social issues.

10.5.4 Farming Systems for Nutrition

Farming System for Nutrition (FS4N) comes under the production pathway through which agriculture can improve nutritional outcomes. It provides agricultural remedies to the nutritional maladies. It considers the farm, the farm household, and off-farm activities in a holistic way to take care not only of farming but also aspects of nutrition, food security, sustainability, risk minimization, income, and employment generation, which make up the multiple objectives of farm households. In order to integrate nutrition and agriculture or livelihood security, the FS4N model should be followed which will target the problem of malnutrition through mainstreaming nutritional criteria in the selection of the components of a farming system involving crops, livestock, poultry, and aquaculture wherever feasible. As agricultural interventions are more likely to be successful if human capital component such as nutrition education and development component such as WASH are incorporated, it strongly advocates their incorporation. Most of the farmers are cultivating same varieties of crop year after year. The need of hour is to diversify the cropping pattern and to include such varieties of crops which are rich in nutrients. The cropping

TABLE 10.2
Nutrient Rich Varieties Developed by ICAR-IARI, New Delhi

Crop	Name of Varieties	Rich in Nutrient
Cereals		
Wheat	HI 8627 (Malavkirti)	Rich in vitamin A
	HI 8663 (Poshan)	High quality for preparation of dalia, suji, and pasta making
		High quality for preparation of dalia and suji
	HD 4672 (Malavratna)	High zinc content
	HD 2932 (Pusa Wheat 111)	Good quality for suji and dalia
	HI 8498 (Malavshakti)	Rich in nutrients like Beta-carotene (precursor of Vitamin A) and
	HI 8713 (Pusa Mangal)	essential micronutrients like iron and zinc
		Rich in micronutrients like iron, zinc, and copper
	HI 1563 (Pusa Prachi)	
Pulses		
Chickpea	Pusa 372 (Desi)	Dal and besan making
	Pusa Chamatkar (BG 1053) (Kabuli)	Good cooking quality
Lentil	Pusa Vaibhav	Rich in iron
Vegetables		
Carrot	Pusa Vasudha	Rice in total carotenoid and lycopene, TSS, and minerals
		Rich source of total carotenoids
	Pusa Rudhira	Roots possess high B-carotene content (7.552 mg/100 g fresh weight
	Pusa Nayanjyoti	
Radish	Pusa Jamuni	Higher anthocyanins (8.04 mg/100 g) and ascorbic acid (44.8 mg/100 g)
	Pusa Gulabi	High total carotenoids (15.6 mg/100 g), anthocyanins (4.41 mg/100 g) and optimal ascorbic acid (39.2 mg/100 g)
Vegetable Mustard	Pusa Sag 1	Higher carotene and ascorbic acid
Oilseeds		
Mustard	Pusa Mustard 29 (LET 36)	low erucic acid
	Pusa mustard 21 (LES-127)	<2% erucic acid
	Pusa Karishma (LES- 39)	<2% erucic acid
	Pusa mustard 30 (LES-43)	Low erucic acid
Fruits		
Mango	Pusa Shreshth	Vitamin C (40.3 mg/100 g pulp) and carotene content (10,964 g/100 g pulp)
	Pusa Pratibha	Vitamin C (34.9 mg/100 g pulp) and carotene content (11,474 g/100 g pulp)
	Pusa Lalima	Vitamin C (34.7 mg/100 g pulp) and high-carotene content (13,028 g/100 g pulp)
	Pusa Peetamber	Rich in vitamin C (39.8 mg/100 g pulp) and carotene content (11,737 g/100 g pulp)
Grape	Pusa Navrang	High antioxidant content

Source: ICAR-IARI (2014), Technological Options for Enhanced Productivity and Profit, Indian Agricultural Research Institute, New Delhi.

pattern at village level needs to be analyzed and according to climatic conditions of a particular village, nutria-rich varieties can be grown. ICAR-IARI, New Delhi has a basket of such varieties which are rich in nutrients as shown in Table 10.2.

10.5.5 Women-Managed Kitchen Gardens

Kitchen gardening does not talk about new concepts and instead emphasizes home food production such as fruits and green vegetables on a small patch of land. It is one of the most cost-effective ways to produce nonstaples, which can significantly contribute to household nutritional security through the supply and consumption of diverse vegetables in rural households. As vegetables and fruits provide vitamins and minerals which are essential for a healthy growth and development, kitchen gardening can act as an invaluable tool in increasing the adoption and consumption of fruits and green leafy vegetables. In addition to the nutritional value of kitchen gardens, it can also help in supplementing household income. Agriculture as a source of food is the most fundamental and direct pathway by which household agricultural production contributes to improved nutrition via consumption of crops cultivated by the household. This makes sense to promote kitchen gardening and good agricultural practices at the household and village level and building capacity of rural women in home food production for food and nutrition security in vulnerable communities. This approach can help in turning around the life of rural people.

10.6 Role of Agricultural Extension in Nutrition Sensitive Agriculture

10.6.1 Agri-Nutri (A2N) Smart Village Model

A2N Smart Village is a model integrating agriculture and nutrition that has been conceptualized by the scientists in the Division of Agricultural Extension, ICAR-IARI, New Delhi. It is a latest approach enriched with highly relevant and modern assets to achieve nutrition security. The focus has been on practical aspects such as the importance and role of local resources in bringing a positive change. Moreover, as gender dynamics play a crucial role in agriculture and how it impacts nutrition, primary focus of the model has been on promoting gender equity and closing the gender gap. Women being the agents of change at the household and community level, we have entered the age of women. Many studies report that economic and political empowerment of women could transform the global landscape over the next 20 years. However, this trend is already evident on the economic front as the explosion in global economic productivity in recent years has been driven by fostering human resources and employment opportunities for women. For instance, in the Southeast Asia, the predominance of women in the export manufacturing sector has been the key driver of the region's economic success. In the agricultural sector also, women account for half of the world's food production even when they don't have the reliable access to land, credit, equipment, and markets. Owing to the global evidence of the role of women in transforming the world, A2N takes gender equality as the strongest stone of the pathway to reach the destination of a smart village. Another research priority in this new agenda is on improving the efficiency and effectiveness of the supply of nutritious foods, particularly fruits and vegetables; animal source foods, such as milk, meat, eggs, and fish; and pulses, which include various types of beans and peas, groundnuts, and lentils, in order to improve their quality and safety and reduce their relative prices.

This envisioned approach integrates agriculture and nutrition in a smart and unique way by targeting different components such as nutri farming system, A2N education, capacity building and Self-Help Groups (SHGs)-based nutri forums for social learning. It is a framework which includes various agriculture interventions, awareness campaigns and capacity building programs, field days and field demonstrations on nutri rich crops and varieties, farmers-scientists interface, exposure visits, streaming of videos on healthy practices as well as on nutri rich varieties, minimal processing techniques of pulses, fruits and vegetables, etc. The interventions target various stakeholders such as farmers, farm women, SHG members, anganwadi workers, school children, etc.

Malnutrition is a multidimensional concept which is determined by a range of problems such as access to diverse and safe diet, WASH, disease burden, health services, environmental factors along with a myriad of socio-political factors. From this perspective, A2N smart village operates on the assumption that the achievement of nutrition goals are more likely if we leverage other development sectors like social protection, food safety and quality assurance, and water and sanitation. This concept has gained attraction, as evidenced by the A2N smart village bulletin that has been introduced under the project of ICAR-IARI as a great initiative to tackle malnutrition and illustrates the importance of understanding and embracing multispectral partnership models if we have to achieve nutrition goals.

10.6.2 Women Empowerment and Education

More than 80% of rural women engaged in the labor force work in the agriculture sector, which provides a significant opportunity to unleash the gender dimensions of agriculture-nutrition linkages. Strengthening women's position both within the agricultural sector and within the household can significantly improve households' nutrition and health. There is evidence that women's employment does have a beneficial effect on household nutrition, female education also has a strong inverse relationship with IMR, educated women have greater roles in household decision-making, particularly those relating to nutrition and feeding practices.

10.6.3 Use of Information and Communication Technology (ICT)

Awareness creation about the problem of nutrition and need for diversified diet is essential for the change in dietary habit of the consumers and production pattern of the farmers. In this scenario, ICT can play an important role right from the production to marketing for the farmers and for the price, information, and awareness among the consumers.

10.6.3.1 Collective Approach

To protect small and marginal farmers in India (about 86%) from national and international price volatility, collective marketing viz. SHG model, cooperative model, contract farming can be used. Through such group actions, farmers can increase their bargaining power, economies of scale, and income from agriculture. They can market their produce collectively, reducing their transaction cost, reducing the price of the commodity to consumers, transferring benefits to the consumers also. Thus, the affordability component of nutrition security can be fulfilled by collectives of farmers.

10.6.4 Farmer Led and Market Led Extension

Agriculture being the primary source of income for farmers and rural people, these extension approaches can impact nutrition and health through the income pathway. In recent years, there has been a reduction in the number of cultivator households implying that the proportion of households for whom agriculture was the source of income has declined and this brings forth the weakness of the pathway which focuses on agriculture as a source of income. Agriculture has not been a viable source of income especially for most smallholders and women, as they lack an effective voice in influencing research and extension priorities. Thus, farmer led extension can be started by organizing farmers into functional groups, such as SHGs, Farmer Interest Groups (FIGs), Commodity Associations (CAs), and/or other types of Farmer Organizations (FOs). These FOs can provide an effective channel for both dissemination of technology to smallholder and feedback to research and extension. Maharashtra State Grape Growers Association (MRDBS) is an excellent example of such an approach. Moreover, in post-WTO regime, marketing extension so far a peripheral issue in the extension scenario needs to be brought to center stage since production is significantly dictated by market requirements and farmers most of the time fails to get a good price for their production.

Income pathway is of huge importance as the effect of income on nutrition is neither direct nor easily predictable. The allocation of income for expenditure on different food and non-food items is always modified by what is available, affordable, and convenient to purchase. The decision is driven by a myriad of other factors like who gets to decide what to purchase and how much to purchase. Thus, increasing farmers' income can help in increasing dietary diversity and social consumption greatly influences nutritional outcomes and health.

10.7 Enhancing Economic Security through Nutrition Sensitive Agriculture

Identifying critical entry point for promoting nutritious crops, varieties, technologies, and practices is very important for successful adoption and integration into the existing agro-food system. Ruel et al. (2013, 2018) identified six pathways through which agricultural interventions can impact nutrition out of which income from the technologies or commodities is the most vital factor beside the dimension of nutrition security. It is because of the fact that income is widely seen as a major motivating factor for adoption of any technology and used as a welfare measure as it is strongly correlated with an improved standard of living, such as food, clothing, shelter, health care, education, and recreation. Therefore, NSA has to be economically viable both to the producer as well as to the consumer for its adoption among the farming communities and consumer society as well. On many occasion, the traditional extension system found to highlight the technological aspect only like yield, nutritive value of crops or varieties only for its promotion ignoring the economic dimension and end up with limited adoption. Several past studies found that nutritious crops and its varieties are not successful in addressing specific nutritional crisis but also helps to raise the income of farmers more than from conventional farming system. Number of studies on role of kitchen gardening in livelihood security admits that it not only ensured the nutritional security but also guaranteed the additional income. Brun et al. (1989), Rowe (2009), Mitchell and Hanstad (2004), and Dilrukshi, Hashini, Galhena et al. (2013) provided ample evidences on how kitchen gardening contributes towards income generation, improved livelihoods, household economic welfare as well as promoting entrepreneurship and rural development.

Dilrukshi, Hashini, Galhena et al. (2013) reported that families in mountain areas of Vietnam were able to generate more than 22% of their cash income through home-gardening activities. A study from Southeastern Nigeria reported that tree crops and livestock produced in home gardens accounted for more than 60% of household income (115). ICAR-IIHR reported that in project areas, out of total vegetable produce of kitchen garden 650 kg was used for household consumption and surplus of 250 kg was sold in local haat to earn gross income to the tune of approximately Rs. 11,500 in a year. The fact that home production is less cost-intensive and requires fewer inputs and investment is extremely important for resource-poor families that have limited access to production inputs. Yet it has been assessed that moderately rigorous crop and livestock production in home gardens can generate as much revenue per unit area as field crop production. Similarly it is found that farmers can earn from revenue from cultivation of nutritious varieties of major crops or nutritious crops. For example, Chetia (2020) reported that cultivation of nutritious black rice which is rich in anthocyanin antioxidant was more remunerative than growing traditional rice. The farmers in Assam earned Rs. 200 to Rs. 250 per kg for nonorganic variety of black rice and Rs. 500/kg for organic black rice. Sharma (2019) found that cost-benefit ratio of black paddy is 1.78 and for rice is 1.87, indicating both the cases is favorable to the farmers, and they can earn more revenue from black rice. Similarly, Deb (2016) in his study calculated the value of return from cultivation of nutritious golden rice under optimistic and conservative scenario. The internal rate of return (IRR) was 70.50% for optimistic scenario and 69.86% for conservative scenario. Under the optimistic scenario, the Benefit Cost Ratio (BCR) is 751 when the benefits and costs are discounted at % indicating that each dollar investment on Golden Rice research and development will generate a social return which wills be 751 dollars. The adoption of nutritious breed of poultry that is Kadaknath chicken proved to be highly remunerative than the conventional breed. It contains more protein (approx 25%) than white chicken meat (approx 18%). Rich in vitamins B1, B2, B6, B12, C and E, niacin, calcium, phosphorus, and iron, the inky fowl has a lower fat content of only 0.73–1.03% as opposed to the approx 25% in other chicken breeds. Kadaknath meat contains 18 amino acids of which, eight are essential for the human body. Black meat is also considered healthier as it contains 24% linoleic acid, as opposed to the 21% in white chicken. Furthermore, the cholesterol content in Kadaknath is only 184.75 mg/100 g as opposed to the white meat. Mishra et al. reported that the chicken sold for Rs. 400–900 a piece, at the farm level against Rs. 150–300 for a regular chicken and farmers were earning additional income of Rs. 90,000–1,00,000 per year by rearing of 100 birds generating employment of 460 man days/unit/year. Gizaki and Madukwe (2019) in a study on Socioeconomic Impact of SG 2000 Quality Protein Maize (QPM) Technology Delivery on Beneficiaries found that the farmers who adopted the technology had more income, better living standard, and accumulated more wealth from cultivating QPM.

Recently, the consumer preferences shifted toward intake of nutria crops specially vegetables and fruits. Therefore, farmers should tap this opportunity for their income security. The increasing health consciousness among the citizen added a new vista for cultivation of nutritional vegetables and fruits among the farming communities. Joosten et al. (2015) in this aspect showed that the profits per hectare are 3–14 times higher in vegetable production than in rice production while profits per labor-day are double. But public and private investments in agriculture are still largely focused on staple crops and oil crops, not on commodities rich in micronutrients (Haddad et al. 2016; Pingali 2015). Ebert (2014) reported that even a small plot of 500 m^2 traditional nutria leafy vegetables of *Amaranthus* can earn a farmer in to a supplementary income of US$250 a year. The acceptance of nutria-farming system also depends on the consumers as they are the demand generator and their preferences, economic capacity, and purchasing behavior are determining factor for the success of nutria-sensitive agriculture. In this regard, Talsma et al. (2017) reported that in India, a survey of rural and urban households (n = 602) as well as students, academicians, and businesspersons (n = 110) showed that 70% and 28%, respectively, were willing to consume genetically engineered nutritious crops and were willing to pay a premium of 19.5% for golden rice. De Steur et al. (2010) found similar results in China with 62% of respondents (n = 944) willing to accept folate-rich rice biofortified through genetic engineering and willing to pay a premium of 34% for it. Factors influencing acceptance of biofortified crops vary largely by type of crop and country but also by characteristics of consumers, such as age, sex, socioeconomic status, and being a disliker or liker of biofortified crops. Hence, the policy makers as well as farming community should keep these facts in mind and accordingly shape their farming system for successful promotion of NSA.

10.8 Conclusions

Competing and complementary theories on NSA have been provided by researchers from different disciplines and literature is growing in recent years. The above essay tries to refresh our understanding on the linkages between agriculture and nutrition and highlights some of the key interventions that can help in improving the synergies between agriculture and nutrition. Furthermore, it brings to notice the need to have a multisectoral and multidisciplinary approach that brings together extension workers, economists, nutritionists, policy makers, and other stakeholders to address key knowledge gaps and drive a change in India's agriculture and nutrition policy and program processes.

REFERENCES

Bhaskar, A. V. V., Nithya, D. J., Raju, S. and Bhavani, R. V. 2017. Establishing agriculture-nutrition programmes to diversify household food and diets in rural India. Food Security 9(5): 981–999.

Bouis, H. and Welch, R. 2010. Biofortification—a sustainable agricultural strategy for reducing micronutrient malnutrition in the Global South. Crop Science 50: 20–32.

Brun, T., Reynaud, J. and Chevaussus-Agnes, S. 1989. Food and nutrition impact of one home garden project in Senegal. Ecology of Food and Nutrition 23: 91–108.

Chetia, P. 2020. Black Rice Cultivation: Success Story of Local Farmers Who Earn Good Income in Assam. https://krishijagran.com/agriculture-world/black-rice-cultivation-success-story-of-local-farmers-who-earn-good-income-in-assam/

Das, P. K., Bhavani, R. V. and Swaminathan, M. S. 2014. A farming system model to leverage agriculture for nutritional outcomes. Agricultural Research 3(3): 193–203.

De Steur, H., Gellynck, X. and Storozhenko, S. 2010. Willingness-to-Accept and Purchase Genetically Modified Rice with High Folate Content in Shanxi Province, China. Appetite 54118–125.

Deb, U. 2016. Returns to Golden Rice Research in Bangladesh: An Ex-ante Analysis. Technical Report. https://www.researchgate.net/publication/344222671

Dixit, A. 2011. Poverty and food security in Gujarat, India. European Journal of Development Research 23 (1): 129–150.

Ebert, AW. 2014. Potential of underutilized traditional vegetables and legume crops to contribute to food and nutritional security, income and more sustainable production systems. Sustainability 6: 319–335; doi:10.3390/su6010319

Economic Survey.2021. Components of Gross Domestic Product at Current Prices (2020-21). Accessed on April 202. https://www.indiabudget.gov.in/economicsurvey/doc/Statistical-Appendix-in-English.pdf

Fan, S. and Pandya-Lorch, R. 2012. Reshaping Agriculture for Nutrition and Health. Washington, DC: International Food Policy Research Institute (IFPRI).

FAO. 2013. The State of Food and Agriculture. Rome: Food and Agriculture Organization of the United Nations.

FAO. 2019, FAO in India. http://www.fao.org/india/fao-in-india/india-at-a-glance/en/ Accessed on: 20 January 2021.

Galhena, D. H., Freed, R. and Maredia, K. 2013. Home gardens: a promising approach to enhance household food security and wellbeing. Agriculture and Food Security 2(8): 1–13. https://doi.org/10.1186/2048-7010-2-8

Gillespie, S., Harris, J. and Kadiyala, S. 2012. The Agriculture Nutrition Disconnect in India: What Do We Know? IFPRI Discussion Paper 1187. Washington, DC: International Food Policy Research Institute.

Gizaki, L. J. and Madukwe, MC. 2019. Socioeconomic impact of SG 2000 quality protein maize (QPM) technology delivery on beneficiaries in Bauchi and Gombe States. European Scientific Journal 15(3). http://dx.doi.org/10.19044/esj.2019.v15n3p169

Gulati, A., Ganesh-Kumar, A., Shreedhar, G. and Nandakumar, T. 2012. Agriculture and malnutrition in India. Food Nutrition Bulletin 33: 74–86.

Haddad, L. and Zeitlyn, S. 2009. Lifting the curse: overcoming persistent undernutrition in India. IDS Bulletin 40(4): 1–8.

Haddad, L., Hawkes, C., Webb, P., Thomas, S., Beddington, J., Waage, J., Flynn, D. 2016. A new global research agenda for food. Nature 540: 30–32.

Headey, D. 2011. Pro-Nutrition Economic Growth: What Is It, and How Do I Achieve It?, Background Paper for the Conference – Leveraging Agriculture for Improving Nutrition and Health. New Delhi: Organised by the International Food Policy Research Institute, 10–12 February.

International Food Policy Research Institute. 2016. Global Nutrition Report 2016: From Promise to Impact: Ending Malnutrition by 2030. Washington, DC. Accessed on 10 Februray 202. DOI:http://dx.doi.org/10.2499/9780896295841

International Food Policy Research Institute. 2018. Global Nutrition Report 2018: Shining a Light to Spur Action on Nutrition. Washington, DC. Accessed on 10 Februray 2021, https://globalnutritionreport.org/resources/nutrition-profiles/asia/southern-asia/india/

Joosten, F., Dijkxhoorn, Y., Sertse, Y., Ruben, R.2015. How does the Fruit and Vegetable Sector contribute to Food and Nutrition Security? LEI Nota 2015–2076, June 2015. Wageningen: LEI Wageningen UR.

Joshi, P. K., Birthal, P. S. and Minot, N. 2006. Sources of Agricultural Growth in India: Role of Diversification towards High-Value Crops: International Food Policy Research Institute Markets, Trade and Institutions Division Discussion Paper 98.

Kadiyala, S., Harris, J., Headey, D., Yosef, S. and Gillespie, S. 2014. Agriculture and nutrition in India: mapping evidence to pathways. Annals of the New York Academy of Sciences 1331: 43–56.

Kadiyala, S., Joshi, P. K., Dev, S. M., Kumar, T. N. and Vyas, V. 2012. A nutrition secure India: role of agriculture. Economic and Political Weekly 47(8): 21–26.

Kotwal, A., Ramaswami, B. and Wadhwa, W. 2011. Economic liberalization and Indian economic growth: what's the evidence?. Journal of Economic Literature 49(4): 1152–1199.

Ministry of Agriculture and Farmers Welfare, Government of India. 2018. Agricultural Statistics at a Glance. India: Department of Agriculture and Cooperation.

Mishra, A., Singh, S. R. K. and Raut, A. A.2020. Kadaknath Farming for Farmers' Livelihood and Empowerment.ICAR-ATARI, Jabalpur Report. pp 1–15 Accessed on 10.3.2021. http://www.zpd7icar.nic.in/download/Kadaknath%20Farming%20for%20Farmers%E2%80%99%20Livelihood%20and%20Empowerment.pdf

Mitchell, R. and Hanstad, T. 2004. Small Homegarden Plots and Sustainable Livelihoods for the Poor. Rome, Italy. LSP Working Paper 11.

Pingali, P. 2015. Agricultural policy and nutrition outcomes – getting beyond the preoccupation with staple grains. Food Security 7: 583–591.

Rowe, W. C. 2009. "Kitchen gardens" in Tajikistan: the economic and cultural importance of small-scale private property in a post-soviet society. Human Ecology 37(6): 691–703.

Ruel, M. T., Quisumbing, A. R. and Balagamwala, M. 2018. Nutrition-sensitive agriculture: what have we learned so far? Global Food Security 17: 128–153.

Ruel, M. T., Alderman, H. and Maternal and Child Nutrition Study Group. 2013. Nutrition-sensitive interventions and programmes: how can they help to accelerate progress in improving maternal and child nutrition? Lancet 382: 536–551.

Sharma, S. K. 2019. Cost benefit analysis of black rice cultivation in Padumani Development Block of Golaghat District of Assam, India. International Journal of Basic and Applied Research 9(6): 5–8.

Subramanyam, M. A., Kawachi, I., Berkman, L. F. and Subramanian, S. V. 2011. Is economic growth associated with reduction in child undernutrition in India?. PLoS Medicine 8(3): e1000424. https://doi.org/10.1371/journal.pmed.1000424

Talsma, E. F., Boonstra, A. M. and Brouwer, I. D. 2017. Acceptance and adoption of biofortified crops in low- and middle-income countries: a systematic review. Nutrition Reviews 75(10): 798–829.

Von Braun, J, Gulati, A., Hazell, P., Ruel, M. and Rosegrant, M. 2005. Indian Agriculture and Rural Development Strategic Issues and Reform Options, IFPRI Strategy Paper.

World Bank. 2007. From Agriculture to Nutrition: Pathways, Synergies and Outcomes. No. 40196- GLB. Agriculture and Rural Development Department. Washington, DC: The World Bank.

11

Biogas Technology for Improving Livelihoods and Agricultural Sustainability

Shiv Prasad[1], Anoop Singh[2], Dhanya MS[3], Dheeraj Rathore[4], and Amitava Rakshit[5]
[1]*Division of Environment Science, ICAR-Indian Agricultural Research Institute, New Delhi, India*
[2]*Department of Scientific and Industrial Research (DSIR), Ministry of Science and Technology, Government of India, New Delhi, India*
[3]*Centre for Environmental Science and Technology, Central University of Punjab, Bathinda, Punjab, India*
[4]*School of Environment and Sustainable Development, Central University of Gujarat, Gandhinagar, Gujarat, India*
[5]*Department of Soil Science & Agricultural Chemistry, Institute of Agricultural Sciences, Banaras Hindu University, Varanasi, Uttar Pradesh, India*

CONTENTS

11.1 Introduction

The current disposal practices for biomass waste and crop residues have caused widespread environmental concern. The pollution from fossil fuel burning has also paved the way for identifying environmentally sound and economically feasible technologies to manage these wastes (Prasad et al., 2017). The nonconventional energy sources, especially biogas from animal waste and biomass residues, attracted developing nation's attention due to the energy crisis and its easy availability. Anaerobic digestion (AD) is a versatile, efficient, and well-established process used extensively to produce biogas energy (Prasad et al., 2017). Agriculture modernization and high-yielding crop varieties lead to produce a large amount of crop residues that can be used for energy generation (Prasad et al., 2020). In India, nearly 32% of its total primary energy usage is still obtained from bio-resources, and more than 70% of the community

DOI: 10.1201/9781003164968-13

depends on it. Organic waste substances of the farm crop residues, industrial, and municipal origin can be anaerobically transformed into biogas by the action of microbes existing in the cow rumen. Ministry of New and Renewable Energy (MNRE) has realized the bio-waste potential to produce biogas energy and also initiated many programs in the Indian context (PIB, 2020).

Biogas is competent in providing substantial energy to assure its maximum benefits. Hence, India's government has launched various programs to encourage effective biogas technologies and their application in multiple areas of the economy, including sustainable alternatives toward affordable transportation (SATAT) initiatives, the Government of India. For effective utilization of biogas, the Indian government intends to assist financial support to compressed biogas (CBG) under priority sector lending to new projects, especially for micro, small, and medium enterprises to develop CBG plants across India. CBG would be deported in cylinders to OMCs' fuel station network for marketing as a green transportation fuel alternative (MoPNG, 2018). Biogas-to-electricity project in India is also executed to advance biogas technologies for the best use of biomass waste for off-grid power generation. Biogas is generated by AD. It has the potential to transition from fossil fuel economy to bio-waste-based improved fuel economy. Biogas is a cost-effective and environmentally sound energy source that helps improve livelihood, promotes sanitation, and a healthier household environment in rural areas.

Biogas technology is also crucial to improving air quality, as it is a cleaner fuel than conventional fuels like wood, coal, and fossil fuel. Moreover, biogas is an important renewable energy source that can be utilized for cooking, lighting, and heating for significantly improving livelihoods (Othman et al., 2017) and agricultural sustainability in rural areas (Prasad et al., 2017). The Indian government also promotes biogas to improve the green energy mix, decrease imports, and minimize pollution. Besides, this can manage the GHG emissions' problem because methane as a principal component of biogas would otherwise be discharged into the air and create a greenhouse effect (Prasad and Dhanya, 2011; Dhanya et al., 2013a,b). This chapter highlights how biogas technology can be used to manage bio-waste from various origins wisely. It also reviewed the progress and advances in this field to provide better opportunities to improve livelihoods and agricultural sustainability. These efforts are in line with the 2015 Paris Climate Agreement and the 'Make in India' initiative (Dabas, 2020).

11.2 Biogas Technology

The biogas technology is attractive and best suited for utilizing bio-degradable organic waste to produce bioenergy in rural areas. In India, biomass resources, i.e., crop residues from farms and fields, are available in plenty. Biodegradable waste from kitchens, agro-processing units, cattle dung from dairy, and municipality, are primary source for biogas production (Prasad et al., 2017). These bio-wastes are subjected to successive steps (i) hydrolysis, (ii) acidogenesis, (iii) acetogenesis, and (iv) methanogenesis, to produce biogas, which is also called AD (Wirth et al., 2012; Luo et al., 2013). The slurry produced from biogas plants after digestion has more nutrient-enriched manure. This digested organic manure can be applied in agriculture. It is not only improving all types of crop yield but also helps in maintaining soil health.

Biogas is a mixture of methane (CH_4) and carbon dioxide gas (CO_2), hydrogen sulfide (H_2S), and many other trace gases, where CH_4 is only a combustible component (Prasad et al., 2017; Sánchez-Muñoz et al., 2020). Therefore, biogas represents interesting and comparable characteristics with natural gas and propane according to its composition (55–65% CH_4). In terms of calorific value, 1 m^3 of biogas equivalent to 5–7.5 kWh energy, 1.63 kWh of electricity, 1–lb of LPG, 0.54 L of petrol, and 0.52 L of diesel are calculated considering 60% CH_4 content (with LHV of CH_4: 9.5–10 kWh Nm^{-3}). According to Mittal et al. (2018), total biogas generation is nearly 2.07 billion m^3 per year from India's installed biogas plants, equivalent to 0.89 billion kg of LPG.

11.2.1 Biogas Plants

In India, the first biogas plant was developed for AD in 1859 at a leper colony in Bombay. Later on, Jashbhai Patel created various small-scale biogas plants. In 1961, the Khadi and Village Industries Commission decided to promote Patel's design named after him in the entire country. Worldwide

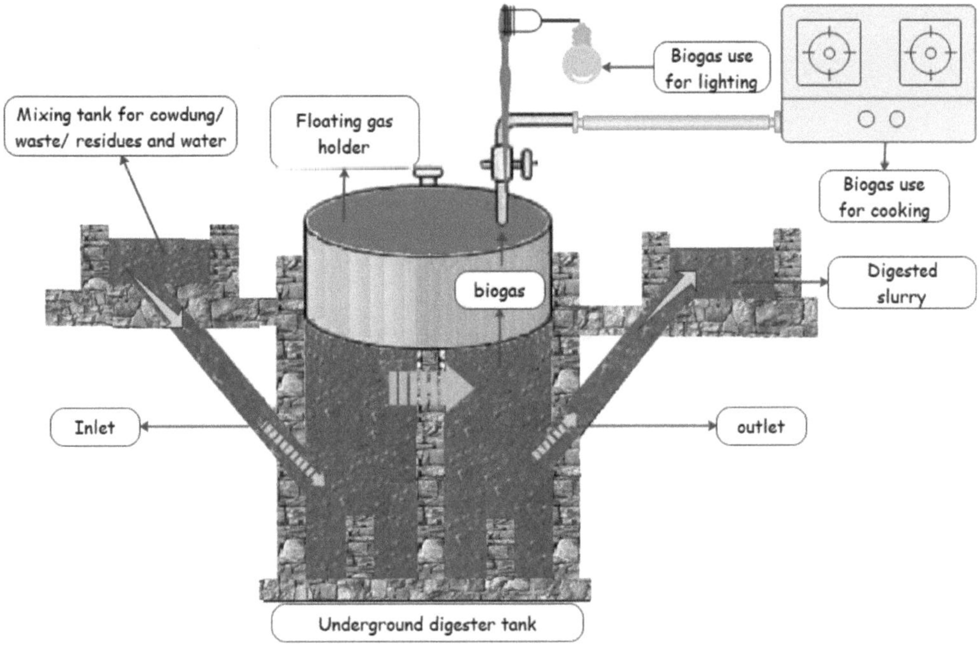

FIGURE 11.1 Small-scale floating type of biogas plant (biogas production and its use).

different efficient biogas plant models suitable for organic waste are available (Nzila et al., 2012; Mittal et al., 2018). However, two basic types, (i) floating gas holder type (Figure 11.1) and (ii) fixed dome type (Figure 11.2), are used (Tiwari et al., 1991). In India, cow dung (gobar) is the traditional primary feedstock utilized in biogas plants to produce CH_4, a significant biogas constituent. Thus, biogas is also called 'gobar gas' (Prasad et al., 2017; Kalyanasundaram et al., 2020).

Biogas plant main parts in both fixed dome and floating drum design are inlet cum mixing tank, digester with a partition wall, gas holder, a conical top placed in the digester, outlet tank, and pipe accessories connecting inlet and outlet tanks to the digester. A floating drum-type biogas model backed by a guide-pipe helps the gas holder move vertically up-down, as shown in Figure 11.1. In this model, the provision of the floating drum's weight is acting as a dead load on biogas accumulated in the drum, leading to the biogas release at constant pressure (Kalyanasundaram et al., 2020).

As shown in Figure 11.2, Khadi Village Industries Commission (KVIC) and Deena Bandhu models are the fixed dome type of biogas model. The Deena Bandhu model constructed using bricks or a ferrocement mixture is widespread in India. The whole plant is made of bricks, cement, and mortar. The digester, gas storage case, and space above the slurry are solely provided in the spherical dome shell. The inlet connects the digester with the slurry mixing tank. All digested slurry is removed from the gas storage chamber and stored in the outlet displacement chamber. Based on the plant size and its establishment, biogas plants are grouped into three subgroups: (i) family-sized biogas plants (<6 m^3), (ii) institutional biogas plants (>15 m^3), and (iii) community biogas plants (>15 m^3) (Kalyanasundaram et al., 2020).

11.2.2 Biogas Production Process and Operational Parameters

Biogas production is an AD process (Singh et al., 2019). Hydrolysis, acidogenesis, acetogenesis, and methanogenesis are significant steps for converting bio-waste to biogas (Lam and Lee, 2011). The entire process is shown in Figure 11.3. Hydrolysis is the beginning stage, where hydrolytic bacteria produce hydrolytic enzymes. These enzymes play a significant role in breaking down the complex molecule into monomers or simpler molecules and creating a suitable environment for acid-forming bacteria. Here, monomers are transformed into fatty acids with a small quantity of hydrogen. Finally, organic acids, especially propionic, butyric, valeric acid, are produced, and it is called acetogenesis.

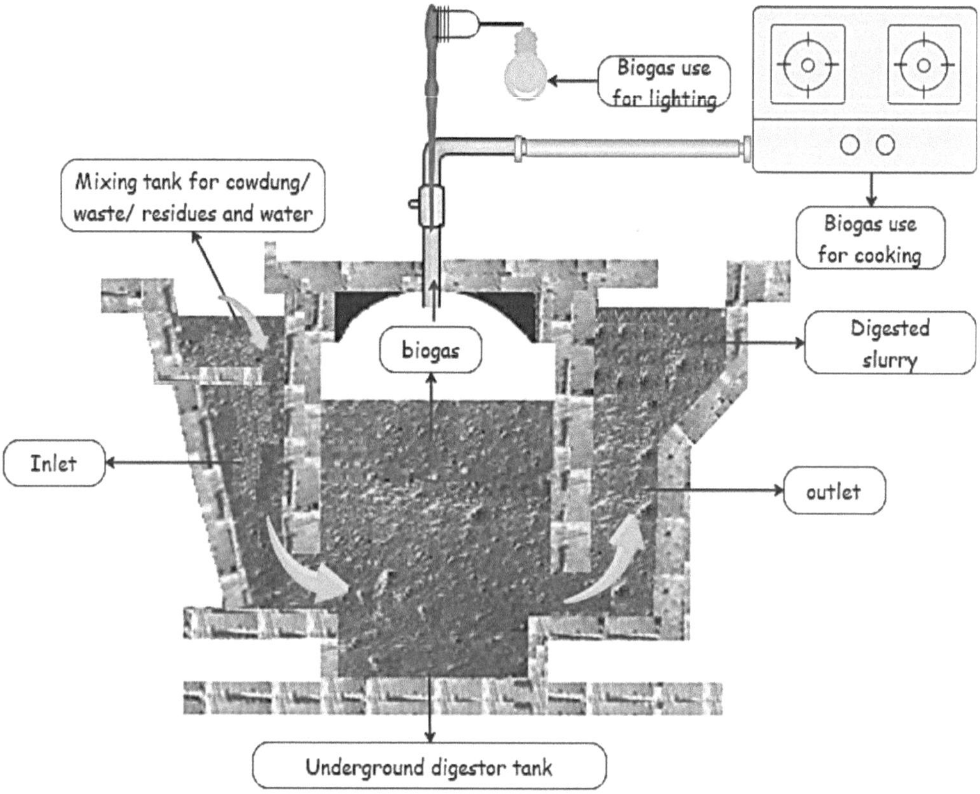

FIGURE 11.2 Small-scale fixed dome-type biogas plant (biogas production and its use).

During AD, in the initial stage of biomass waste hydrolysis, bacterial strains *Bacteroides*, *Bifidobacterium*, *Megasphaera*, *Lactobacillus*, *Propionibacterium*, *Sphingomonas*, *Sporobacterium* play a significant role in converting those complex biomass components to simple sugars, peptides, and fatty acids (Dassa et al., 2014; Prasad et al., 2017; Reddy et al., 2019). In the next step, syntropic acid-forming bacteria (acidogens), both facultative and obligate microbial species like *Ruminococcus*, *Paenibacillus*, and *Clostridium*, help convert simple sugars peptides and fatty acids into volatile fatty acids (VFA). Acetogenic bacteria, especially *Desulfovibrio*, *Aminobacterium*, and *Acidaminococcus*,

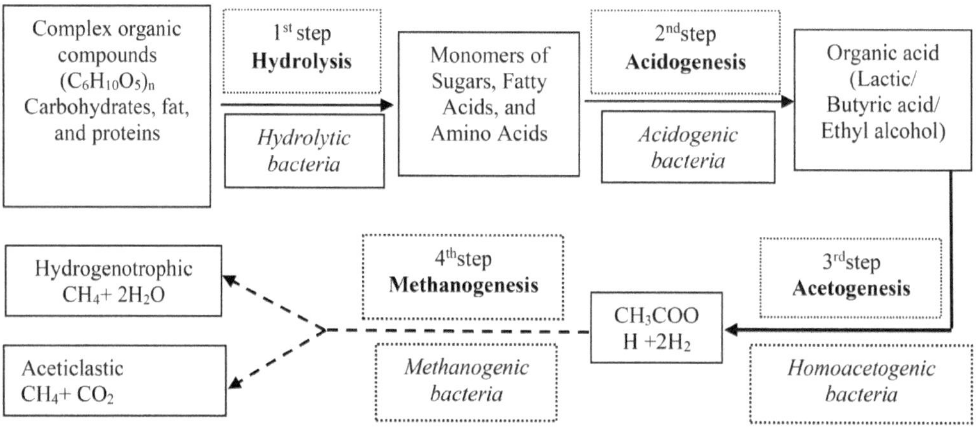

FIGURE 11.3 Biogas production process (Prasad et al., 2017).

convert VFA to acetate (CH_3COOH). Finally, during methanogenesis, methanogens (Archaea), particularly *Methanosarcina, methanosalsus, Methanohalobium, Halomethanococcus*, and *Methanosaeta*, convert acetate into CH_4, a central component of biogas (Dassa et al., 2014; Prasad et al., 2017).

Various factors, such as pH, alkalinity, volatile acid concentration, temperature, and nutrient availability, influence organic substrates' anaerobic fermentation. Composition of the organic substrate, retention time, substrate concentration, and carbon nitrogen ratio, organic loading rate, and degree of mixing also play a significant role during the plant operation. The AD process's efficiency can be improved by modifying the factors mentioned above, such as (i) temperature, which can affect microbial growth. The ideal temperature for mesophilic digestion is reported to be 37°C, and thermophilic digestion is 55–70°C (Feng et al., 2018); (ii) pH interferes with microbial enzymatic activities, a variation in pH derives from metabolites generation in biogas production in the range of 5.5–8.5 are optimum for substrate degradation (Divya et al., 2015); (iii) carbon/nitrogen (C/N) ratio describes the nutritional demands of microbes associated with the AD process. Enhancing the C/N ratio induces a lower protein solubilization rate, creating a reduction in total ammonium nitrogen and fatty acid.

Hence, optimal C/N ratio in the AD process may avoid ammonia inhibition (Divya et al., 2015; Mao et al., 2015); (iv) VFA need to be adjusted to keep the efficiency of the digester better. Liu et al. (2018) reported that the increase of additional VFA of 1,541 mg L^{-1} generates a fast reduction in pH and breaks the digestion process, which negatively influences methanogens activity; (v) digester configuration, especially substrate loading is also significant during AD process (Wu et al., 2015; Baldi et al., 2019); (vi) mixing time must be reduced to cut operational costs because the decrease in mixing period improved the substrate's viscosity and do not let biogas escape (Lebranchu et al., 2017; Kress et al., 2018). Likewise, the organic Rankine cycle (Baccioli et al., 2019) and methanation units (Guilera et al., 2020) increase biogas plants' efficiency.

11.3 Scope and Potential of Biogas Production in India

There is the enormous potential of biogas plant setting and its commercialization in India. According to MNRE, nearly 300 million livestock populations comprising cattle, buffalo, mithun, and yak may contribute directly to the biogas production program. Manure/cow dung, municipal solid wastes (MSWs), other agro-industrial waste, and crop residues are readily available at nominal charges that can be used for the AD process to produce biogas. They have much scope for integrating agro-processing waste and other biomass waste to generate renewable power and heat, reduce GHG emissions, and contribute to the rural socio-economy. The biogas generation from the various potential bio-waste is given in Table 11.1.

TABLE 11.1

Potential of Biogas Production from Different Feedstocks

Substrate	Dry Matter (%)	Biogas Yield Normal L kg^{-1} Fresh Matter	Methane Content Normal L kg^{-1} Fresh Matter
Pig slurry	4–19	20–35	10–21
Cattle slurry	6–11	20–30	11–19
Cattle manure	20–25	60–120	33–36
Poultry manure	34–50	130–270	70–140
Maize silage	28–39	170–230	68–120
Grass silage	15–50	102–200	46–109
Sugar beets	13–23	120–140	65–113
Olive pomace	57–90	92–147	65–104
Wheat straw	91–94	–	135–237
Corn (corn stover)	66–89	–	261–402
Rye	62–93	130	70

Source: Modified and Adopted from Nsair et al. (2020).

TABLE 11.2
Composition of Biogas Produced from Household, Agricultural Waste, and Treated Sludge

Component of Biogas	Bio-Waste from Household	Wastewater from Sewage-Sludge Treatment Plants	Agricultural Crop Waste
		(% by volume)	
CH_4	50–60	60–75	40–70
Carbon dioxide	38–34	33–19	30–60
Nitrogen	0–5	0–1	0–1
Hydrogen	0–1	0–1	0–1
Hydrogen sulfide	0.01–0.09	0.1–0.4	0.3–1

Source: Wilkie et al. (2015), Kumara and Varma (2016), Prasad et al. (2017).

Various researchers studied the potential of many biomass substrates for biogas production. The biogas production potential of water hyacinth, *Lantana camara*, castor, neem, groundnut, and coconut was reported. The coir pith mixed with cattle dung at a 3:2 ratio gave a better gas output, and 80–85% CH_4 content was reported by Radhika et al. (1983). The retention time taken for co-digestion of cattle dung and plant biomass may vary. Considering mesophilic conditions, the retention time for efficient anaerobic fermentation for cattle dung, piggery waste, and poultry waste is 20–30 days, 15–25 days, and 20–40 days, respectively.

The co-digestion of biomass usually increases the materials' digestibility, and higher biogas yields faster than cattle dung alone. Several types of biomass waste and crop straw, stalks-trash from harvesting and processing important crops like maize, cotton, rice, wheat, and water hyacinth anaerobically digested with cattle dung enhanced biogas production in the range of 10–80% (Prasad et al., 2017; Nsair et al., 2020). In addition, several inorganic amendments, like iron salts at varying concentrations (50-mM $FeSO_4$) (70-lM $FeCl_3$), have been found to improve the biogas generation rate. Dhanya et al. (2013a,b) investigated the role of cellulolytic strains Actinomycetes and mixed consortia and reported improve biogas production rate in the range of 8–44% from the admixture of cattle dung.

MSW consists of food waste, paper waste, grass, and branches or leaves that are also a good source for biogas generation. The agro-industrial wastes, including hotel wastes, bio-sludge from the paper and pulp industry, rejected the animal food, fishery by-products, harvest residues, and critical crops, are another potential source for biogas. Food processing waste and wastewaters from various produce processing, especially fruit canning unit waste, animal and dairy product finishing, vegetable canning and processing, breweries, and sugar production, can also be used for biogas production (Wilkie et al., 2015; Kumara and Varma, 2016; Prasad et al., 2017). Biogas from various alternate sources and their composition are given in Table 11.2.

11.4 Policies for Adoption of Biogas Technology

The policies have a crucial role in the successful adoption of any technology. In India, many biogas technology support programs have been launched. They focus on biogas production from animal waste and other waste feedstock utilization to reduce firewood use, improve soil fertility, and reduce indoor air pollution. India's government supports renewable energy technologies (RETs) to provide technologies that utilize energy sources known as environmentally benign as possible. A significant effort has been made to mobilize the resources to realize this potential for rural power development.

Reducing rural poverty through agricultural development is the critical requirement for achieving the Millennium Development Goals (MDGs) of the United Nations General Assembly (UNGA) by 2015, connecting clean energy sources and rural energy access. MDG-ensuring environmental sustainability promotes RETs to expand access to energy services and offers solutions to existing environmental issues and many unexpected benefits. The government of India also launched National Project on Biogas Development (NPBD) in 1981 (Nahar et al., 2017). Under this program, around 50.3 lakh family-sized biogas plants were installed in India in 2018–2019.

Indian government launched a biogas-based off-grid power generation program to promote power generation in rural and remote areas of the country. More emphasis is given to the efficient utilization of available wastes resources from dairy, fisheries, poultry, forestry, agroforestry, waste from food processing units, waste food, and vegetables from the kitchen. Waste-to-energy projects, especially for households' programs, were implemented and financial assistance through India's government and nodal agencies of the states,/Union Territories, KVIC, and private and nongovernmental institutions. Since up to 2018–2019, a total of 389 power projects with 87,990 m^3 biogas and 8,951.5 kW power capacity have been commissioned.

To provide biofertilizers to reduce chemical fertilizers, mitigate climate change, reduce rural women's drudgery and pressure on forests, improve sanitation, control black carbon and air pollution, the National Biogas and Manure Management Program (NBMMP) was launched. In this program, the Indian government mainly focuses on setting up family-type biogas plants. In addition to this, under GOBAR-DHAN (Galvanizing Organic Bio-Agro Resources) scheme, 700 projects across the country have been established to convert cattle dung and solid waste to biogas-based CBG and also produce a sufficient amount of nutrient-rich compost. NBMMP provides subsidies, free maintenance, financial support, training, and service charges. State-level Biogas Development and Training Centres (BDTC) also offer financial assistance to cattle-dung-based power generation facilities. According to NITI Aayog, the Ministry of Road Transport and Highways has permitted the use of BioCNG for motor vehicles. The government of India is also promoting bio-waste technologies to improve BioCNG quality (MoPNG, 2018).

11.5 India Moves to Include Compressed Biogas under Priority Sector Lending

Biogas has a calorific value of 5,000 kcal per m^3 and widely used energy source. CH_4 is a combustible component. However, it also contains CO_2, H_2S with other traces of gases as impurities. These impurities are removed and purified to meet specifications, and then it is compressed for their delivery to use for transport and cooking, and it is called CBG. It is emerging as a substitute for natural gas like LPG and CNG (Spitoni et al., 2019). Under the SATAT initiatives, India's Government has targeted to set up 5,000 CBG plants by 2023–2024 with 15 million tons CBG. According to the India MoPNG, in the SATAT program, Rs. 2 lakh crore investment in the CBG program will provide affordable transport (PIB, 2020).

SATAT has a clear-cut roadmap for implementing this program. A letter of intent (LoI) and MoU for 600 CBG plants have already been signed with the Indian government and leading oil marketing agencies for setting up 900 CBG plants. Under the Flagship program, 5,000 CBG bio-waste-based biogas plants will be installed by 2023–2024. It would also drive several advantages, such as reducing oil imports, decreasing GHG discharges, minimizing crop biomass burning, and boosting farmers' income. SATAT intends to produce 15 million tonnes of CBG from bio-degradable waste. Simultaneously, 50 million tonnes of bio-slurry as manure will be made and used as plant nutrients in crop production (Dabas, 2020).

11.6 Biogas-to-Electricity Project in India

India has excellent expertise in biogas production, keeping this in view, the biogas-to-electricity generation project was launched to use abandoned biomass resources for off-grid power generation efficiently. Bloom Energy, United State, and Energy Power, India, are collaborated to produce electricity from biogas. Energy Power is deploying a new set of modern digesters to efficiently utilize field crop residues and MSW combined with Bloom Energy's solid oxide fuel cell technology in the Shirala district, Maharashtra. USA-Bloom Energy establishes a 4-MW power servers' plant in Shirala that will supply reliable, renewable energy to consumers to cater to their power needs domestically. In addition, bloom technology has developed the advanced process for reaching net-zero carbon emissions from the biogas-to-electricity creation project (TPRMG, 2020; PIB, 2020).

USA-Bloom Energy provides technology for biogas-to-electricity generation projects to supply clean, cost-effective energy in partnership with many countries. India-Energy Power and USA-Bloom Energy are working to implement biogas-to-electricity generation projects and provide employment opportunities and businesses to local peoples (Bloom Energy, 2019). They have installed a 3.5-MW power project to provide a reliable electric supply in Mumbai and Bengaluru's selected townships. It would reduce GHG emissions by 65%. In addition, Tata Power Renewable Microgrid (TPRMG) declared cow-dung-based biogas plants with the capacity of 5-kW power generation nearby villages in Muzaffarpur, Bihar. The cow-dung-fed biogas-plant-digested slurry is processed to create good quality organic manure for agriculture and horticultural and crop production (TPRMG, 2020).

11.7 Achievement of Family/Small-Sized Biogas Plants in India

NPBD was started in 1981. In this program, the emphasis is given to promote family/small-sized biogas plants. The Government of India provides financial support to produce alternative fuel to the rural masses. Its implementation emphasizes providing incentives to link toilets to biogas digesters to enriched organic manure for growing agriculture crops and use in gardening. This program also fulfills the objective to reduce the use of nonrenewable fuels and solve the health issues associated with improper biowaste and sewage disposals (Kurian, 2004; Surendra et al., 2014). According to MNRE (2020), under the NPBD program, 50.3 lakh family/small-sized biogas plants have been installed until 2018–2019. India-state-wise potential and achievements for family/small biogas plants are shown in Table 11.3. India's government also provides financial and technical support to develop a new concept, design, and setting up new biogas plants (Abhishek, 2019; TPRMG, 2020).

TABLE 11.3
India-State-Wise Potential and Achievements for Family/Small Biogas Plants

Indian State/Union Territories	Potential (No. of Biogas Plants)	Achieved up to 2018–2019 (No. of Biogas Plants)
Andhra Pradesh	1,065,000	555,294
Maharashtra	897,000	918,201
Manipur	38,000	2,128
Meghalaya	24,000	10,659
Mizoram	5,000	5,838
Nagaland	6,700	7,953
Arunachal Pradesh	7,500	3,591
Assam	307,000	138,423
Bihar	733,000	129,905
Chhattisgarh	400,000	58,908
Goa	8,000	4,226
Odisha	605,000	271,656
Punjab	411,000	183,835
Rajasthan	915,000	72,132
Gujarat	554,000	434,995
Haryana	300,000	62,825
Himachal Pradesh	125,000	47,680
Jammu and Kashmir	128,000	3,195
Jharkhand	100,000	7,823
Karnataka	680,000	503,935
Kerala	150,000	152,019
Madhya Pradesh	1,491,000	373,037
Sikkim	7,300	9,044
Tamil Nadu	615,000	223,618
Telangana	–	19,694

(Continued)

TABLE 11.3 ***(Continued)***

India-State-Wise Potential and Achievements for Family/Small Biogas Plants

Indian State/Union Territories	Potential (No. of Biogas Plants)	Achieved up to 2018–2019 (No. of Biogas Plants)
Tripura	28,000	3,688
Uttar Pradesh	1,938,000	440,385
Uttarakhand	83,000	363,615
West Bengal	695,000	972
A&N Islands	2,200	97
Chandigarh	1,400	169
Dadra and Nagar Haveli	2,000	681
Delhi/New Delhi	12,900	578
Puducherry	4,300	17,541
Daman and Diu	–	–
Lakshadweep	–	–
Total	12,339,300	50,28,340

Source: MNRE (2020).

11.8 Achievement of Biogas-Based Power Generation (Off-Grid)

This scheme was launched in 2006 for biogas-based power generation (off-grid). Three hundred eighty-nine power projects completed with 87,990 m^3 biogas generation capacity and 8,951.5-kW power have been commissioned since then. Cumulative achievement of biogas-based power generation (off-grid) state-wise is shown in Table 11.4. After successfully implementing the community/institutional biogas plants program for off-grid power generation, the MNRE focused on setting up medium-sized biogas plants to generate electricity and use various thermal and cooling applications.

TABLE 11.4

India-State-Wise Achievement of Biogas-Based Power Generation (Off-Grid)

	Installed/Finalized as on 31.03.2019		
Name of the State	Nos	m^3	kW
Andhra Pradesh	34	4,320	481
Karnataka	70	15,670	1,582
Maharashtra	68	11,690	1,258
Punjab	41	9,980	1,035
Haryana	3	2,540	155
Tamil Nadu	52	30,360	2,854
Uttarakhand	17	1,070	124
Uttar Pradesh	30	4,400	591
Kerala	36	1,010	118
Telangana	25	5,410	574
Madhya Pradesh	6	735	70
Rajasthan	2	120	15
West Bengal	1	340	60
Gujarat	2	285	30
Odisha	2	60	10
Total	389	87,990	8,952

Source: MNRE (2020), Bioenergy Overview.

11.9 Biogas Technology for Improving Livelihoods and Agricultural Sustainability

Biogas technology provides a versatile use of biomass resources to produce renewable energy. Their application can play a crucial role in sinking pollutant gases into the air and mitigating GHG emissions. Thus, it would help save the environment with cleaner and greener approaches and ultimately meet the sustainability goals of livelihoods and agricultural production.

11.9.1 Improving Livelihoods

Biogas can be converted into natural gas quality fuel for use in vehicles. As CH4, it can be used as a supplement to fossil fuel, clean energy for cooking, lighting, generating electricity, and various other household applications (Scarlat et al., 2018). Biogas is used in specially designed stove burners for cooking (Biogas Digest, 2010; Sooch and Gautam, 2013). It is safe for indoor use and does not produce unhealthy smoke, reduces health-care costs, and increases productivity. A biogas plant of 2 m^3 capacity is sufficient for providing the cooking fuel needs of a family of about five persons. Biogas is also used for lighting (MoPNG, 2018). For powering a 60-W equivalent lamp, 0.13 m^3 per hour biogas is required. Biogas lamps allow households without electricity to light their homes after dark, enabling children to study and family members to increase their working hours. Biogas lamps are brighter than kerosene lanterns and safer for domestic use (MNES, 2006).

Biogas can generate electricity, especially useful in remote areas that are not connected to the electricity grid. Biogas can operate a dual-fuel engine to replace up to 80% of diesel oil and diesel engines. Petrol and CNG engines can be easily modified to run 100% on biogas. With the anaerobic treatment of animal waste through the controlled capture of CH_4 and its energetic use, about 13.24 MMTCH_4 per year can be eliminated worldwide. In total, about 4% of the anthropogenic CH_4e missions could be reduced by biogas technology (Bajgain and Shakya, 2005; Jahangirian et al., 2009). Moreover, its use can significantly improve livelihoods and environmental quality (Prasad and Dhanya, 2011; Singh et al., 2014).

Human and animal waste is loaded with several types of pathogens like *Salmonella*, *Escherichia coli*, *Yersinia enterocolitica*, *Giardia lamblia*, etc. Biogas technology helps to improve hygienic conditions through the reduction of pathogens, worm eggs, and flies (Brown, 2006). Besides these health and sanitary benefits, biogas technology provides several indirect social advantages, such as reducing the struggle to collect fuelwood from long distances. Consequently, women and children spend more time on their education when they do not have to collect firewood from long distances (Bajgain and Shakya, 2005). In addition, biogas has been reported to drop lung and eye diseases from cooking in smoky kitchens and improve living standards.

11.9.2 Improving Agricultural Sustainability

Biogas technology provides a comprehensive utilization of biomass waste from agroforestry, farm animals, fishery, and other agricultural activities. Its uses not only protect the environment but also realize agricultural sustainability through recycling those organic wastes wisely. It is known that open exposure of animal dung heaps resulted in the loss of nutrients, mainly nitrogen, potassium, phosphorus, and others, through volatilization and leaching (Matsumoto et al., 1997). Biogas technology helps in biomass waste management, which reduces nutrient losses from this resource-based manure. The digested slurry obtained as a by-product after AD is rich due to its manurial value than compost and FYM (Prasad et al., 2017). It has not only environmental and economic benefits but also very relevant input to sustainable agriculture. Its use surpasses the energy-intensive chemical-based agriculture's adverse effects like global warming and soil erosion. The recovery of plant nutrients from the biogas slurry represents an additional economic return to the farmer using this technology.

Biogas-digested slurry in appropriate combination with chemical fertilizer is a significant step in integrated nutrient management for sustainable agriculture (Singh et al., 2014; Prasad et al., 2017). After digestion, the nutrients composition in the spent slurry residue is higher on the percent dry weight basis than in the undigested material. It is mainly due to the conservation of nutrients during waste digestion and mechanization. In digested biogas, slurry carbon, nitrogen, phosphorous, and other plant micronutrients are readily available for growing plants. Their agricultural land application acts as manure/fertilizer, improves soil – physical and chemical properties medium, and increases crop yields (Prasad et al., 2017). It has been found that the use of biodigester slurry can replace 20–100% chemical fertilizers for different crops. The slurry that has been digested is also a high-grade fertilizer.

Biogas slurry from 1-kg digested dung can yield more N up to 0.5 kg than fresh manure (Sasse et al., 1991). The anaerobically treated biosolids are often easier to handle and are known to inactivate weed seeds, plant pathogens, and pests and decrease the number of phytotoxic compounds in manure. The sludge production by anaerobic treatment is 80% less compared to aerobic treatment processes. By this technology, the plant nutrients, nitrogen, potassium, and phosphorus are not removed. Instead, its enrichment occurs, which is an advantage, and the effluent can also be applied in the agricultural field, which can replace chemical fertilizer. This can solve soil degradation problems in areas where earlier dung has been used as a burning fuel and save costs in purchasing chemical fertilizers, bringing revenue to the household (Li et al., 2005).

11.10 Conclusions

Biogas or gobar gas is a renewable energy source produced from organic waste. It is used for cooking, lighting, and heating extensively. Biogas technology is a well-proven technique for waste recycling and sanitation. This practice offers integrated and sustainable management and safe disposal of waste. India's government is promoting the effective use of biomass and other bio-waste for biogas production and implementing several programs, especially the biogas-to-electricity project, CBG, SATAT, and Bio-CNG for transport. The slurry produced from biogas plants after digestion has more nutrient-enriched manure. It can be used either alone as organic manure or in an appropriate combination with chemical fertilizer in agriculture. It is not only increasing crop yield but also helps in maintaining soil health. Biogas used as energy will improve the green energy mix, decrease imports, and minimize pollution with many other environmental benefits. India's biogas-to-electricity project is also implemented to provide electricity in remote areas by reaching off-grid power generation. It is expected that this new program would enable new job opportunities, enhancing farmers' income, and simultaneously improving rural and semi-urban peoples' livelihood.

REFERENCES

Abhishek, S., 2019. Biogas Production in India is equivalent to 5% of the total LPG consumption https://factly.in/biogas-production-in-india-is-about-5-percent-of-the-total-lpg-consumption.

Baccioli, A., Ferrari, L., Vizza, F. and Desideri, U., 2019. Feasibility analysis of coupling an ORC to a mGT in a biogas plant. *Energy Procedia*, *158*, pp. 2311–2316.

Bajgain, S. and Shakya, I. S., 2005. *A Successful Model of Public-Private Partnership for Rural Household Energy Supply*. SNV, Kigali, Rwanda.

Baldi, F., Pecorini, I. and Iannelli, R., 2019. Comparison of single-stage and two-stage anaerobic co-digestion of food waste and activated sludge for hydrogen and methane production. *Renewable Energy*, *143*, pp. 1755–1765.

Biogas Digest 2010. Basics Information and Advisory Service on Appropriate Technology. Available from http://www.gtz.de/de/dokumente/enbiogas-volume1.pdf.

Bloom Energy, 2019. Bloom Energy announces biogas-to-electricity project in India http://biomassmagazine.com/articles/16517/bloom-energy-announces-biogas-to-electricity-project-in-india

Brown, V. J., 2006. Biogas: a bright idea for Africa. Environmental Health Perspectives, 114, 5 (nih.gov).

Dabas, C., 2020. India moves to include compressed biogas under Priority Sector lending https://www.energylivenews.com/2020/06/30/india-moves-to-include-compressed-biogas-under-priority-sector-lending/

Dassa, B., Borovok, I., Ruimy-Israeli, V., Lamed, R., Flint, H. J., Duncan, S. H., Henrissat, B., Coutinho, P., Morrison, M., Mosoni, P. and Yeoman, C. J., 2014. Rumen cellulosomics: Divergent fiber-degrading strategies revealed by comparative genome-wide analysis of six ruminococcal strains. *PLoS One*, *9*(7), p. 99221.

Dhanya, M. S., Prasad, S. and Singh, A., 2013a. Biogas technology for developing countries: An approach to sustainable development. In *Bioenergy Production by Anaerobic Digestion: Using Agricultural Biomass and Organic Wastes*, N. E. Korress, et al., Eds. (pp. 30–45). Routledge (Taylor and Francis Group), New York, NY.

Dhanya, M. S., Prasad, S. and Singh, A., 2013b. Biogas technology for developing countries. In Bioenergy Production by Anaerobic Digestion, N. E. Korres, P. O'Kiely, J. A. H. Benzie, J. S. West, Eds.Routledge (Taylor & Francis Group), New York (pp. 397–421).

Divya, D., Gopinath, L. R. and Christy, P. M., 2015. A review on current aspects and diverse prospects for enhancing biogas production in sustainable means. *Renewable and Sustainable Energy Reviews*, *42*, pp. 690–699.

Feng, L., Ward, A. J., Guixé, P. G., Moset, V. and Møller, H. B., 2018. Flexible biogas production by pulse feeding maize silage or briquetted meadow grass into continuous stirred tank reactors. *Biosystems Engineering*, *174*, pp. 239–248.

Guilera, J., Andreu, T., Basset, N., Boeltken, T., Timm, F., Mallol, I. and Morante, J. R., 2020. Synthetic natural gas production from biogas in a waste water treatment plant. *Renewable Energy*, *146*, pp. 1301–1308.

Jahangirian, S., Engeda, A. and Wichman, I. S., 2009. Thermal and chemical structure of biogas counterflow diffusion flames. *Energy & Fuels*, *23*(11), pp. 5312–5321.

Kalyanasundaram, G. T., Ramasamy, A., Godwin, B., Desikan, R. and Subburamu, K., 2020. Prospects and challenges in biogas technology: Indian scenario. In *Biogas Production* (pp. 19–37). Springer, Cham.

Kress, P., Nägele, H. J., Oechsner, H. and Ruile, S., 2018. Effect of agitation time on nutrient distribution in full-scale CSTR biogas digesters. *Bioresource Technology*, *247*, pp. 1–6.

Kumara, B. B. and Varma A., 2016. Biomethanization. In *Microbial Resources for Sustainable Energy*. Springer, Cham. https://doi.org/10.1007/978-3-319-33778-4_2

Kurian, P., 2004. *Socio-economic and Environmental Impact of Biogas Program*. Kerala Research Program on Local Level Development, Centre for Development Studies, Thiruvananthapuram. www.cds.ac.in/krpcds/report/pkkurian.pdf

Lam, M. K. and Lee, K. T., 2011. Renewable and sustainable bioenergies production from palm oil mill effluent (POME): Win-win strategies toward better environmental protection. *Biotechnology Advances*, *29*(1), pp. 124–141.

Lebranchu, A., Delaunay, S., Marchal, P., Blanchard, F., Pacaud, S., Fick, M. and Olmos, E., 2017. Impact of shear stress and impeller design on the production of biogas in anaerobic digesters. *Bioresource Technology*, *245*, pp. 1139–1147.

Li, Z., Tang, R., Xia, C., Luo, H. and Zhong, H., 2005. Towards green rural energy in Yunnan, China. *Renewable Energy*, *30*(2), pp. 99–108.

Liu, X. F., Song, D. Z., He, X. Q. and Nie, B.S., 2018. Effect of microstructures on methane adsorption characteristics of soft and hard coal. *Journal of China University of Mining and Technology*, *47*(1), pp. 155–161.

Luo, G., Wang, W. and Angelidaki, I., 2013. Anaerobic digestion for simultaneous sewage sludge treatment and CO biomethanation: Process performance and microbial ecology. *Environmental Science & Technology*, *47*(18), pp. 10685–10693.

Mao, C., Feng, Y., Wang, X. and Ren, G., 2015. Review on research achievements of biogas from anaerobic digestion. *Renewable and Sustainable Energy Reviews*, *45*, pp. 540–555.

Matsumoto, T., Noshiro, M. and Hojito, M., 1997. The effect of barnyard manure of different degradation levels on grass production. In *Plant Nutrition for Sustainable Food Production and Environment* (pp. 591–592). Springer, Dordrecht.

Mittal, S., Ahlgren, E. O. and Shukla, P. R., 2018. Barriers to biogas dissemination in India: A review. *Energy Policy*, *112*, pp. 361–370.

MNES, 2006. Renewable energy for rural applications. Ministry of Non-conventional Energy Sources, Government of India.

MNRE, 2020. Bioenergy Overview. Current Status | Ministry of New and Renewable Energy, Government of India (mnre.gov.in). MoPNG, 2018. http://petroleum.nic.in/sites/default/files/biofuelpolicy2018_1.pdf

Nahar, G., Mote, D. and Dupont, V., 2017. Hydrogen production from reforming of biogas: Review of technological advances and an Indian perspective. *Renewable and Sustainable Energy Reviews*, *76*, pp. 1032–1052.

Nsair, A., OnenCinar, S., Alassali, A., Abu Qdais, H. and Kuchta, K., 2020. Operational parameters of biogas plants: A review and evaluation study. *Energies*, *13*(15), p. 3761.

Nzila, C., Dewulf, J., Spanjers, H., Tuigong, D., Kiriamiti, H. and Van Langenhove, H., 2012. Multi criteria sustainability assessment of biogas production in Kenya. *Applied Energy*, *93*, pp. 496–506.
Othman, M. N., Lim, J. S., Theo, W. L., Hashim, H. and Ho, W. S., 2017. Optimisation and targeting of supply-demand of biogas system through gas system cascade analysis (GASCA) framework. *Journal of Cleaner Production*, *146*, pp. 101–115.
PIB, 2020. Rs. 2 lakh crores to be invested for setting up 5000 Compressed biogas in the country, says Petroleum Minister at MoU signing event for setting up 900 CBG plants; https://pib.gov.in/Pressreleaseshare.aspx?PRID=1674428
Prasad, S. and Dhanya, M. S., 2011. Air quality and biofuels. In *Environmental Impact of Biofuels*, M. A. D. S. Bernardes, Ed. InTech. ISBN: 978-953-307-479-5, DOI: 10.5772/17889.
Prasad, S., Rathore, D. and Singh, A., 2017. Recent advances in biogas production. *Chemical Engineering and Process Technology*, *3*(2), p. 1038.
Prasad, S., Singh, A., Korres, N. E., Rathore, D., Sevda, S. and Pant, D., 2020. Sustainable utilization of crop residues for energy generation: A life cycle assessment (LCA) perspective. *Bioresource Technology*, *303*, p. 122964.
Radhika, L. G., Seshadri, S. K., Mohandas, P. N., 1983. A study of biogas generation from coconut pith. *Journal of Chemical Technology and Biotechnology*, *33*(3), pp. 189–194.
Reddy, K. P., Ajeet, I.A., Denver Lawrence, S., Chandrashekar, 2019. A novel method for kitchen anaerobic and aerobic digestion. *International Journal of Engineering Research & Technology*, *08*(08), pp. 119–128.
Sánchez-Muñoz, S., Barbosa, F. G., Jiménez-Ascencio, J., Mier-Alba, E., Singh, A. K., dos Santos, J. C., Balagurusamy, N., da Silva, S. S. and Chandel, A. K., 2020. Technological routes for biogas production: Current status and future perspectives. In *Biogas Production* (pp. 3–17). Springer, Cham.
Sasse, L., Kellner, C. and Kimaro, A., 1991. *Improved Biogas Unit for Developing Countries*. GATE Publication, Eschborn, *2*.
Scarlat, N., Dallemand, J. F. and Fahl, F., 2018. Biogas: Developments and perspectives in Europe. *Renewable Energy*, *129*, pp. 457–472.
Singh, N. B., Kumar, A. and Rai, S., 2014. Potential production of bioenergy from biomass in an Indian perspective. *Renewable and Sustainable Energy Reviews*, *39*, pp. 65–78.
Singh, B., Szamosi, Z. and Siménfalvi, Z., 2019. State of the art on mixing in an anaerobic digester: A review. *Renewable Energy*, *141*, pp. 922–936.
Sooch, S. S. and Gautam, A., 2013. Present status of renewable energy sources in Punjab. *International Journal of Agriculture, Environment and Biotechnology*, *6*(2), p. 317.
Spitoni, M., Pierantozzi, M., Comodi, G., Polonara, F. and Arteconi, A., 2019. Theoretical evaluation and optimization of a cryogenic technology for carbon dioxide separation and methane liquefaction from biogas. *Journal of Natural Gas Science and Engineering*, *62*, pp. 132–143.
Surendra, K, Takara, D, Hashimoto, A., et al., 2014. Biogas as a sustainable energy source for developing countries: Opportunities and challenges. *Renewable & Sustainable Energy Reviews*, *31*, pp. 846–859.
Tiwari, G. N., Singh, S. K. and Srivastava, L. M., 1991. A comparison of active heating of biogas systems for higher production. *International Journal of Solar Energy*, *10*(1–2), pp. 115–126.
TPRMG, 2020. TP Renewable Microgrid, a Tata Power subsidiary, inaugurates 5 kilowatts (kW) biogas plant in Bihar https://www.business-standard.com/article/companies/tata-power-subsidiary-inaugurates-5-kilowatt-biogas-plant-in-bihar-120122800766_1.html
Wilkie, A. C., Graunke, R. E. and Cornejo, C., 2015. Food waste auditing at three Florida schools. *Sustainability*, *7*(2), pp. 1370–1387.
Wirth, R., Kovács, E., Maróti, G., Bagi, Z., Rákhely, G. and Kovács, K. L., 2012. Characterization of a biogas-producing microbial community by short-read next-generation DNA sequencing. *Biotechnology for Biofuels*, *5*(1), pp. 1–16.
Wu, C., Wang, Q., Xiang, J., Yu, M., Chang, Q., Gao, M. and Sonomoto, K., 2015. Enhanced productions and recoveries of ethanol and methane from food waste by a three-stage process. *Energy & Fuels*, *29*(10), pp. 6494–6500.

12

Emerging and Remerging Diseases and Their Innovative Management in Jute and Allied Fiber Crops under Small Farm

A. N. Tripathi
Division of Crop Protection, ICAR-Indian Institute of Vegetable Research, Varanasi, India

CONTENTS

DOI: 10.1201/9781003164968-14

12.1 Fungal Diseases of Jute and Allied Fiber Crops

In India, jute mostly cultivated in the Eastern-Indian States, namely, Assam, Bihar, Eastern Utter Pradesh, Orissa, and West Bengal. West Bengal alone contributes more than 80% of the total production (110 lakh bales) of Indian jute with the highest acreage (0.9 million ha) with productivity of 23 q/ha. Acreage, productivity and production of under jute and allied fiber crops started declining over the years due to incidence and emergence of new pests and diseases under changing climatic scenario (Anonymos, 1949–1956; Tripathi *et al.* 2015a–h). Due to such change in climate, the disease earlier reported to be minor and sporadic in nature in some pockets is now becoming more and more severe and causing more damage to the crops (Selvaraj *et al.* 2015). Integrated disease management strategy is a holistic approach/strategy for developing location specific cost-effective modules to manage diseases and prevent economic losses caused by diseases under abrupt weather (Sarkar *et al.* 2014; Tripathi *et al.* 2014a–h).

Root knot caused by nematode (*Meloidogyne javanica/Meloidogyne incognita*) on jute and mesta.

12.2 Fungal Diseases of Jute

Stem rot of jute caused by *Macrophomina phaseolina* (Tassi) Goid. is the most destructive wide host range pathogen. In jute, the pathogen incites disease complex, viz., seed rot, collar rot, seedling blight, stem rot, and root rot in various growth stages of the crop. Incidence of stem rot on jute crop results in fiber yield loss around 11–20% and affects the fiber quality to its commercial value (Ghosh *et al.* 2008; Kumar *et al.* 2015a,b; Tripathi 2015; Bhattacharya *et al.* 2016) (Table 12.1).

TABLE 12.1
Fungal Diseases of Jute and Allied Fiber Crops

Jute	
Stem rot	*Macrophomina phaseolina*
Black band	*Botryodiplodia theobromae*
Anthracnose	*Colletotrichum corchori*
Soft rot	*Sclerotium rolfsii*
Powdery mildew	*Oidium* sp.
Tip blight	*Curvularia subulata*
Stem gall	*Physoderma corchori*
Mesta	
Foot and stem rot	*Phytophthora parasitica var. sabdariffae*
Leaf blight and leaf spot	*Phoma sabdariffae*
White stem rot	*Sclerotinia sclerotiorum*
Collar rot and wilt	*S. rolfsii*
Sunhemp	
Anthracnose	*Colletotrichum crotalariae*
Wilt	*Fusarium udum* f. sp. *crotalariae*
Ramie	
Leaf spot	*Cercospora* spp.
Sisal	
Zebra disease	*Phytophthora* spp.
Flax	
Wilt	*Fusarium oxysporum* f. sp. *Lini*

12.2.1 Diagnostic Symptoms

The *M. phaseolina* hyphae initially invade the cortical tissue of jute plants, followed by microsclerotia formation, causing stem rot (Figure 12.1(1 and 2)). Gray black mycelia and microsclerotia are produced in the infected area of the plant. Diseased capsules/pods discolored black and seeds become discolored and small.

12.2.2 Disease Cycle

The pathogen is soil and seed-borne in nature. The hyphae initially invade the cortical tissue of jute plants, followed by microsclerotia formation, causing stem rot disease. It is transmitted from seed to seedling/and seedling/plant to seeds (Akanda and Fakir 1985). Islam (1987) has reported that disease development in jute plants due to *M. phaseolina* increases gradually with the increase of seed-borne infections.

12.2.3 Management

Prophylactic measures of disease management are always better than curative application. Integrated Disease Management is a sustainable approach for eco-friendly management of soil and seed-borne; fungal diseases of jute under changing climatic scenario (Gotyal *et al.* 2014; Meena *et al.* 2014; De and Tripathi 2015; De *et al.* 2016; Bhattacherjee, 2016). IDM modules are described here:

(1) Use of disease-free certified seed; (2) deep plowing during the summer season; (3) field sanitation/clean cultivation; (4) liming of soil with 2–4 t/ha applied in field three to four weeks before sowing; (5) judicious use of fertilizer @ 60–30–30 kg, NPK/ha; (6) field sanitation, viz., rouging of diseased plants, timely weeding, and thinning of the field.

(7) Prefer stem-rot-resistant/tolerant varieties for cultivation, viz., JRO-524 (Navin), JRO-8432, JRO-128, JRO-204 (Suren), JRC-212 (Basudev), S-19 (Subala), and IRA-2003. (8) Seed treatment with

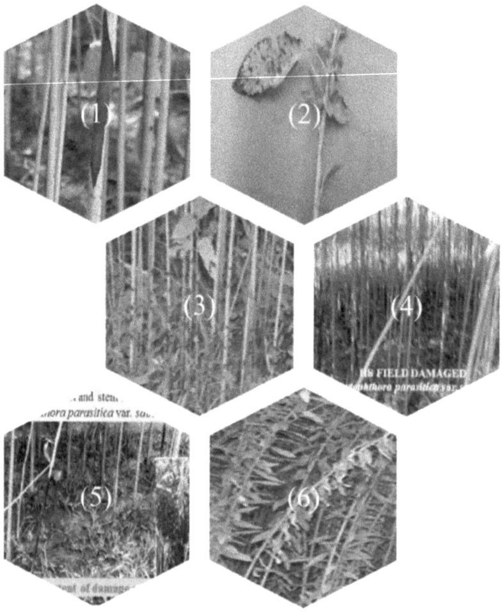

FIGURE 12.1 (1) Symptom of stem rot of jute; (2) leaf rot of jute; (3) anthracnose of jute; (4) foot and stem rot of mesta; (5) foot and stem rot incidence on mesta; and (6) symptom of wilt on sunhemp.

fungicide carbendazim 50 WP @ 2.0 g/kg. If it exceeds 2% spray, Dithane M-45 @ 5 g/l of water or @ copper oxychloride @ 5 g/l of water is used. Seed treatment with *Trichoderma viride* @ 10 g/kg or *Pseudomonas fluorescens* @ 10 g/kg of seed or *T. viride* @ 1 kg/ha mix with FYM @ 100 kg is applied in the field three to four weeks before sowing.

12.3 Anthracnose of Jute

Anthracnose of jute caused by *Colletotrichum corchori* is considered as one of the destructive disease of fiber and seed crop of jute in Assam, North Bengal, Bihar, and Uttar Pradesh.

12.3.1 Symptomatology

Diseased plants produces black and brown depressed dot-like spots that form asexual fruiting body acervuli on stem, which crack the bast fiber bundle (Figure 12.1(3)).

12.3.2 Management

Seed treatment with carbendazim 50% WP @ 2 g/kg of seed eliminates primary source of inoculum/infection. Seed lots having 15 percent or more infection should not be used as seed. Foliar spray of carbendazim 50% @ 0.5% or copper oxychloride 50% WP @ 0.75% is recommended for the management of anthracnose.

12.4 Foot and Stem Rot of Mesta

Foot and stem rot is one of the most economically important diseases of mesta caused by fungal pathogen *Phytophthora parasitica* var. *sabdariffae*. It causes more damage to roselle (*Hibiscus sabdariffa*) than to kenaf (*Hibiscus cannabinus*) (Tripathi *et al.* 2015a–h; Biswas *et al.* 2011). *Phytophthora* of oomycetes are devastating plant pathogens, causing damages to both agricultural and horticultural ecosystems specially malvaceous and solanaceous crops. High to epiphytotic outbreak of foot and stem rot was observed at Barrackpore (WB), Kendrapara (Orissa), and Amadalavalasa (Tamil Nadu) in India.

12.4.1 Diagnostic Symptoms

Initially, lesions are light brown in color and cover the stem. A black basal rot spreading over the stem up to 90 cm above the ground level is the diagnostic symptom of this disease. The plants may topple down by breaking at the point of infection. The tip of the plant collapses leaves' wilt and finally dies. The plants wilt, defoliate, stand leafless for some time and finally dry up prematurely (Figure 12.1(4 and 5)).

12.4.2 Management

(i) Pre-sowing seed treatment with metalaxyl 72 WP @ 2 g per kg seed is highly effective. (ii) A foliar spray of metalaxyl 35 WS @ 0.2% or metalaxyl 72 WP + mancozeb 8 WP @ 0.2% or copper oxychloride 50 WP @ 0.5% may check the disease to a great extent.

12.5 Sunnhemp Wilt

Wilt in sunhemp caused by *Fusarium udum* f. sp. *crotalariae* is an important disease of sunhemp. Incidence of the disease is about 10–12% and average loss due to this disease is 11–15%. *Fusarium* spp. has a wide host range and is a cosmopolitan fungal plant pathogen. The pathogen is widespread throughout the world, i.e., tropical, subtropical, and temperate geographic region of the world. The incidence increases with decreasing temperature, prolonged drought, low soil moisture, and increasing crop age.

12.5.1 Symptoms

Characteristic symptoms of the disease are the plant gradually withers, droops down, and ultimately dies within a day (Figure 12.1(6)). It attacks the young plants causing wilt and necrosis and in the older plants causing yellowing of the leaves, which leads to eventual necrosis.

12.5.2 Disease Cycle

It causes infection in the plant through roots and even through the cracking in the basal portion of the stem with germ tubes arising from micro- or macroconidia or chlamydospores reaching the vascular tissues to establish and multiply rapidly, causing wilting of parts or all the plant.

12.5.3 Management

Seed treatment with carbendazim @ 3 g/kg of seed and spraying with carbendazim @ 2 g/l of water is found most effective. Showing early disease-tolerant variety like K12 yellow and SH4 reduces incidence of the disease.

12.6 Anthracnose of Sunnhemp

Anthracnose sunnhemp is caused by *Colletotrichum crotalariae* (Kundu 1964). It occurs in all sunnhemp growing areas of India.

12.6.1 Diagnostic Symptoms

Initially, disease is appeared on cotyledons and followed by stem and growing point. Soft discolored areas on the cotyledons are the characteristic symptoms of this disease. Later, brownish spot is formed on all parts of host except underground parts. The affected seedling drops from the point below cotyledon.

12.6.2 Disease Cycle

The infection was severe during the seedling stage. Rain splash helps the spore of the fungus within the field. The pathogen required high rainfall, higher temperature (30–32°C), and high relative humidity favor infection.

12.6.3 Management

Seed treatment with carbendazim @ 2 g/kg seed checks the seed-borne infection. Secondarily, infection is checked by foliar spray with carbendazim @ 1.5%.

12.7 Minor and Other New Emerging Diseases of Jute and Allied Fiber Crops

12.7.1 Stem Gall of Jute

The disease is caused by *Physoderma corchori*. The disease is found on late sown *olitorius* crop subjected to submergence in low-lying areas during monsoon. Gall may be initially green but gradually turns pink to brown and finally rusty brown. Each infected plant may have innumerable galls, some of which coalesce to form a large blister with lacerated fiber (Figure 12.2(7)). Finally, the dead plants turn into black sticks (De and Tripathi 2012). Stem gall of jute is more prevalent during heavy rain and waterlogged condition of the cropping season. Gall (knot)-like structure is formed on the basal portion of the plant, which later turns brown and withers bark/bast fiber. The causal agent of this disease produces brown sporangia that are packed inside infected cells. Each sporangium releases motile zoospores that require both light and water to germinate and infect the plant. Infection most commonly occurs in the whorl where water tends to accumulate during periods of rain and irrigation, which is why lesions tend to develop in galls.

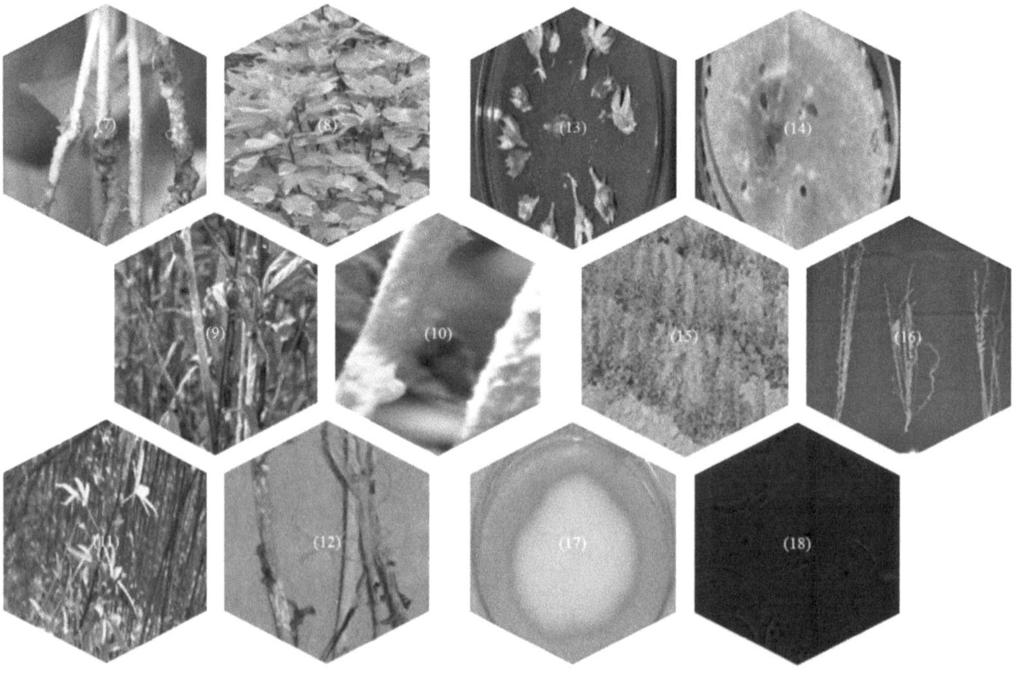

FIGURE 12.2 (7) Stem gall of jute, (8) phoma leaf spot of mesta, (9) white stem rot of mesta, (10) sclerotial mat, (11) stem break of mesta, (12) exposed fiber of mesta, (13) infected capsule of mesta, (14) PDA plate of sclerotinia with sclerotia, (15) wilted flax field, (16) wilted flax plant, (17) PDA culture plate of fusarium, and (18) macroconidia of fusarium.

12.7.1.1 Management

Maintaining proper field drainage and preventing waterlogged condition, the incidence of disease can be prevented by (i) avoiding submerged conditions by arranging proper drainage system and (ii) by growing resistant varieties.

12.7.2 Leaf Blight and Leaf Spot of Mesta

Phoma sabdariffae causes leaf spot on mesta. Symptoms can be observed on the leaf apex in one-week-old seedlings as a small spot that gradually spreads toward the petiole along the midrib. In some cases, the infection initiates from the base of the lamina and moves along the sides of the midrib toward the apex and it is often confined to one or two lobes (Figure 12.2(8)).

12.7.2.1 Management

Spray of mancozeb 72 WP @ 3 g or copper oxychloride 50 WP 3 g/l of water twice is done at an interval of seven days.

12.7.3 White Stem Rot of Mesta

Stem rot of roselle was a minor disease, but during the last two years (2012–2014), the incidence of the disease was recorded first time as high as 50% in seed crops at CRIJAF Farm, Barrackpore. The disease appears in severe form during the cooler months of December and January. White stem rot of mesta is caused by *Sclerotinia sclerotiorum*. The pathogen infects number of economically important crops, but very little information is available on mesta. Pathogen produces typical large irregular black colored rat dropping like sclerotia (0.5–1.5 cm), on diseased stems and bolls/capsules and culture plates (Tripathi *et al.* 2013a,b, 2014a–h, 2015a–h, 2016).

12.7.3.1 Symptom

The disease appears in water-soaked areas on any parts of stem, which turn into brown patches that are visible from distance. Initially, the portion of the stem above or below the patches looks healthy. Finally, the infection girdles the stem completely, which extends as much as a foot or more. The rot causes the tissues to become soft and easily peeled off into shreds. The portions above the affected part may ultimately wilt, die back, and break away (Figure 12.2(9–12)). The surface of the affected parts is covered with white stands of fungus mycelium, which form cushion in the axils of the branches. Black colored sclerotia may be observed on this mycelial mat (Figure 12.2(13 and 14)). The sclerotia were also noticed in bolls. The infected plant neither yields any fiber nor seeds.

12.7.3.2 Management

Spray of fluopyrum 17.7 + tebuconazole 17.7 SC or tebuconazole 50% + trifloxystrobin 25% or tebuconazole 25.9% or azoxystrobin 23% SC 1 g/l of water twice is done at an interval of 15 days.

12.7.4 Wilt of Flax

Linum usitatissimum L is an important bast fiber-producing cash crop grown in temperate as well as tropical countries, including India. So far, flax wilt [(*Fusarium oxysporum* f. sp. *lini* (Schlecht)] was rarely noticed at CRIJAF research farm, Barrackpore. But due to continuous monoculture of flax during the last five years, localized outbreak of the disease was noticed at research plot during December 2012–January 2013. Out of seven genotypes, JRF-2 was the most susceptible (80%) by JRF-1 (60%) and JRF-3 (40%). However, wilt incidence was least on genotypes, viz., JRF-4, FT-850, FT-896, and FT 897 (Tripathi *et al.* 2013a,b; Tripathi 2016).

12.7.4.1 Symptoms

Initially, wilt-infected plants appear sickly yellowing, start wilting at the top followed by whole plant wilting, and eventually the plant dries with brown discoloration both on stem and leaves (Figure 12.2(15 and 16)). The incidence was drastically increased with the age of the crop. The pathogen was isolated from the stem section of the infected plant but not from root, pure culture was established, and the pathogen was identified as *F. oxysporum* f. sp. *lini* (Figure 12.2(17 and 18)).

12.8 Zebra Diseases of Sisal

Zebra disease is caused by different species of *Phytophthora*, viz., *P. nicotianae*, *P. aracae*, and *P. palmivora*. Dark black, water-soaked lesions with concentric wavy rings appear on the leaves. The older leaves touching soil line during moist soil condition are more susceptible to infection. Two foliar spray of metalaxyl 72 WP @ 2.5 g/l of water at 15-day interval is recommended to manage this disease.

12.9 Diseases Caused by Bacteria in Jute and Allied Fiber Crops

12.9.1 Bacterial Wilt or Hooghly Wilt of Jute

Bacterial wilt is one of the most important diseases of jute. The disease is most prevalent in the areas where jute is followed by potato or other solanaceous crops. Bacterial wilt is first time recorded in West Bengal by Hutchinson in 1913 (Kelman 1953). During 1950, typical wilt disease in jute (*Corchorus olitorius*) was observed in Tarkeshwar area in Hooghly district of west Bengal. During the late 1980s and early 1990s, 5–37% of disease was recorded in Kamarkundu area of Hooghly district and 2–20% in some area of Nadia and North 24 Parganas districts (Mandal and Mishra 2001). *Ralstonia solanacearum* has an extended host range that includes hundreds of plant species in 54 botanical families (Allen *et al.* 2005; Sood and Tripathi 2005). In India, the economically important hosts include potato, tomato, brinjal, chillies, groundnut, fennel, cumin, ajwain, cotton, jute, ginger, and broad bean (Mondal *et al.* 2011). Bacterial wilt has been regarded as a problem primarily of solanaceous crops. However, it is also causing losses in non-solanaceous crops such as jute and mesta (Table 12.2).

In India, the disease was observed in Burdwan, Hooghly, Jalpaiguri, Murshidabad, North 24 Parganas, Nadia, and Pundibari district of West Bengal. *Olitorius* jute cv. is more prone to bacterial wilt than *capsularis* jute (Tripathi *et al.* 2014a–h).

12.9.1.1 Diagnostic Symptoms

In jute bacterial wilt starts with drooping of the leaves at the base and proceeds upward (Figure 12.3(19)). The stem on pressing produces slimy, turbid fluid (bacterial ooze) and causes wilt. And it started with wilting and drooping of the leaves from the base to upward and brown discoloration of the vascular tissues.

TABLE 12.2

Bacterial Diseases of Jute and Allied Fiber Crops

Jute	
Bacterial wilt	*Ralstonia solanacearum*
Bacterial leaf spot and blight	*Xanthomonas axonopodis* pv. *olitorii*
Mesta	
Bacterial wilt	*R. solanacearum*

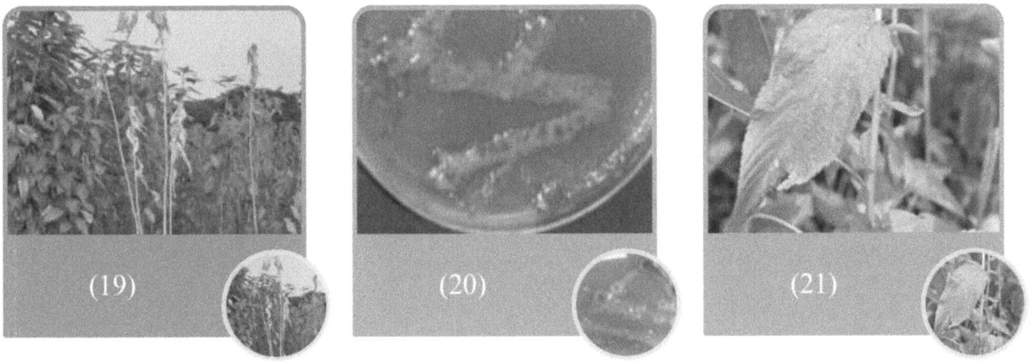

FIGURES 12.3 (19) Incidence of bacterial wilt in jute, (20) virulent colonies of *Ralstonia Solanacearum*, and (21) bacterial leaf spot and blight of jute.

12.9.1.2 Biology

Bacterial wilt pathogen *R. solanacearum* is a gram-negative, rod-shaped, mesophilic, nonfluorescent pigment-producing second topmost economically important plant pathogenic bacteria. It is reflecting a high level of phenotypic and genotypic plasticity. Yabuuchi et al. (1995) proposed new genus from *Pseudomonas* to *Burkholderia* to accommodate RNA homology group II, including *Pseudomonas solanacearum* with *Pseudomonas cepacia* as type species. There are considerable differences in host range of individual strains of *R. solanacearum* isolates that are classified into five races based on host range, geographic distribution, and ability to survive under different environmental conditions (Quinon *et al.* 1964; Buddenhagen *et al.* 1962) and six biovars on the basis of utilization/oxidization of disaccharides and sugar alcohols (Hayward and Hartman 1994; Opina *et al.* 1997). Virulent isolates of *R. solanacearum* forms white and irregular fluidal/mucoid colonies with pinkish center and produces abundant extracellular polysaccharide (EPS) on Kelman's triphenyl tetrazolium chloride (TZC) medium (Tripathi and Sood 2005, 2009). Avirulent types produce round, non fluidal/nonmucoid dark red colonies and are deficient in production of EPS (Figure 12.3(20)). The colony morphology changes, and its associated collective alterations are called "phenotype conversions (PC)." The range of generic variation in *R. solanacearum* and the basis of this variation are poorly understood. The jute bacterial wilt pathogen identified as race 1 biovar III of *R. solanacearum* (Tripathi *et al.* 2014a–h).

12.9.1.3 Disease Cycle

R. solanacearum is primarily a root inhabitant soil-borne bacterial pathogen. From infected plants, secondary spread over short, medium, and long distances takes place by root contacts, flood/irrigation water, farm implements, storage, and packaging material.

12.9.1.4 Management

Crop rotation with potato and other solanaceous crops avoids this disease. Break cropping system "jute after potato" followed crop rotation with non-host crop, viz., paddy and wheat for two years. The diseased plants or plant parts should be rogue out and destroy them, burning the solanaceous plants and rejecting rotten potato tubers are important cultural practices to control the disease. Applied organic manure @ 5 t/ha and Spraying copper oxychloride or copper hydroxide @ 2 g/liter of water helps to reduce root rot incidence, which favors the entry of the bacteria.

12.9.2 Bacterial Leaf Spot/Blight of Jute

Symptoms of bacterial leaf spot (*Xanthomonas campestris* pv. *olitorii*) in some capsularis jute fields were noticed as water-soaked lesions on leaves, which later turn into brown colored spots with halo blight

TABLE 12.3
Viral and Phytoplasmal Diseases of Jute and Allied Fiber Crops

Jute	
Jute leaf mosaic	*Jute Leaf Mosaic Virus* (JLMV)
Yellow vein virus	*Corchorus Yellow Vein Virus* (CYVV)
Mesta	
Yellow mosaic	*Mesta Yellow Vein Mosaic* (MYVMV)
Phyllody	Phytoplasmas
Sunhemp	
Sunhemp mosaic	*Sunhemp mosaic virus* (SMV)
Sunhemp leaf curl	*Indian Tomato Leaf Curl Virus* (ITLCV)
Flax	
Phyllody	Phytoplasmas

appearance (Figure 12.3(21)). During severe infection, the leaf rapidly becomes necrotic and defoliated. Pure culture of the bacterial pathogen was established on nutrient agar medium from infected young leaves virulent colonies of the isolates that were *Xanthomonas* – like pale yellow cream in color.

12.10 Phytoplasmal and Viral Diseases

Among phytoplasmal diseases, phyllody of mesta, sunnhemp, and flax limits the cultivation and decreases fiber production and productivity of these bast fiber crops. Better understanding of diagnostics, virus-vector relationship, etiology, and epidemiology of the viruses play important roles in the development of cost-effective, eco-friendly, integrated management strategy. (Lee et al., 2012) Important phytoplasmal and viral pathosystem of jute, mesta, sunhemp, and flax are described below (Table 12.3).

12.10.1 Jute Leaf Curl and Mosaic (JYLM)

Jute Yellow Leaf Mosaic is a virus belonging to member of begomovirus under the family of Geminiviridae. The incidence of the disease has been found to be around 50% on some of the leading *Corchorus capsularis* cultivars. It is distributed in West Bengal, Orissa, Assam, Bihar, and Eastern Uttar Pradesh.

12.10.2 Diagnostic Symptoms

The disease is characterized by symptoms such as small yellow flakes on the lamina during the initial infection stage, which gradually increase in size to form green and chlorotic intermingled patches producing a yellow and light-yellow patches and mosaic appearance. Leaves may be reduced in size and may be curled (Figure 12.4(22 and 24)).

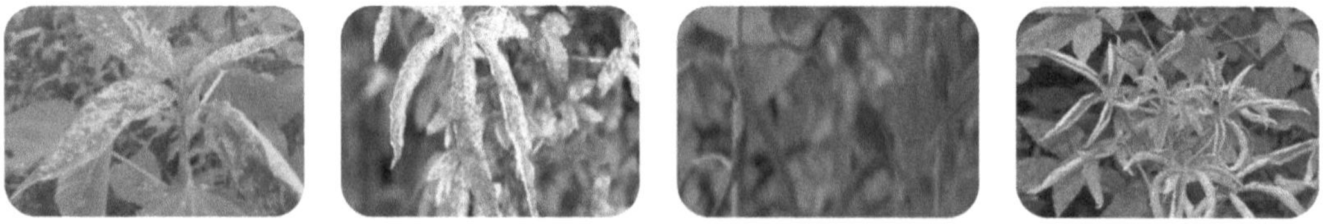

(22) (23) (24) (25)

FIGURE 12.4 (22) Mosaic of jute, (23) mosaic of mesta, (24) leaf curl of jute, and (25) leaf curl of mesta.

12.10.3 Disease Cycle

Whitefly (*Bemisia tabaci*) transmission of the disease has been reported by many workers. The whitefly has a minimum acquisition feeding period (AFP) and minimum inoculation feeding period (IFP) of 30 minutes for successful transmission of JLYMV.

12.10.4 Management

Application of neem cake @ 250 kg/ha before sowing was found effective in controlling whitefly. Yellow sticky traps can be used to detect and monitor whitefly activity. Around 3–5 traps should be placed in a block of 2–3 ha level with the tops of the plants. Seed treatment with imidacloprid 70% WS @ 1 g/kg of jute seeds and use of imidacloprid @ 0.25 ml/l as spray are used for insect vector control.

12.10.5 Jute Yellow Vein Mosaic (JYVM)

This is newly occurring minor viral disease of jute noticed in *C. capsularis* cultivars from all jute-growing areas of India.

12.10.5.1 Diagnostic Symptoms

The first visual symptom of the disease is the clearing bright-yellow network of veins, which usually starts at various points near the margins of top leaves. In severe cases, chlorosis extends to interveinal areas resulting in complete yellowing of leaves. The leaves become reduced in size.

12.10.5.2 Management

Seed treatment with imidacloprid 70% WS @ 1 g/kg of jute seeds and use of neem cake @ 250 kg/ha before sowing of the crop was much effective against whiteflies. Application of insecticides like imidacloprid @ 0.25 ml/l was found effective in controlling whitefly.

12.10.6 Mesta Yellow Vein Mosaic (MYVM)

Among viral diseases, most importantly Mesta Yellow Vein Mosaic (MYVM) was first observed in Bahraich district in Uttar Pradesh.

12.10.6.1 Diagnostic Symptoms

Leaves and stems are completely chlorotic or yellow and plant defoliates. Affected leaves become leathery, smooth, and reduced in size (Islam *et al.* 1993). Diseased plants mature late and bear very few flowers and capsules (Figure 12.4(23 and 25)).

12.10.6.2 Management

Sowing of mesta in the first fortnight of June and spraying with thiamethoxam (0.1 g/l) or imidacloprid (0.25 ml/l) at 50 days after sowing to check the whitefly is a vector for the disease.

12.11 Sunnhemp Mosaic and Phyllody

12.11.1 Diagnostic Symptoms

Characteristic symptoms of the disease are mottling, severe mosaic, puckering, and malformation with enations on the undersides of the leaves. As the disease progressed, patches of light- and dark-green areas become more prominent. Infected plants are shorter in height. Diseased leaves were smaller than the normal (Figure 12.5(26)).

FIGURE 12.5 (26) Phyllody of sunnhemp, (27) phyllody of mesta, and (28) phyllody of flax.

12.11.2 Management

Sanitation is one of the common control methods for SMV, which includes removing the infected plants and weed host from field. April sowing of tolerant variety like K-12 yellow and SH-4 largely escapes the disease.

12.12 Little Leaf and Bunchy Top of Jute

In June 2012, symptoms suggestive of phytoplasma infection were noticed on tossa jute in different experimental fields of the CRIJAF research farm, Barrackpore, and the incidence of the disease varied from 5 to 20%. The infected plants showed profuse lateral branching with a bushy appearance.

12.13 Phyllody and Reddening of Mesta

Stems were malformed and proliferated into many brooms like small and thick secondary branches (Biswas *et al.* 2013) (Figure 12.5(27)). The phytoplasmas (formerly known as plant-pathogenic mycoplasma-like organism) are a group of plant pathogenic wall-less, phloem-inhabiting prokaryotes (bacteria) in the class mollicutes. They have been found in more than 1,000 plant species worldwide, including several economically important crops, vegetables, fruit and fiber crops, and ornamental plants

12.14 Phyllody of Flax

The phyllody symptom includes stem flattening with abnormal green-structure virescence and yellowing, internode shortening, intense proliferation of leaves, and flower buds (Figure 12.5(28)). Infected plants showed reproductive malformation in which stamen and ovary transformed into leaf or shoot-like structure and commonly exhibited symptoms such as virescence (green pigmentation in plant tissues that are not normally green) and witches' broom (wild, erratic, broom-like growth at the ends of shoots, stems, or branches).

12.14.1 Management

However, some phytoplasma diseases could be managed by spraying systemic insecticides to control the leafhopper vectors. The disease symptoms can also be suppressed temporarily by the foliar application of antibiotics such as oxytetracycline hydrochloride, but the use of resistant germ plasm is the best option.

12.15 Diseases Caused by Nematodes on Jute and Mesta

Root-knot is caused by *Meloidogyne* and *Heterodera* spp. It is one of the most important diseases of jute and mesta affected at various growth stages of crop during hot and humid climate. Root-knot nematode-infested plants produce knot-like globular swelling in roots. As a result of this, translocation of water and nutrients is blocked. Infected plants show yellowing and stunting of the plant. Roselle plants have been seriously attacked by root-knot nematodes such as *Meloidogyne arenaria*, *M. incognita*, and *M. javanica*. Another nematode, *Heterodera radicicola*, has been recognized as a major pest affecting the roots of roselle plants.

12.15.1 Management

Various soil amendments, namely, cakes of karanj, mahua, groundnut castor cowdung manure, and Nemagon, were important for checking the nematode infestation and crop rotation with paddy and wheat for two years reduced nematode population in jute field.

12.16 Infestation of Phanerogamic Parasitic Weed (*Cuscuta* spp.) on Jute and Allied Fiber Crops

Infestation of *Cuscuta* spp. on seed crop of jute, sunhemp, and flax fields were recorded at CRIJAF, research farm during cropping season 2012–2013 and 2013–2014. The genus *Cuscuta* (known as Dodder) is a dicotyledonous holostem phanerogamic plant parasite, which belongs to the family of convolvulaceae. It was the first time observed in seed crop of flax; later, it gradually spreads to seed crop of sunhemp and jute in the following year. In spite of bast fiber crops being dicotyledonous and seed propagated, they are heavily infested with this growing menace. If left unchecked, it has potential to soon become a serious threat to production of seed crop of all bast fiber crops, including jute, flax, and sunhemp. Parasite appeared in the form of yellow vine strands, which grow around the stem and other above the ground part of the plants. The infected seed crops of jute, flax, and sunhemp were found to be encircled and entangled with the yellow slender and thin vines of *Cuscuta*, which in advance stage has drawn nourishment from jute by producing haustoria-like structure piercing into the vascular portion of stem (Figure 12.6(29–31)). Its serious infestation suppressed the crop growth and yield (Tripathi *et al.* 2014a–h). It developed flowers and produced seeds, which mixed with seed of crop plants and contaminated seed lots. Contaminated seeds play important role in its long-distance dissemination as well as survival under seed/germplasm/gene bank facility.

FIGURE 12.6 (29) Infestation of *Cuscuta* spp. on jute, (30) infestation of *Cuscuta* spp. on sunnhemp, and (31) infestation of *Cuscuta* spp. on flax.

India has comprised a varied agrometerological parameters and diverse agroclimatic conditions from tropical, subtropical to temperate. Jute, mesta, sunhemp, and flax are seed propagated bast fiber crops. Therefore, infected seed/germplasm-borne pathogens act as vital source for short-, medium-, and long-distance dissemination of seed-borne and seed-transmitted pathogens. An international standard (http://www.ippc.org) for phytosanitary measures and other national regulatory measures to protect agriculture have been (Khetarpal and Gupta 2006) promulgated. Accurate and quick diagnosis of viral pathogens is imperative for the success of mapping of epidemics, breeding for resistance, and developing quarantine and other control measures (Nene 2010 and Rishi 2006). The chemical pesticides when applied in higher doses or frequently in field for management of diseases give residual effects, which have adverse effect on soil microbiota and below underground diversity. Therefore, in perspective to changing climatic scenario, it is very important to develop greener and compatible innovative management strategies for bast fiber crops.

REFERENCES

Akanda, M. A. M. and Fakir, G. A. 1985. Prevalence of major seed borne pathogens of jute. *The Journal of Plant Pathology* 1(1): 76.

Allen, C., Prior, P. and Hayward, A. C. 2005. *Bacterial wilt disease and the Ralstonia solanacearum species complex*. Saint Paul, MN: American Phytopathological Society (APS Press).

Anonymous 1949–56. Annual Report, Jute Agriculture Research Institute (JARI), Barrackpore.

Bhattacherjee, D., Tripathi, A. N. and Singh, S. R. 2016. Characterization of PGPR strains for their plant growth promoting activities and evaluation of *in vitro* biocontrol efficacy against *Macrophomina phaseolina* (Tossi) Goid. In: *Book of souvenir & abstract*. p. 357. Varanasi, Uttar Pradesh: ICAR-IIVR. National Symposium on Vegetable legumes for soil and human health 2016.

Biswas, C., Dey, P., Mandal, K. and Satpathy, S. 2013. Phyllody and stem fasciation: a new phytoplasmal disease of flax in India. *JafNews* 11(2): 16.

Biswas, C., Sarkar, S. K. and De, R. K. 2011. Diseases of jute and mesta: present status and management options. In S. K. Biswas and S. R. Singh (Eds.), *Sustainable disease management of agricultural crops*. Pp. 62–79. New Delhi: Daya Publishing House.

Buddenhagen, I. W., Sequeria, L. and Kelman, A. 1962. Designation of races in *Pseudomonas solanacearum*. *Phytopathology* 52: 726.

De, R. K. and Tripathi, A. N. 2012. Incidence of stem gall in jute. *JafNews* 10(2): 10.

De, R. K. and Tripathi, A. N. 2015. Effect of plant population and sowing methods on stem rot of jute. *JafNews* 13 (1): 16.

De, R. K., Ghorai, A. K. and Tripathi, A. N. 2016. Potential of organic materials against stem rot of jute caused by *Macrophomina phaseolina* (Tassi) Goid. In R. Peshin, A. K. Dhawan, F. Bano and K. S. Risam (Eds.), *Proceeding of the Indian ecological society: international conference on natural resource management ecological prospectives*. Pp. 826–827. Jammu and Kashmir: Sher-e-Kashmir University of Agricultural Sciences and Technology of Jammu, Vol (2). ISBN: 978-93-5254-337-3.

Ghosh, R., Paul, S., Das, S., Palit, P., Acharyya, S., Das, A., Mir, J. I., Ghosh, S. K. and Roy, A. 2008. Molecular evidence for existence of a New World begomovirus associated with yellow mosaic disease of *Corchorus capsularis* in India. *Australasian Plant Disease Notes* 3(1): 59–62.

Gotyal, B. S., Tripathi, A. N., Selvaraj, K., Ramash Babu, V., Meena, P. N. and Satpathy, S. 2014. Screening of some jute (*Corchorus* spp.) germplasms against stem rot caused by *Macrophomina phaseolina* (Tassi) Goid. *Journal of Mycopathological Research* 52(2): 363–365.

Hayward, A. C. and Hartman, G. L. 1994. Bacterial wilt: the disease and its causative agent, *Pseudomonas solanacearum*. Wallingford: CAB International.

Islam, F. 1987. *Study of transmission of seed-borne fungal pathogens from seed to plant to seed (Corchorus capsularis L.)*.M.Sc. Thesis, Department Plant Pathology, Bangladesh Agriculture University, Mymensingh, Bangladesh.

Islam, H. R., Bhardwaj, M. L., Sharma, I. M. and Sharma, N. K. 1993. Performance of commercial okra (*Hibiscus esculentus*) varieties in relation to diseases and insect-pests. *Indian Journal of Agricultural Science* 63(11): 747–748.

Kelman, A. 1953. The bacterial wilt caused by *Pseudomonas solanacearum*. In *Technical bulletin of North Carolina agricultural experiment station*. Vol. 99. Salisbury, NC: North Carolina Agricultural Experiment Station.

Khetarpal, R. K. and Gupta, K. 2006. Policy issues in plant protection. In S. S. Chahal, R. K. Khetarpal, T. S. Thind (Eds.), *One hundred years of plant pathology in India: an overview.* Pp. 401–421. Jodhpur: ISMPP, Scientific Publishers.

Kumar, S., Shamna, A., Jha, S. K., Sarkar, S., Majumder, B., Tripathi, A. N., Naik, M. R. Naik, R. K. and Roshan, R.K. August, 2015a. *Unnat Taknikon Dwara Patsan Utpadan: Ek Sankshipt Vivran.* Extension folder published, published in August. Barrackpore, Kolkata: ICAR-CRIJAF.

Kumar, S., Singh, V. K., Tripathi, A. N. and Kumar, P. 2015b. Diseases of jute and sunhemp crops and their management. In G. Chand and S. Kumar (Eds.), *Crop diseases and their management: integrated approaches.* Pp. 555–563. New York: Published by Apple Academic Publication, CRC Press, Publication, Taylor and Francis. Hard-Book-ISBN: 9781771882705; E-Book -ISBN: 9781771882613.

Kundu, B. C. 1964. Sunnhemp in India. *Proceedings of the Soil and Crop Society of Florida* 24: 396–404.

Lee, S., Kim, C. E. and Cha, B. 2012. Migration and distribution of graft-inoculated jujube witches broom phytoplasma within a *Catharanthus roseus* plant. *Plant Pathology Journal* 28: 191–196.

Mandal, R. K. and Mishra, C. D. 2001. Role of different organisms in inducing Hooghly wilt symptom in jute. *Environment and Ecology* 19(4): 969–972.

Meena, P. N., Tripathi, A. N., Gotyal, B. S., and Satpathy S. 2014. Bio-efficacy of phytoextracts and oil cakes on *Macrophomina phaseolina* (Tassi) causing stem rot disease of jute, *Corchorus* spp. *Journal of Applied and Natural Science* 6(2): 530–533.

Mondal, B., Bhattacharya, I. and Khatua, D. C. 2011. Crop and weed host of *Ralstonia solanacearum* in West Bengal. *Journal of Crop and Weed* 7(2): 195–199.

Nene, Y. L. 2010. Prologue: status of plant pathology in India before 1905. *Plant Pathology in India* 1.

Opina, N., Tavner, F., Hollway, G., Wang, J. F., Li, T. H., Maghirang, R., Fegan, M., Hayward, A., Krishnapillai, V., Hong, W. and Holloway, B. 1997. A novel method for development of species and strain-specific DNA probes and PCR primers for identifying *Burkholderia solanacearum* (formerly *Pseudomonas solanacearum*). *Asia Pacific Journal of Molecule Biology and Biotechnology* 5: 19–30.

Quinon, V.L., Aragaki, M. and Ishi, M. 1964. Pathogenicity and serological relationships of three strains of *Pseudomonas solanacearum* in Hawaii. *Phytopathology* 54: 1096–1099.

Rishi, N. 2006. Significant achievements and current status: virology. In S. S. Chahal, R. K. Khetarpal, T. S. Thind (Eds), *One hundred years of plant pathology in India: an overview.* Pp. 401–421. Jodhpur: ISMPP, Scientific Publishers.

Sarkar, S., Shamna, A., Jha, S. K., Kumar, S., Majumder, B., Ghorai, A. K., Gotyal, B. S., Tripathi, A. N., Naik, R. K., Sharma, H. K., Chaudhary, S. B. and Roshan, R. K. 2014. *Improved jute production technology: a guide to farmers.* Barrackpore, Kolkata: CRIJAF.

Selvaraj. K., Gotyal, B. S., Gawande, S. P., Tripathi, A. N. and Satpathy, S. 2015. Emerging insect pests and diseases in jute and allied fibres. In S. K. Pandey, K. Selvaraj, R. T. Maruthi and S. P. Mazumdar (Eds), *Training manual on "technological advances in production of jute and allied fibre crops".* Pp. 100–106. Kolkata: Published by All India Network Project on Jute and Allied Fibres, ICAR-Central Research Institute for Jute & Allied Fibres Barrackpore.

Sood, A. K. and Tripathi, A. N. 2005. Current status of bacterial wilt of vegetable crops of Himachal Pradesh. *Indian Phytopathology* 58(3): 346.

Tripathi, A. N. 2015. Patsan ki fasal me adhik resha upaj hetu samanvit rog prabandhan: ek samadhan. *Agroxy Fasal Kranti* 2(4): 29–32.

Tripathi, A. N. 2016. Occurrence of vascular wilt incidence in fibre crop of flax caused by *F. oxysporum* f. sp. *lini* Schlecht. *Journal of Mycopathological Research* 54(1): 41–44.

Tripathi, A. N. and De, R. K. 2015. Characterization of isolates of *Ralstonia solanacearum* (E.F. Smith) Yabuuchi *et al.* causing bacterial (Hooghly) wilt of jute (*Corchorus* spp.) in west. Bengal book of abstracts, National seminar on "Poverty, inequality and health in India with special reference to North East India", 8–10 October, 2015, North-Eastern Hill University, Shillong, India, p. A-056.

Tripathi, A. N. and Sood, A. K. 2005. Variability studies on *Ralstonia solanacearum* causing bacterial wilt in Himachal Pradesh. *Indian Phytopathology* 58(3): 350.

Tripathi, A. N. and Sood, A. K. 2009. Cultural and physiological variability in *Ralstonia solanacearum* causing bacterial wilt in Himacal Pradesh. 5th International Conf. on Plant Pathology in the Globalized Era, Nov. 10–13, 2009, New Delhi, 134.

Tripathi, A. N., Babu, V. R. and Sharma, H. K. 2015f. Current status of diseases of mesta and their management. In P.C. Trivedi (Ed.), *Plant pathology: problems and progress.* Pp. 129–138. Jaipur: Pointer Publishers.

Tripathi, A. N., Babu, V. R., De, R. K. and Satpathy, S. 2014e. An emerging threat of holostem parasite (*Cuscuta* spp.) on seed crop of jute, sunhemp and flax. *JafNews* 12(1): 21.

Tripathi, A. N., Babu, V. R., Sharma, H. K., Kumar, S., Pandey, S. K. and Satpathy, S. 2014d. *Mesta: upaj badhane hetu unnat krishi karyamala*. Resha Kiran: Ank. Pp. 39–41. Barrackpore, Kolkata: Published by ICAR-Central Research Institute for Jute & Allied Fibres, Vol (1).

Tripathi, A. N., De, R. K. and Babu, V. R. 2015e. Mesta me kiton evm rogon ka samanvit prabandhan. *MAC Krishi Jagran* 20(4): 40–41.

Tripathi, A. N., De, R. K., Meena, P. N. and Sharma, H. K. 2015a. Emergence of leaf blight in mesta (*Hibiscus* spp.) caused by *Phoma exigua* (Desm). *JafNews* 13(1): 12.

Tripathi, A. N., De, R. K., Meena, P. N., Babu, V. R., Meena, Y. R., Kumar, S., Singh, V. K. 2015b. Mesta pathosystems: an overview. In G. Chand and S. Kumar (Eds.), *Crop diseases and their management: integrated approaches*. Pp. 555–563. New York: Apple Academic Publication, CRC Press, Publication, Taylor and Francis. Hard ISBN: 9781771882705; E-Book ISBN: 9781771882613.

Tripathi, A. N., De, R. K., Sharma, H. K. and Karmakar, P. G. 2016. Morphometric characterization of *Sclerotinia sclerotiorum* (Lib.) de Bary causing stem rot on seed crop of mesta and its sensitivity towards bioagents. In R. Peshin, A. K. Dhawan, F. Bano and K. S. Risam (Eds.), *Proceeding of the Indian Ecological Society: International Conference on Natural Resource Management Ecological Prospectives*. Vol (2). Pp. 745–746. Jammu and Kashmir: Sher-e-Kashmir University of Agricultural Sciences and Technology of Jammu. ISBN: 978-93-5254-337-3.

Tripathi, A. N., De, R. K., Sharma, H. K., Karmakar, P. G. 2015c. Emerging threat of *Sclerotinia sclerotiorum* causing white/cottony stem rot of mesta in India. *New Disease Reports* 32: 19. http://dx.doi.org/10.5197/j.2044-0588.2015.032.019.

Tripathi, A. N., Gotyal, B. S., Meena, P. N., Kumar, S., Sarkar, S. K., Pandey, S. K. and Satpathy, S. 2014a. *Patsan evmsamvargiya resha faslonme samanvit rog prabandhan: badhayen evm nidan*. Resha Kiran: Ank. Vol (1). Pp. 29–32. Barrackpore, Kolkata: Published by ICAR-Central Research Institute for Jute & Allied Fibres.

Tripathi, A. N., Kumar, M., Pandey, S. K., Satpathy, S., Kar, C. S., Bera, A. and Shil, S. 2014b. *Tosha Patsan (JRO-204 (Suren) Ki Unnat Krishi Karyamala*. Barrackpore, Kolkata: ICAR-CRIJAF.

Tripathi, A. N., Meena, P. N. and Kumar, S. 2015d. Diseases of phytoplasma and plant viruses of bast fibre crops. In P.C. Trivedi (Ed.), *Plant diseases: biocontrol and management*. Pp. 160–167. Jaipur: Aavishkar Publishers, Distributors.

Tripathi, A. N., Meena, P. N., De, R. K., Kar, C. S. and Satpathy, S. 2014c. Prevalence of bacterial wilt/Hooghly wilt on jute caused by *Ralstonia solanacearum* (E. F. Smith) Yabuuchi *et al*. *JafNews* 12(2): 29–30.

Tripathi, A. N., Sarkar, S. K. and Meena, Y. 2013a. Outbreak of vascular wilt of flax. *JafNews* 11(1): 15.

Tripathi, A. N., Sarkar, S. K., Kar, C. S. and Satpathy, S. 2014f. Characterization of *Ralstonia solanacearum* (Yabuuchi et *al*.) causing bacterial wilt of jute (*Corchorus olitorius* and *C. capsularis*). In *Book of abstracts, International conference on "India biodiversity meet 2014"*. p. 68. 21–23rd November, 2014. Kolkata, West Bengal: Agricultural and Ecological Research Unit Indian Statistical Institute.

Tripathi, A. N., Sarkar, S. K., Sharma, H. K. and Karmakar, P. G. 2013b. Stem rot of roselle: a major limitation for seed production. *JafNews* 11(1): 14.

Tripathi, A. N., Sarkar, S. K., Sharma, H. K. and Karmakar, P. G. 2014g. Detection and characterization of roselle stem rot pathogen, *Sclerotinia sclerotiorum* (Lib.) de Bary and its sensitivity towards bioagents. In *Book of abstract cum souvenir, national symposium on "plant pathology in genomic era"*. Pp. 8–9. 26–28th May, 2014. Raipur, Chhattisgarh: Department of Plant Pathology, Indira Gandhi Krishi Vishwavidyalaya.

Tripathi, A. N., Sarkar, S. K., Sharma, H. K. and Karmakar, P. G. 2014h. First report of White/Cottony stem rot caused by *Sclerotinia sclerotiorum* (Lib.) de Bary on mesta (*Hibiscus sabdariffa* L.) seed crop in west Bengal, India. In *Book of abstracts, International conference on "India biodiversity meet 2014"*. p. 28. 21–23rd November, 2014, Kolkata, West Bengal: Agricultural and Ecological Research Unit Indian Statistical Institute.

Tripathi, A. N., Sharma, H. K, Meena, P. N. and Babu, V. R. 2015g. Mesta (Patua) ki vaigyanik kheti. *Krishi World* 8 (12): 23–24.

Tripathi, A. N., Singh, L. N. K., Tripathi, S. K., Gupta, A. K., Tripathi, R., Salgotra, S. K. and Sharma, P. K. 2015h. Diversity, biology, epidemiology and management of bacterial wilt: an overview. In P. C. Trivedi (Ed.), *Plant diseases: biocontrol and management*. Pp. 151–159. Jaipur: Aavishkar Publishers, Distributors.

Tripathi, S. K., Sarkar, S. K. and Agrawal, P. C. 2012. In vitro sensitivity of fungal pathogens isolated from seed samples of jute (*Corchorus* sp.) and mesta (*Hibiscus* sp.) germplasm (abstract). *Journal of Mycology and Plant Pathology* 42(4): 539.

Yabuuchi, E., Kosako, Y., Yano, I., Hotta, H. and Nishiuchi, Y. 1995. Transfer of two *Burkholderia* and an *Alcaligenes* species to *Ralstonia* gen. nov.: proposal of *Ralstonia pickettii* (Ralston, Palleroni and Doudoroff 1973) comb. nov., *Ralstonia solanacearum* (Smith 1896) comb. nov. and *Ralstonia eutropha* (Davis 1969) comb. nov. *Microbiology and Immunology* 39(11): 897–904.

13

Transition toward Agroecology among Family Farmers: Crop Protection Practices

C. A. Costa[1], A. Aguiar[2], C. Parente[3], J. Neto[1], E. Valério[4], M. C. Godinho[4], and E. Figueiredo[5]
[1]*Research Centre for Natural Resources, Environment and Society (CERNAS-IPV), Polytechnic Institute of Viseu, Viseu, Portugal*
[2]*GreenUPorto & DGAOT, Faculty of Sciences, University of Porto, Porto, Portugal*
[3]*Department of Sociology and Institute of Sociology, Faculty of Arts and Humanities, University of Porto, Porto, Portugal*
[4]*Life Quality Research Centre (CIEQV), Polytechnic Institute of Santarém, Santarém, Portugal*
[5]*Linking Landscape, Environment, Agriculture and Food (LEAF), Associated Laboratory TERRA Instituto Superior de Agronomia, Universidade de Lisboa, Lisboa, Portugal*

CONTENTS

13.1 Introduction

Due to its worldwide relevance, family farming transcends the mere productive or economic concern to become a lifestyle for rural populations, preserving and transmitting local knowledge, promoting the protection of natural resources, and ensuring food security for families. Unlike conventional agriculture, through the use of agricultural practices such as crop diversity with a preference for regional varieties, cultural associations, crop rotation, addition of organic materials of animal origin, green manure, or manual defoliation to control pests and diseases, family farming generally adopts agroecological practices.

13.2 Family Farmers: Who, What, How Different They Are?

Although the concept of family farming varies between countries, it is recognized as an agricultural system based on smallholdings, mostly dependent on family labor, and rooted in traditional practices.

DOI: 10.1201/9781003164968-15

Family farming is the dominant type of agricultural production in capitalist countries despite the restructuring that agriculture underwent in the post-war period with the so-called Green Revolution and Agricultural Policies in developed countries. These two circumstances foresaw a process of modernization of agriculture that would transform "small family farmers into entrepreneurs" (Dinis 2019).

This vision, based on economic considerations like efficiency, productivity, and profitability, ignores other aspects and functions of family farming. In fact, the debate about family farmers needs a more holistic notion, which articulates the dynamics between farm and family and their strong relationship with the social environment and the territory. As Djurfeldt (1996) highlights "The notional family farm is characterized by an overlapping between three functional units: (a) the unit of production (the farm), (b) the unit of consumption (the household) and (c) the unit of kinship (the family)."

In summary, family farming is rooted in the existence of a farm managed by a family that owns or leases it, and worked mainly by family labor, not always remunerated. The economic income derived from this is complementary to other sources of income resulting from other professional activities, often carried out primarily by one of the members of the couple. This plurality of activities and the corresponding multi-income is one of the aspects identified to explain the resilience of family farmers. Family farming is far beyond a mere professional occupation and a mere economic rationality. It is a lifestyle associated with values, beliefs, and traditions and about living and working based on a fusion between home and work (Brandth 2002).

13.3 Crop Protection Practices in Family Farming

The sustainability and productivity of a farm depend, among other aspects, on a rational organization of work (Aguiar et al. 2020), a correct management of crop protection and of other ecosystem services such as pollination, soil fertility, or water quality (van Zanten et al. 2014). For almost a century, crop protection on family farms has added to traditional practices, curative practices to control pests, diseases and weeds, particularly using pesticides. Crop protection in sustainable agriculture systems is based on a sequential approach focused, in a first phase, on the use of indirect control measures before any intervention followed by, if necessary, direct control (or curative) measures. Indirect control measures include cultural control (e.g. intercropping, plant consociation, crop rotation), genetic control (e.g. resistant species/varieties/cultivars), and conservation biological control. Direct control measures, applied after risk assessment and decision-making, include cultural control (e.g. manual weed removal), biological control, biotechnical control (e.g. mating disruption), and chemical control.

A survey carried out on family farms in Portugal showed that 98% use regional varieties and more than 80% practice crop rotation and 90% practice manual weeding (Costa et al. 2018, 2020), practices that are at the core of agroecology. Even though family farmers traditionally use some agroecological practices transmitted through the generations, small and medium family farmers are less likely to innovate or adopt new approaches to renounce some of the unsustainable practices they have implemented in the last seventy years, such as the intensive use of synthetic pesticides. This change-aversion may be due to a lack of education, training, and technical support closely related to a mostly aged work class (Ngowi et al. 2007). These social-demographic characteristics may limit the options to consider in crop protection.

Family farmers commonly use several cultural practices: the use of nets as a barrier for insects, defoliation to achieve better aeration of the canopy or removal of fruits. Weed control is achieved either through (i) cultural practices such as adapting soil conditions, soil preparation, use of false sowing, crop rotation, ground cover, and use of suitable cultivars or by (ii) mechanical or thermal intervention such as weeding, interrow cultivation, and heat treatment.

Biological control is the use of living organisms to suppress the population density of a specific pest. Although family farmers are not considering biological control as a crop protection solution, they usually adopt practices to protect and enhance natural enemies (namely preserving ecological infrastructures [EISs] or cover crops), or in other words they practice conservation biological control, even without knowing it. Due to the high landscape complexity and diversity of crops that characterize family farms, they contribute significantly to the promotion of natural control and ecosystem services (Steward et al. 2014).

In fact, the diversity and abundance of different groups of living beings (birds, spiders, and other predators and parasitoids) enhance ecosystem services such as pollination and biological control. In a survey of Portuguese family farmers, 17% stated that they maintain EISs that provide shelter or food to beneficials and promote their presence in crops (Costa et al. 2018). Studies carried out by Burgio et al. (2006) on small farms in northern Italy showed that the green borders of the fields accommodate high populations of coccinellids and other beneficials.

Biotechnical control refers to a set of techniques based on advanced scientific and/or technical knowledge based on insect physiological processes or environmental behaviors negatively affecting the organisms' survival. Examples are insect mating disruption used mainly in large farms for some key pests, and mass capture with attractive traps, a traditional technique used in small farms. Family farmers use these traps, with attractive liquid or pheromone, for capturing pests such as the olive fly *Bactrocera oleae* or the Mediterranean fruit fly *Ceratitis capitata*.

The use of chemical control, meaning the application of synthetic or naturally occurring pesticides is strongly regulated: in organic farming, only pesticides of natural origin (biopesticides) are allowed; on the other hand, in integrated production and conventional agriculture, they can only be used if registered by authorities for that pair pest-crop (Chandler et al. 2011). Regarding its use by small farmers, some authors refer to the nonuse of pesticides, particularly in developed countries (Gurr et al. 2016), and others claim that the use of pesticides by family farmers is widespread and unproperly used (Abreu & Alonzo 2016).

The protection practices mentioned are commonly used by family farmers; some are part of the collective and deep knowledge at the local level and regional tradition, which may allow family farmers to hold most of the solutions needed to maintain resilient and sustainable production systems, and thus overcome market difficulties and meet environmental demands (Walpole et al. 2013). This is the case of cultural practices such as the choice of regional varieties/cultivars and crop rotation, conservation biological control through the maintenance of EISs ecological infrastructures, or the use of traps to capture pests.

13.4 Crop Protection and Agroecology

Curative approach and chemical control are dominant in agricultural crop protection yet but regulations in European Union (EU) are moving toward an ecological attitude. Integrated pest management (IPM) has represented a great advance in crop protection, trying to change the control paradigm to a management approach. However, despite the intentions of the policy makers, the IPM strategy has not changed deep-rooted habits, namely those related to pesticide use, and is being subverted, mainly, to integrated pesticide management (Parsa et al. 2014). So, due to health, economic and environmental reasons (consumers' demand for healthier food, citizens' demand for the reduction of water, soil and air pollution, collective concern about climate changes, and the increasement of pesticide resistance in important pests, diseases, and weeds), agroecological solutions are gaining importance in crop protection strategies. Instead of focusing on reducing damage through tactics to suppress crop enemies, as preconized in IPM, agroecology focuses on improving the resilience of the agroecosystems, changing from a pest-by-pest approach to a more holistic one, including agricultural practices thought to maintain soil health and fertility (Altieri & Nicholls 2005).

Agroecological crop protection is being discussed since the 1980s and 1990s. Development of agroecological crop protection draws on knowledge of several sciences, as agronomy, biology, ecology, and, particularly, the use of tools and techniques of IPM, landscape management, and conservation biology (Deguine 2017). The transposition of the two more important principles of agroecology (preserving local biodiversity and soil health) to crop protection is evident: incorporation of plant diversity in time and space through habitat and landscape management, and improvement of soil quality (organic matter and biodiversity.

The agroecosystems are deeply changed ecosystems. Decreased landscape and on-farm plant diversity, horticultural simplicity, phenological uniformity, changes in plant defenses to herbivorous brought by breeding (e.g. decrease in secondary compounds), inorganic fertilization, and pesticide application inducing insect outbreaks (resistance development and/or secondary pest emergence due to decrease in their natural enemies) tend to make agroecosystems more vulnerable to pest outbreaks (Altieri &

Nicholls 2005). Consequently, agroecological crop protection relies, firstly, on preventive measures to reduce crop enemies' population, as the cultivation of healthy plants, maintenance of good soil health, crop rotation and consociations, resistant or less susceptible cultivars, minimal or no tillage, sustainable fertilization and irrigation, push-and-pull techniques, conservation biological control, and preventive biotechnical control measures. If necessary, and only after risk evaluation and decision-making, curative measures can be adopted (augmentative or inundative biological control and biotechnical control) (Deguine 2017) and as a very last resort, the use of active ingredients of natural origin with the least environmental impact might be considered. To achieve this, species, ecosystem, and landscape diversity are essential (Boller et al. 2004).

Annual crops as monocultures and perennial crops in extensive areas are easier to find by the pests and diseases since they are cultivated in a higher density. Monocultures (same species, same cultivar, same seed/plantation date) turn the plants more apparent to the pests. Those see a monoculture field as a dense and pure mass of well nourish – fertilized and watered – and easy to find food, that have not evolved to defend themselves in this scenario - a real *fast-food* for insects and pathogens. Inversely, monocultures are less attractive to the natural enemies, as they do not normally provide the resources they need to survive. In fact, Boller et al. (2004) separated arthropod species in three groups regarding their habitat requirements: (i) species that require only field crops; (ii) species that need refuge outside crops to overwintering; (iii) species that require resources other than crop to complete their development. These authors found that more pest species and fewer beneficial arthropods laid on the first group. In contrast, almost all the antagonists belong to groups (ii) and (iii). Most of the biological control agents need pollen and nectar to feed, to increase longevity, or even to develop their eggs when adult; alternative hosts or prey as feed when they are absent from the crop, and shelter when they are facing adverse conditions (Boller et al. 2004; Franco et al. 2006; Wäckers & van Rijn 2012).

More complex landscapes with a mosaic of natural and cultivated habitats, and at farm level, through temporal and spatial plant diversification, such as crop rotation, cover crop, no tillage or reduced tillage, crop mixtures, intercrop or relay crops, and plant mixtures, and creating or improving EISs enhance natural enemies' populations. The impact of some of these practices on conservation biological control is site dependent.

An EIS is any infrastructure at the farm or within a radius of about 150 m that increases farm functional biodiversity (Boller et al. 2004), including nutrient cycling, pollination, and the one related with pest regulation, providing food, alternative prey or hosts, and shelter for natural enemies. The EIS network should be composed of three basic elements: (i) large permanent habitats as large low-intensity grasslands, forests, large ruderal areas, high-stem fruit tree orchards; (ii) stepping stones or temporary habitats – concentrated and small sized structures – as woodland patches, stone, or wood heaps; (iii) ecological corridors – linear or strip structures – as hedges, grass and wildflowers strips, road/trail margins, ditches and their riparian vegetation, and stone-walls (Boller et al. 2004) (Figure 13.1).

The effective contribution of the EIS to the functional biodiversity is heavily dependent on their ecological quality, distribution, and interconnection among the EISs existing inside the farm and to other EIS located outside (Franco et al. 2006). The area of all the EIS should be at least 5% of the farm surface, excluding forest (Boller et al. 2004). The size, distance, and form of the fields should be taken into account when designing the EIS, as colonization depends on the size and dispersion capacity of the natural enemies. Generally, EIS long and narrow are easier colonized by natural enemies than square fields, after a disturbance (e.g. annual crop planted or sown, pesticide application). Being most of them small arthropods and several of them not good flyers, field width should be inferior to 100 m. Different natural enemies have different needs for food and shelter (Table 13.1). Parasitoid adults need flower or extrafloral nectar. But the majority, due to the small size of their mouthparts, is only able to feed if the nectary is sufficiently exposed (Wäckers & van Rijn 2012). It is crucial to have plants that bloom when the adults are flying, that is, the bloom period and phenology of natural enemies should be synchronized.

The plant species included in the EIS should be carefully selected to have the best impact in biological control and pollination. For an adequate selection, it is crucial to know the biology and ecology of the beneficial arthropod and the plant species (Wäckers & van Rijn 2012). This choice will depend on the purpose of the EIS, the region, and the site (soil type and moisture, and sun exposition) (Boller et al. 2004). Several criteria for selecting plants should be considered (Table 13.2) (Boller et al. 2004; Franco et al. 2006; Rodríguez & González 2014). One of the most important criteria is the selection

FIGURE 13.1 Different examples of EIS: (a) mosaic landscape (forest and woodland patches) and natural vegetation between the rows in a pear orchard; (b) pear orchard with flower strip inside the field to improve wild bees' population and pollination, sown vegetation between rows and contiguous woodland patches; (c) outside, and (d) inside a tomato greenhouse crop.

(Photos by (a) and (b) Catarina Reis, (c) and (d) Gonçalo Duarte.)

of autochthonous plant species as they are easy to maintain, and less susceptible to pests and diseases than crops or ornamentals. Plant species selected must provide nectar and pollen throughout the season (bloom sequence should be ensured), with some of the flowers with floral nectar depth <0.2 mm (exposed), and refuge for hibernation, estivation, and/or nesting. If possible, decision about plant species to choose should consider the available knowledge about alternative prey/host and crop diseases or pests for each plant species.

Plant species and guidelines to design of EIS were presented, for example, for vineyards, fruit orchards and arable crops in temperate regions of Europe (Boller et al. 2004), citrus orchards (Franco et al. 2006), protected crops (Rodríguez & González 2014), and vineyards (Carlos et al. 2014) in Mediterranean regions.

As a corner of agroecology, soil health is of major importance, especially soil microbial community and its relationship with organic matter. The evolution of crop protection and the soil fertility management have developed separately. However, the reliability of plant health relies, not only on plant diversity, but also on good and equilibrated fertilization. High-nitrogen fertilization relates with high populations of picking-sucking insect pests as aphids, psyllids, and whiteflies. In contrast, a good fertilization with calcium and potassium helps plants to defend themselves from several diseases (Magdoff & van Es 2010). There suggests that the ability to resist or tolerate attack of pests and diseases is related to the soil properties, especially biological properties due to an active and diverse microbial community in a complex food web.

TABLE 13.1

Main Groups of Natural Enemies and Food Sources for Ecological Infrastructures (Boller et al. 2004; Eubanks & Styrsky 2005; Franco et al. 2006; Wäckers & van Rijn 2012; Klecka et al. 2018)

Natural Enemies	Refuge	Mobility	Food Source	Plants
Phytoseiidae	H: adult female, hedges and tree bark	+ Wind-passive transport	Pollen, mites, thrips, other small insects, honeydew	Apiaceae, Asteraceae, Brassicaceae, Chenopodiaceae, Oxalidaceae, Papaveraceae, Poaceae, Rosaceae
Anthocoridae	H: adult, near prey, mainly outside crop fields but nearby, hedges, vegetation, forest (bark)	++	Pollen, several prey	Amaranthaceae, Asteraceae (e.g. *Chamaemelum mixtum*), Fabaceae, Poaceae (e.g. *Zea mays*), Urticaceae (e.g. *Urtica* spp.), extrafloral nectar on beans (at least, some species prefer deeper and wider corollas in which they can hide)
Miridae		++	Several prey, pollen, phytophagy	Asteraceae (e.g. *Calendula arvensis*), Boraginaceae (e.g. *Phacelia tanacetifolia*, *Borago officinalis*), Pedaliaceae (*Sesamum indicum*), Solanaceae, Verbenaceae (*Verbena* sp.)
Coccinellidae (S)	H: adult, near prey, mainly outside crop fields but nearby, hedges, vegetation, forest (bark)	++	Mainly prey (ex. aphids), sometimes pollen and nectar	Apiaceae (e.g. *Daucus carota*), Asteraceae (e.g. *Tripleurospermum inodorum*, *Sonchus* sp.), Boraginaceae (e.g. *Myosotis arvensis*), Brassicaceae (e.g. *Brassica* sp.), Lamiaceae (e.g. *Lamium purpureum*), Plantaginaceae (e.g. *Plantago major*), Scrophulariaceae (e.g. *Verbena* sp.), Urticaceae (*Urtica* spp.)
Chrysopidae and Hemerobiidae (S)	H: adult, hedges	++/+++	Pollen, nectar and honeydew (some species also on prey)	Opportunist species – flowers that are available open at twilight, mainly Apiaceae (e.g. *Ammi majus*, *Conopodium majus*, *D. carota*, *Foeniculum vulgare*[b], *Pastinaca sativa*[b]); Asparagaceae (e.g. *Asparagus acutifolius*), Asteraceae (e.g. *Matricaria chamomilla*, *Centaurea* spp., *Helianthus annuus*), Boraginaceae (e.g. *Borago officinalis*, *P. tanacetifolia*), Brassicaceae, Caprifoliaceae (e.g. *Lonicera periclymenum*), Caryophylaceae, Chenopodiaceae, Liliaceae, Poaceae, Polygonaceae (e.g. *Fagopyrum esculentum*), and shrubs from Rosaceae, and *Sambucus* sp. and *Corylus* sp.
Syrphidae (S)	H: adult, mainly outside crop field: vegetation strips, hedges, tree patches, forest	+++	Pollen, nectar, and honeydew	Apiaceae (e.g. *A. majus*, *Coriandrum sativum*[b], *D. carota*[b]; *F. vulgare*[b], *P. sativa*[b]; Asteraceae (e.g. *Achillea millefolium*, *Centaurea* spp., *H. annuus*), Boraginaceae (e.g. *P. tanacetifolia*), Brassicaceae (e.g. *Lobularia maritima*, *Diplotaxis muralis*), Polygonaceae (e.g. *F. esculentum*), Ranunculaceae (e.g. *Ranunculus acris*), Rubiaceae (*Galium* spp.)
Hymenoptera Parasitica (parasitoids)		+	Pollen[c], nectar, and honeydew (some females also host feeding)	Apiaceae (e.g. *C. sativum*[b], *D. carota*[b], *F. vulgare*[b]), Asteraceae (e.g. *Centaurea* spp., *Cosmos bipinnatus*), Brassicaceae (e.g. *L. maritima*), Euphorbiaceae, Fabaceae (*Vicia sativa*[d]), Malvaceae[d], Passifloraceae[d], Polygonaceae (e.g. *F. esculentum*)

S – Synovigenic species; [a]the parasitoids whose hosts do not produce honeydew (as aphids, whiteflies, and mealybugs do) may have no sugar sources available near hosts and, consequently, may need to look for carbohydrates; if they are not available nearby, they will go out the field; [b]floral nectar depth <0.2 mm; [c]observations of direct pollen feeding by parasitoids are scarce (Wäckers and van Rijn 2012); [d]extrafloral nectar.

TABLE 13.2

Criteria for Selection of Plant Species for Hedges, Edges, and Margins

Criteria	Importance
Autochthonous species commercially available adapted to the ecosystem	6
Risk of being repository of crop pathogens and/or crop pests	6
Risk of being a weed for the crop	5
Price of seeds/plants	5
Source of pollen and/or nectar (or extrafloral nectary)	5
Flowers with exposed nectaries	5
Shelter	5
Toxic nectar, extrafloral nectar or pollen	5
Capacity to compete with weeds	4
Bloom period (sequential; continuous flower source, not coincident with crop)	4
Easy and cheap management	4
Stratum (include shrubs and herbaceous plants, in some cases trees)	3
Equilibrium perennial/deciduous species	2
Color of the flowers (avoid homochromous)	2

Source: Adapted after Boller et al. (2004), Franco et al. (2006), Rodríguez & González (2014).

Pathogenic organisms and even nonpathogenic soil microorganisms can promote above or below ground production of salicylic acid. The increase of this phytohormone induces the plant to produce defense proteins that protect the plant from a wide range of pests. Moreover, roots exposed to specific plant growth-promoting rhizobacteria (PGPR), *Trichoderma* spp., or mycorrhizae are stimulated to produce jasmonic acid, another defensive phytohormone (Magdoff & van Es 2010; Ferlian et al. 2018). Salicylic and jasmonic acids protect plants from herbivorous insects. Mycorrhizal fungi abundant in a soil with equilibrated microbioma protect plants from the attack of pathogenic soil fungi and nematodes. Soil organic fertilization induces resistance in plants. Cover crops increase protection to nematodes and the diversity and abundance of soil predators. A few studies support that mycorrhizal root colonization has implications in above-ground plant protection attracting predators and parasitoids (Ferlian et al. 2018; Prieto et al. 2017).

13.5 Family Farmers and Agroecology: Near or Distant?

Family farmers do not always present environmental or health concerns (Vaidya et al. 2017). However, society seeks for their traditional products, in local markets and fairs, as they associated them with agricultural practices with lower impacts on health and the environment.

However, family farmers are, in general, more exposed to the risks associated with the use of pesticides, in particular due to the fact that they have less education and training, and because they constitute a group of agents for whom there are no technical support and supervision programs (Ngowi et al. 2007; Abreu and Alonzo 2016). On the other hand, farm productivity depends, among other things, on proper management of crop protection and other ecosystem services, such as pollination or water quality. Current knowledge about ecosystem services in landscape systems consisting of small family farms is limited and has been neglected (Walpole et al. 2013; Steward et al. 2014); however, due to the high complexity of these landscapes, it is estimated that small family farms can contribute to a large extent to the promotion of natural limitation and associated ecosystem services. It is, also for this reason, that agroecology has the potential to reduce vulnerabilities and bring resilience to family farms, using techniques and strategies that focus primarily on strengthening natural processes.

The management of crop enemies (pests, diseases, and weeds) emerges as the main differentiating area in relation to conventional agriculture, presenting high levels of technical demand and increased risks for farmers, and therefore constituting the most difficult group of cultural practices to be adopted by family farmers. Family farmers differ in terms of agroecological practices adopted (Costa et al. 2020). In fact, it is possible to find farmers that are market oriented, concerned with productivity and using pesticides intensively aiming to reduce production risks and that ignore or are unaware of the negative effects of their use. Other farmers tend to have a greater perception of the risks associated with the use of pesticides and still use farming techniques considered "traditional," saving seeds from one year to the other, using regional varieties, allowing the presence of weeds on the farm and with production oriented to self-consumption. They are very open and adopt agroecological practices, such as biological and biotechnical control. The use of pesticides seems to be a widespread practice in some regions, without awareness of the risks that may be associated with (Hanafi 2005). Even though its small dimension, altogether the small farmers represent a large portion of the agriculture practiced worldwide, so more studies aimed at small farms are necessary along with a committed investment in the upgrade of farms skills and knowledge. Education is determinant for an informed decision-making process regarding crop protection practices, so the knowledge of control options and protocols, risk assessment, and self-protection are not a limiting factor for the adoption of more sustainable practices.

13.6 Innovation in Crop Protection through Agroecology

There are numerous examples of innovation in crop protection through agroecology among family farmers, with direct impact on the increase in farm and household income. Successful examples around the world include Southeastern African farmers and push-and-pull technique, agroecological management of vegetable flies in Réunion and Kenya through biotechnical control, riparian IES to reduce chemical runoff into water bodies in Lake Naivasha, Kenya, among many others.

13.6.1 Southeastern African Farmers and Push-and-Pull Technique

Stemborer causes maize between 20 and 40% yield losses but can achieve 80% of the potential grain production. Preventing these losses is sufficient to feed more 27 million people in the eastern Africa (ICIPE 2015). In the last 20 years, several improvements on the control of cereal stemborers in sub-Saharan Africa have been achieved. In 1994, researchers from ICIPE (Kenya) and Rothamsted Research Station (UK) began to study the stemborers, key pests of maize, and sorghum in eastern and southern Africa, namely the native moths *Busseola fusca* (Fuller) and *Sesamia calamistis* (Hampson) (Noctuidae), and the exotic *Chilo partellus* (C. Swinhoe) (Crambidae) (ICIPE 2015). These species were pests of autochthonous weeds but when maize began to be cultivated, they adapted to feed on this crop. The push-and-pull technique consists in intercropping a plant that repels the pest (push), and placing a plant that attracts the pests (pull) in the border of the field. In this case, the push-and-pull system is implemented by using desmodium (*Desmodium* spp.) (Fabaceae) for pushing and elephant grass (=Napier grass), *Cenchrus purpureum* (Schumach.) Morrone (=*Pennisetum purpureum*) (Poaceae) for pulling (Khan et al. 2014). Desmodium improves soil fertility through nitrogen fixation and increasing organic matter. Furthermore, desmodium and elephant grass provide high-quality animal food, increasing milk production, and farmers income (Khan et al. 2014; ICIPE 2015). This system has proved to be also effective to protect maize and sorghum from other key enemies, such as striga (Scrophulariaceae), a parasitic weed and the noctuid invasive species *Spodoptera frugiperda* (Khan et al. 2014; ICIPE 2015; Midega et al. 2018). In Kenya, the grain production of maize or sorghum can duplicate or triplicate with this system compared to monocrops (Khan et al. 2014).

13.6.2 Agroecological Management of Vegetable and Fruit Flies in Réunion and Kenya, through Biotechnical Control

In Réunion, vegetable crops, especially Cucurbitaceae, are heavily damaged by three Tephritidae fruit flies: *Bactrocera cucurbitae* (Coquillett), *Dacus ciliatus* Loew, and *Dacus demmerezi* (Bezzi). Crop losses can reach as 90–100% even applying chemical insecticides (Ryckewaert et al. 2010). The program

GAMOUR aimed to decrease the population of these pest species and losses in vegetable crops. Several approaches were used: sanitation measures, mass trapping, trapping crops, augmentation biological control, increase plant diversity, including crop association and vegetable cover (Deguine et al. 2017). The combination of all these measures concurs to the decrease of these tephritid flies' populations. Sanitation comprises collecting infested fruits and closing them in a device that traps the adult flies but allows the parasitoid to escape through a net. Additionally, maize rows are planted around the fields, and together with a bait based on liquefied protein and a very low concentration of spinosad on very small spots. Mass trapping is performed with sexual parapheromone, and augmentation releases of the parasitoid wasp *Psyttalia fletcheri* (Silvestri) contribute to decrease the population. Plant diversity increased around and inside the fields as well as the organic matter by using the compost from the sanitation device (Deguine et al. 2017).

In Kenya, Tephritidae fruit flies are major pests of several tree and non-tree crops, infesting fruits at ripeness, when pesticide can no longer be used. *Bactrocera invadens* Drew, Tsura & White was first reported in coastal Kenya, in 2003 (Lux et al. 2003) in mango orchards, but completing its cycle through avocado fruits. Because of that, and as it poses major constraints to international trade of these fruits, the pest was declared as a quarantine one in 2008. At that time, Kenya promoted a strategy, based on the use of a pheromone trap (biotechnical control) laced with an insecticide, empowering family farmers with an agroecological solution.

13.6.3 Riparian IES to Reduce Chemical Runoff into Water Bodies in Lake Naivasha, Kenya

In Kenya, the availability of freshwater from Lake Naivasha, together with the closeness to Nairobi, makes the area attractive for horticulture. The concentration of small family farmers in this area increases the vulnerability of the lake to agricultural residues, from fertilizers and pesticides. To avoid such impacts, farmers adopt highly innovative ecological measures such as water treatment systems that use natural processes involving riparian IES, soils organic content and their associated microbial assemblages. This combination protects the lake, the associated fishing industry and sustains a rich landscape (Wanjala et al. 2018).

Despite the success of the examples presented, technological and innovative agroecological farming interventions face challenges and threats. Agroecological crop protection practices require its deeper understanding when compared to other solutions, such as the use of pesticides. Challenges start with the lack of governmental policies supporting agroecological practices but continue with the need of knowledge, research, and training on pest management agroecological solutions. Also, a proper identification of organisms (and the lack of experts) is a major impediment to an adequate decision-making process. Because of globalization and climate change, new pests and diseases are emerging and the environment is facing a growing degradation process (from soil and water degradation to biodiversity loss). Extension services are limited in most countries, little funding is available for farmer training, and basic information is still needed, especially among family farmers.

REFERENCES

Abreu PHB, Alonzo HGA (2016) Family farmer and (un)safe use of pesticides in Lavras, Minas Gerais, Brazil. *Rev Bras Saude Ocup* 41: e18, doi:10.1590/2317-6369000130015.

Aguiar A, Silva L, Parente C, Costa CA (2020) Organizational innovation in the context of family farms: lean diagnosis. *J Innov Manage* 8: 68–86, doi:10.24840/2183-0606_008.002_0006.

Altieri MA, Nicholls CI (2005) *Agroecology and the search for a truly sustainable agriculture* (Mexico DF: United Nations Environmental Programme – Environmental Training Network Latin America and the Caribbean). http://www.agroeco.org/doc/agroecology-engl-PNUMA.pdf

Boller EF, Häni F, Poehling H-M, eds. (2004) *Ecological infrastructures: ideabook on functional biodiversity at the farm level: Temperate zones of Europe* (Lindau: IOBCwprs).

Brandth B (2002) Gender identity in European family farming: a literature review. *Sociol Ruralis* 42: 181–200, doi:10.1111/1467-9523.00210.

Burgio G, Ferrari R, Boriani L, Pozzati M, van Lenteren J (2006) The role of ecological infrastructures on Coccinellidae (Coleoptera) and other predators in weedy field margins within Northern Italy agroecosystems. *Bull Insectol* 59: 59–67.

Carlos C, Gonçalves I, Guenser J, Maillard B, Porte B (2014) *Technical guide: conservation actions to promote functional biodiversity in viticulture. Douro - Portugal* (Vila Real: ADVID), http://www.advid.pt/imagens/workshops/technicalguide_ADVID_EN.pdf

Chandler D, Bailey AS, Tatchell GM, Davidson G, Greaves J, Grant WP (2011) The development, regulation and use of biopesticides for integrated pest management. *Philos Trans R Soc Lond B Biol Sci* 366: 1987–1998, doi:10.1098/rstb.2010.0390.

Costa CA, Guiné R, Correia HE, Costa DT, Costa T, Parente C, et al. (2018) Agricultura familiar e proteção das culturas: abordagens tradicionais e proximidade com práticas de agricultura biológica. *Rev Ciênc Agrár* 41(spe): 164–173, doi:10.19084/rca.17086.

Costa CA, Parente C, Aguiar A, Guiné R, Costa D, Correia HE, et al., eds. (2020) *Pontes Entre Agricultura Familiar e Agricultura Biológica*, Iberografias 37 (Guarda: Centro de Estudos Ibéricos).

Deguine J-P, Nurbel T, Gloanec C, Laurent P (2017) Application of agroecological crop protection to vegetable crops: the GAMOUR experience. In *Agroecological crop protection*, J-P Deguine, C Gloanec, P Laurent, Ratnadass A, and J-N Aubertot, eds. (Dordrecht: Springer Netherlands), pp. 47–75, doi:10.1007/978-94-024-1185-0_2.

Deguine J-P (2017) Applying agroecological principles to crop protection. In *Agroecological crop protection*, J-P Deguine, C Gloanec, P Laurent, Ratnadass A, and J-N Aubertot, eds. (Dordrecht: Springer Netherlands), pp. 1–46, doi:10.1007/978-94-024-1185-0_1.

Dinis, I (2019) The concept of family farming in the Portuguese political discourse. *Social Sci* 8(7): 213, doi:10.3390/socsci8070213.

Djurfeldt G (1996) Defining and operationalizing family farming from a sociological perspective. *Sociol Ruralis* 36: 340–351, doi:https://doi.org/10.1111/j.1467-9523.1996.tb00026.x.

Eubanks MD, Styrsky JD (2005) Effects of plant feeding on the performance of omnivorous "predators". In *Plant-provided food for carnivorous insects: a protective mutualism and its applications*, FL Wäckers, J Bruin, PCJ van Rijn, eds. (Cambridge: Cambridge University Press), pp. 148–177, doi:10.1017/CBO9780511542220.007.

Ferlian O, Biere A, Bonfante P, Buscot F, Eisenhauer N, Fernandez I, Hause B (2018) Growing research networks on mycorrhizae for mutual benefits. *Trends Plant Sci* 23 (11): 975–984, doi:10.1016/j.tplants.2018.08.008.

Franco JC, Ramos AP, Moreira I (2006) *Infra-estruturas ecológicas e protecção biológica – Caso dos citrinos* (Lisboa: ISAPress).

Gurr GM, Lu Z, Zheng X, Xu H, Zhu P, Chen G, Yao X, et al. (2016) Multi-country evidence that crop diversification promotes ecological intensification of agriculture. *Nature Plants* 2: 16014, doi:10.1038/nplants.2016.14.

Hanafi A (2005) Invasive pests and diseases: a challenge to IPM in greenhouse crops. *Phytoparasitica* 33: 423–426, doi:10.1007/BF02981390.

ICIPE (2015) *The 'push–pull' farming system: Climate-smart, sustainable agriculture for Africa* (Nairobi, Kenya: ICIPE—African Insect Science for Food and Health).

Khan ZR, Midega CAO, Pittchar JO, Murage AW, Birkett MA, Bruce TJA, Pickett JA (2014) Achieving food security for one million sub-Saharan African poor through push–pull innovation by 2020. *Philos Trans R Soc Lond B Biol Sci* 369: 20120284, doi:10.1098/rstb.2012.0284.

Klecka J, Hadrava J, Biella P, Akter A (2018) Flower visitation by hoverflies (Diptera: Syrphidae) in a temperate plant-pollinator network. *PeerJ* 6: e6025, doi:10.7717/peerj.6025.

Lux S, Copeland R, White I, Manrakhan A, Billah M (2003) A new invasive fruit fly species from the *Bactrocera dorsalis* (Hendel) group detected in East Africa. *Insect Sci Appl* 23: 355–361, doi:10.1017/S174275840001242X.

Magdoff F, van Es H (2010) *Building soils for better crops*, 3rd ed. (Washington, DC: SARE).

Midega CAO, Pittchar JO, Pickett JA, Hailu GW, Khan ZR (2018) A climate-adapted push-pull system effectively controls fall armyworm, *Spodoptera frugiperda* (J E Smith), in maize in East Africa. *Crop Prot* 105: 10–15, doi:10.1016/j.cropro.2017.11.003.

Ngowi AVF, Mbise TJ, Ijani ASM, London L, Ajayi OC (2007) Pesticides use by smallholder farmers in vegetable production in Northern Tanzania. *Crop Prot* 26: 1617–1624, doi:10.1016/j.cropro.2007.01.008.

Parsa S, Morse S, Bonifacio A, Chancellor TCB, Condori B, Crespo-Pérez V, Hobbs SLA, et al. (2014) Obstacles to integrated pest management adoption in developing countries. *Proc Natl Acad Sci USA* 111(10): 3889–3894, doi:10.1073/pnas.1312693111.

Prieto JD, Castañé C, Calvet C, Camprubi A, Battaglia D, Trotta V, Fanti P (2017) Tomato belowground–aboveground interactions: *Rhizophagus irregularis* affects foraging behavior and life history traits of the predator *Macrolophus pygmaeus* (Hemiptera: Miridae). *Arthropod-Plant Interact* 11: 15–22, doi:10.1007/s11829-016-9465-5.

Rodríguez E, González M (2014) Vegetación autóctona y control biológico: Diseñando una horticultura intensiva sostenible. *Fichas de Transferencia Caja Mar Caja Rural* 4, https://www.grupocooperativocajamar.es/storage/documents/004-vegetacion-autoctona-1403862179-ccc2c.pdf

Ryckewaert P, Deguine J-P, Brévault T, Vayssières J-F (2010) Fruit flies (Diptera: Tephritidae) on vegetable crops in Reunion Island (Indian Ocean): state of knowledge, control methods and prospects for management. *Fruits* 65(2): 113–130, doi:10.1051/fruits/20010006.

Steward PR, Shackelford G, Carvalheiro LG, Benton TG, Garibaldi LA, Sait SM (2014) Pollination and biological control research: are we neglecting two billion smallholders. *Agric Food Secur* 3(1): 5, doi:10.1186/2048-7010-3-5.

Vaidya A, Gyenwali D, Tiwari S, Pande BR, Jørs E (2017) Changes in perceptions and practices of farmers and pesticide retailers on safer pesticide use and alternatives: impacts of a community intervention in Chitwan, Nepal. *Environ Health Insights* 11: 1–12, doi:10.1177/1178630217719270.

van Zanten BT, Verburg PH, Espinosa M, Gomez-y-Paloma S, Galimberti G, Kantelhardt J, Kapfer M, et al. (2014) European agricultural landscapes, common agricultural policy and ecosystem services: a review. *Agron Sustain Dev* 34: 309–325, doi:10.1007/s13593-013-0183-4.

Wäckers FL, van Rijn PCJ (2012) Pick and mix: selecting flowering plants to meet the requirements of target biological control insects. In *Biodiversity and insect pests*, GM Gurr, SD Wratten, WE Snyder, eds. (Chichester: John Wiley & Sons), pp. 139–165, doi:10.1002/9781118231838.ch9.

Walpole M, Smith J, Rosser A, Brown C, Schulte-Herbruggen B, Booth H, Sassen M, et al. (2013) *Smallholders, food security, and the environment* (Rome: IFAD).

Wanjala S, Mwinami T, Harper DM, Morrison EHJ, Pacini N (2018) Ecohydrological tools for the preservation and enhancement of ecosystem services in the Naivasha Basin, Kenya. *Ecohydrol Hydrobiol* 18: 155–173, doi:10.1016/j.ecohyd.2017.09.004.

14

Does Conservation Agriculture Work for Small-Scale Farmers in Developing Nations? A Mini-Review

Anandkumar Naorem[1], Shiva Kumar Udayana[2], and Somasundaram Jayaraman[3]
[1]*ICAR-Central Arid Zone Research Institute, Regional Research Station, Kukma-Bhuj, Gujarat, India*
[2]*MSSSOA, Centurion University of Technology and Management, Paralakhemundi, Odisha, India*
[3]*ICAR–Indian Institute of Soil Science, Bhopal, Madhya Pradesh, India*

CONTENTS

14.1 Introduction: Background

Smallholder agriculture is the most prominent component of food production in developing nations and could ensure long-term food security in the world. Around 80% of the total food produced in developing countries is generated through 500 million global smallholder farms (Sims and Heney 2017). Moreover, the burden of increasing food production is already upon them to feed the rising human population. Soil degradation is one of the most devastating problems and may be accelerated due to anthropogenic activities. It is more prevalent in resource-poor countries due to misuse of land to feed the growing human population bust in these countries (Mortimore and Harris 2005). The crop residues are generally burnt in the fields, causing environmental pollution such as smog. The intensive cultivation leads to excessive nutrient mining and overgrazing, failing to support enough crop productivity (Thierfelder and Wall 2009). It will be more difficult for small-scale farmers to raise their crop production with an increasing rate of soil degradation. Therefore, conservation agriculture (hereafter CA) is recommended in these developing nations with an aim to promote soil health and increase crop productivity in a sustainable and environment-friendly manner.

14.2 Conservation Agriculture: Principles and Benefits

CA is a worldwide known concept that has been implemented over 160 million hectares. CA has been actively supported by several international organizations, including Food and Agriculture Organization (FAO), the European Conservation Agriculture Federation, and the Global Forum on Agricultural

DOI: 10.1201/9781003164968-16

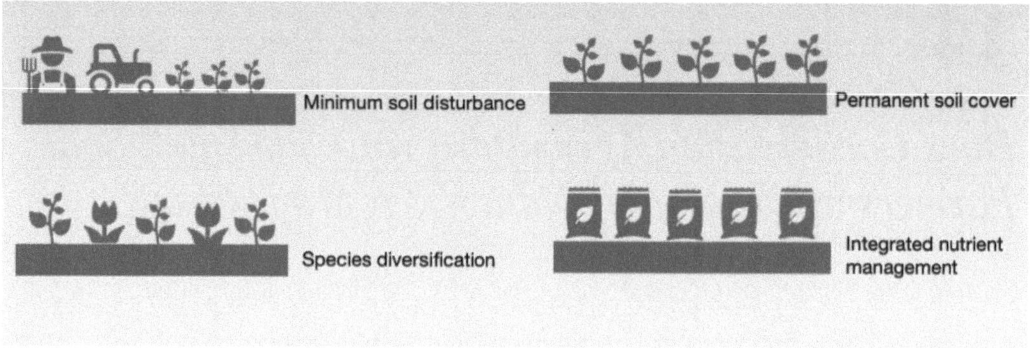

FIGURE 14.1 Main principles of conservation agriculture.

Research (Lahmar 2009). FAO introduced CA as a holistic approach to crop production that efficiently utilizes the available resources (FAO 2008). The CA is based on the integrated management of soil, water, and biological resources. It is simply an opposite trend of what has been perceived as conventional agriculture. The need for a clean seedbed without crop residues is replaced with the idea of leaving crop residues in the soil.

CA encompasses three basic principles (Ngwira et al. 2013):

a. zero or minimum soil disturbance

   It includes zero tillage or reduced tillage done to produce a narrow seeding slot in the soil.

b. providing permanent soil cover through mulching or cover crops

   It includes avoiding the burning of crop residues while it can be left in the soil surface as mulch or incorporated in the soil to increase soil organic matter and better soil aggregation. It also includes protecting the soil surface from erosion agents by using cover crops such as legumes.

c. crop diversification through intercropping or crop rotation

   Planting legumes followed by nonlegumes after the harvest of the earlier crop has been an old practice to reduce pest and disease attack. Intercropping increases soil biodiversity and benefits the farmers in case one crop fails to provide sufficient yield.

In addition to these three agronomic principles, a fourth principle "Integrated nutrient management" has been added, especially for developing countries. Figure 14.1 shows the three main principles of CA with an additional fourth principle for developing nations.

The benefits of CA have been reported in a large body of literature that includes an increase in soil biodiversity, soil and water conservation, reduction of energy consumption, higher water infiltration, lesser soil runoff and erosion, lesser greenhouse gas emissions, higher soil organic matter, and increased yield (Six et al. 2002; Gowing and Palmer 2008; FAO 2010). However, it is reported that CA will only work when all these principles are carried out simultaneously and followed strictly, which might not be easy for smallholders. CA recognizes the judicious use of land and water resources for sustainable crop productivity as well as to increase the carbon sink of the soil and mitigate climate change. Therefore, it is expected that a generic switch from conventional intensive tillage to CA among small-scale holders might solve the problem of global land degradation and increasing climate change. Figure 14.2 depicts the overall benefits and the ecosystem services provided by CA.

14.3 Adoption of Conservation Agriculture by Smallholders

One of the main aims of introducing CA is to reverse soil degradation and promote soil health on a global scale. Moreover, this aim was successfully implemented in North and South America where large-scale agriculture is predominant (Triplett and Warren 2008). The high success rate of CA in these regions

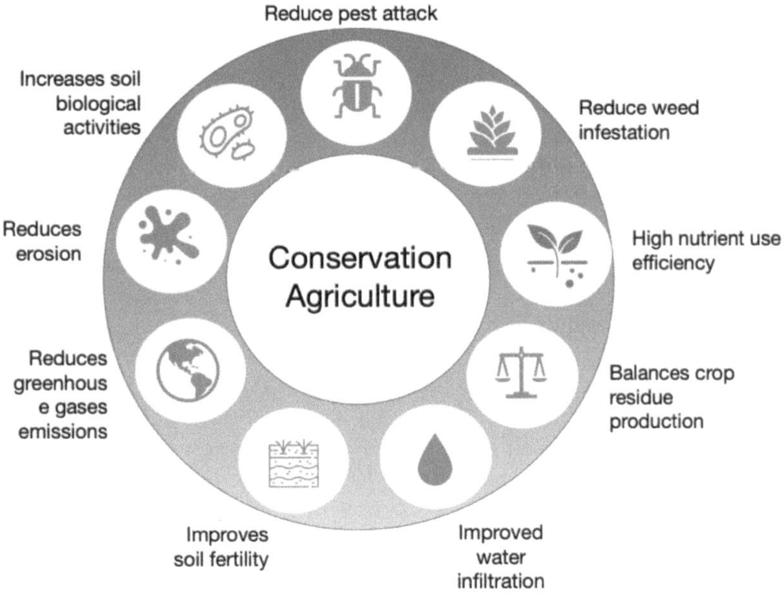

FIGURE 14.2 A series of ecosystem services provided by conservation agriculture.

fueled the assumption that CA will also solve other problems. For example, since CA practices involve minimum soil disturbance, less labor requirement is obvious and expected. On the other hand, CA has shown significant positive results in soil and water conservation, especially in dry soils (Thierfelder et al. 2017). Umar et al. (2011) reported that in Zambia, out of 129 households, only one was found to implement CA in which they were trained by a nongovernmental organization and the inputs were also provided to them through the organization. After it took a few years to observe the benefits of CA, they have converted all their farms to CA.

14.3.1 The Controversial Yield Effect and Adoption Rate

The promotion of CA for sustainable intensification must be addressed in several aspects because the impacts of CA on crop yield have been controversial. The short-term yield benefits of CA are found to be variable in various studies, indicating either positive, negative, or neutral response (Gill and Aulakh 1990). The short-term yield advantage is the main attraction for most small-scale farmers. The null and negative results of CA on yield responses discourage the farmers to adopt CA in the long term. The yield differences are due to the interaction of several variables that affects soil, crop, and climate. In drier soils, the implementation of CA decreases soil water evaporation losses and improves soil water infiltration, thus increasing crop yield. On the other hand, humid regions might face an opposite result as it will increase waterlogging in low-lying areas and depress crop yield (Rockström and Barron 2007).

Corbeels et al. (2020) conducted a meta-analysis of CA trials in sub-Saharan Africa in which 79 experiments from 16 countries were included. CA was compared with the conventional practices that include crop removal and cereal monocropping. CA could register a slightly higher average crop yield (+3.7%) and be more effective with the adoption of all three CA principles. Descheemaeker (2020) emphasized that CA could not overcome the increasing problems of crop productivity of the small-scale holders in sub-Saharan Africa in the short term. CA has been recognized and widely studied in various parts of the world whether it can increase crop yield without posing any environmental drawbacks. This issue is more serious and challenging in developing countries where the human population is continuously increasing and leading to more challenges in achieving food security. Most of these developing nations are predominated by small-scale holders. Therefore, to recommend and advocate a new system is always a risky situation in these nations. So, the big question arises whether CA can fulfill the food demands of small-scale farmers in developing nations. Few possible reasons could explain the discrepancies in crop yields due to CA practices (Pittelkow et al. 2015):

a. CA involves a series of practices that could vary between farmers based on their resources, cropping system, climate, and other influencing variables. For CA to be successfully profitable, all the components must be implemented, which might not be possible for smallholders.
b. Improving soil health through CA is a time-taking process. Therefore, increasing crop yields under CA might be observable only after a few years of CA implementation.

Although the farmers are adopting CA components based on their conditions, the adoption rate is much lesser than the efforts given to promote CA in these developing countries (Corbeels et al. 2014). Corbeels et al. (2020) contradicted the claims that CA could improve the food security of small-scale holders in developing countries. It is also added that a small increase in crop yield seems meaningless because these farmers have small land areas. Most of the farmers are demotivated by the time-taking aspect of CA to deliver the full benefits to these farmers. Giller et al. (2009) highlighted that the benefits of CA in addressing the problems of soil fertility in the resource-poor developing countries seem to show better results under controlled conditions than field conditions. It is not attributed to the technical problems related to CA implementation. It is rather related to the poor resource of the smallholders that constrain the applicability of CA components in their fields. Trials conducted on CA and its benefits include the systematic supply of inputs such as fertilizers, pesticides, improved seeds at the right time and right amount. However, in the case of small-scale farmers, these inputs might vary and are lacking which might affect the performance of CA (Haggblade and Tembo 2003). Due to labor constraints or less resource, small-scale farmers are often found not adopting all the three principles of CA (Baudron et al. 2007).

14.3.2 Weed Menace under Conservation Agriculture

Few constraints hinder the high adoption rate of CA in these nations. For example, minimum soil tillage leads to higher weed growth in field conditions. The lack of market access in some areas limits the use of pesticides in CA. Moreover, weed management through manual weeding increases the labor cost. The poor markets in the developing countries also limit the inclusion of legume crops in CA as intercrop or crop rotation because many small-scale farmers rely on sustenance type of agriculture due to which they prefer to cultivate energy-dense cereal crops.

The most recognized problem in CA practice is the control of weed infestation in the field (Vogel 1995). Weed control requires labor inputs and is not affordable for most small-scale farmers (Wall 2007). On the other hand, proponents of CA claimed that with a sufficient permanent ground cover, there will be less burden of weeds under CA farm. The increased labor requirement in weed control might outweigh the labor-saving through no-tillage (Siziba 2008). In addition to it, hand tillage or plowing is mainly carried out by men while hand weeding is performed by the women section. Eliminating the tillage requirement while increasing the need for weed control leads to an unacceptable load of the burden of labor on women. In the short term and without the use of herbicides, CA could not increase the net income of the small-scale holders in the developing nations. Rather, it might increase the labor requirements for women.

14.3.3 Crop Residues: More Competition and Less Availability

Another point to address in CA is the incorporation of crop residues in the soil. Arid zones have less vegetative cover owing to their low annual precipitation. Therefore, most of the arid agriculture is livestock-based type, generating more income through livestock. In such cases, mulching with crop residues might not be completely possible. Most of these crop residues are used as livestock feed and a small percentage

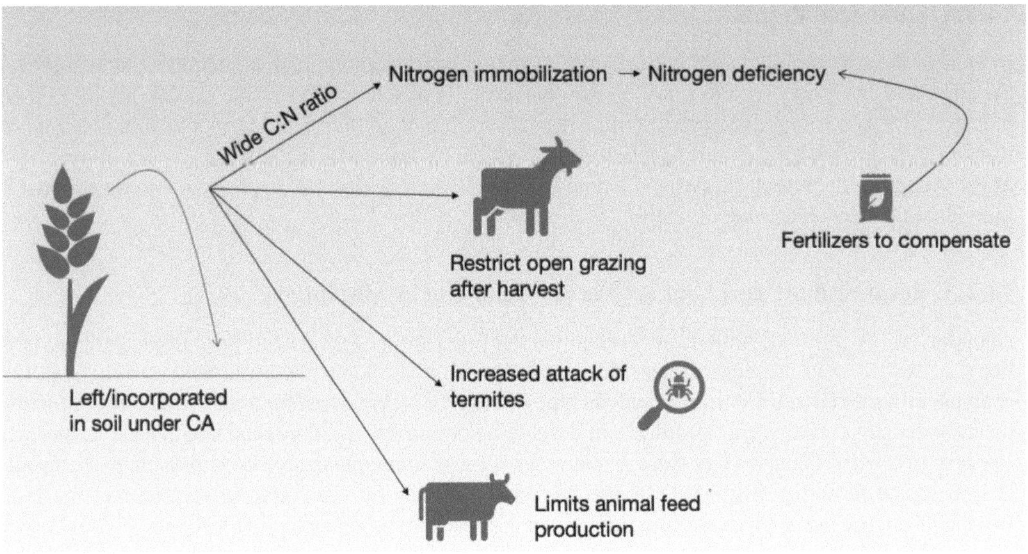

FIGURE 14.3 Challenges for smallholders related to crop residue management under conservation agriculture.

of this crop residue goes to the mulching process. The limited availability of crop residues in many farming systems limits the adoption of CA in smallholding farming. Small-scale farmers often use the crop residues for livestock feed because of few reasons:

a. Livestock is a form of investment and insurance against any unpredictable risk.
b. Livestock is used in traction and intercultural practices.
c. Livestock produces inputs for making farmyard manure, compost, vermicompost, etc.
d. Livestock also generates additional income from milk and meat.
e. Improvement of soil health by incorporating crop residues into the soil takes a long time.

Figure 14.3 shows the crop residue-related challenges faced by smallholders. Therefore, prioritizing the short-term monetary gains over the long-term advantages of CA in resource-poor countries is not surprising. The remaining crop residues are used for mulching which is not sufficient to increase the crop yield under CA (Wezel and Rath 2002). Recommending using limited crop residues in CA rather than feeding to livestock might not be accepted by most small-scale holding farmers. In some cases where farmers have a few livestock, tillage is often carried out with hand hoes and the crop residues are burnt since it is the easiest way to clear the field before sowing of the next crop. Other reasons for burning crop residues are soil fertility enhancement, clearing weeds, killing soilborne pest and controlling rodent infestation (Kanmegne 2004).

Another secondary problem of retaining crop residues in the field under CA is observed in Mozambique where it was being removed by termites within a few weeks (Giller et al. 2009). The crop residues initially retained in the soil to protect the soil from erosion agents are susceptible to surface runoff or heat due to termite attacks, diminishing the effects of CA on soil health. Crop residues with a wide C:N ratio, when a large amount is left in the soil, create a temporary immobilization of nitrogen (Abiven and Recous 2007). It increases the input requirement of fertilizers to compensate for the nitrogen deficiency due to immobilization, failing to do so might suffer yield loss. However, it is reported that such immobilization does not always occur (Lal 1979). But, if soil nitrogen availability decreases due to the abovementioned issue, a larger amount of nitrogen fertilizer needs to be applied to achieve the yield comparable to that without incorporating crop residues (Giller et al. 2009).

14.3.4 Land-Use Rights

In most of the smallholding communities, open grazing is a common practice. After the harvest season, the goats and cattle are freed to graze on the open crop fields due to the relative availability of feed and better quality as compared to open grazing areas. Therefore, if an individual farmer wishes to leave crop residues as a crop mulch, he must negotiate with the local by-laws, which challenges the traditional rights of the villagers. Therefore, targeting a whole community to practice CA is preferable to be adopted by individual farmers (Martin et al. 2004).

14.3.5 Requirement and Lack of Accessibility to CA Machinery

Adoption of CA by small-scale farmers requires the provision of specialized farm implements to practice CA. No-till seeding requires specific instruments to make a narrow seeding slot in the soil. One example of a specialized CA implement is Happy Seeder. The big question arises whether these implements are easily available and affordable in developing countries. An individual small-scale farmer with low resource will hesitate to buy these implements if these implements are not affordable to the farmer. A well-equipped and well-trained CA mechanization service provider may solve a part of the issue. Instead of buying the implements, the farmer can approach these service providers and get the far work done without investing in buying the implements. This will also benefit the service provider and can make a living by deploying this equipment over several CA farms. Interested farmers can later purchase these CA machines by gathering the technical knowledge from these service providers. For the connection between the service providers and the farmers to successfully be made, meetings and discussion programs must be organized involving a large group of farmers and the service providers. Through the interactions, both the parties will be able to understand each other's needs and requirements. The service provider must mapped an area that could be provided with these CA implements. The selection of the location must consider the transportation issues, location of the market and the village. The service provider must be equipped with internet services to understand how long it might take to deliver the equipment to the individual farmer because sowing, intercultural practices, and harvesting are season dependent. Clusters of interested farmers can be identified based on their requirements. Figure 14.4 depicts an overall concept of CA adoption-related issues among smallholders.

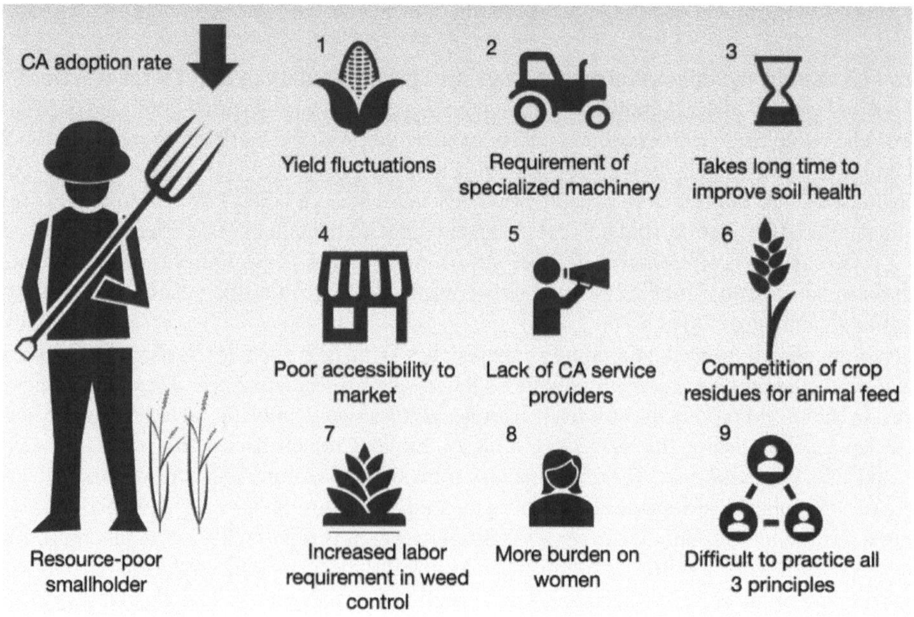

FIGURE 14.4 Main constraints in adoption of conservation agriculture among small-scale farmers.

14.4 Conclusions

Technology such as CA can be successfully implemented in developing countries only when it can be embedded into the economic, social, and cultural context. Many of the CA adoption cases in developing countries transpired after a few years due to the temporary influence and monitoring of the project. As the support given to the farmers stops, the small-scale farmers revert to their conventional crop management practice. Farmers must be made aware of the long-term benefits of CA and the ill effects of soil degradation. Ecosystem services provided through CA must be well discussed with the farmers in association with social workers, policymakers, researchers, etc. However, lack of specialized machinery and technical know-how on CA will discourage the farmers from adopting CA. It must be implemented after choosing the right cropping system and should be modified based on the climatic conditions, soil, and other socioeconomic factors. Overall, the constraints identified that limits the adoption of CA in small-scale holding dominated agriculture are:

a. lack of accessibility to specialized implements used in CA
b. unavailability of crop residues
c. competition of crop residues between animal feed and incorporating in the soil
d. time taking effects of CA on crop yield
e. difficult to carry out all the three principles of CA in resource-poor farm
f. lack of appropriate technical knowledge
g. blanket recommendations
h. increased problem of weed control
i. more burden on agricultural women
j. lack of access to credit

REFERENCES

Abiven, S. and Recous. S. 2007. Mineralisation of crop residues on the soil surface or incorporated in the soil under controlled conditions. *Biology and Fertility of Soils* 43: 849–852.

Baudron, F., Mwanza, H. M., Triomphe, B. and Bwalya. M. 2007. Conservation Agriculture in Zambia: A Case Study of Southern Province. *African Conservation Tillage Network, Centre de Coopération Internationale de Recherche Agronomique pour le Développement*, Food and Agricultural Organization of the United Nations, Nairobi.

Corbeels, M., De Graaff, J., Ndah, T. H., Penot, E., Baudron, F., Naudin, K., Andrieu, N., Chirat, G., Schuler, J., Nyagumbo, I. and Rusinamhodzi. L. 2014. Understanding the impact and adoption of conservation agriculture in Africa: A multiscale analysis. *Agriculture, Ecosystems, and Environment* 187: 155–170.

Corbeels, M., Naudin, K., Whitbread, A. M., Kühne, R. and Letourmy. P. 2020. Limits of conservation agriculture to overcome low crop yields in sub-Saharan Africa. *Nature Food* 1(7): 447–454. https://doi.org/10.1038/s43016-020-0114-x.

Descheemaeker, K. 2020. Limits of conservation agriculture in Africa. *Nature Food* 1(7): 402–402. https://doi.org/10.1038/s43016-020-0119-5.

FAO. 2008. Investing in Sustainable Agricultural Intensification. The Role of Conservation Agriculture. A Framework for Action. Food and Agriculture Organization of the United Nations, Rome.

FAO. 2010. What is Conservation Agriculture? [Online] Available: http://www.fao.org/ag/ca. Retrieved 14th August 2010.

Gill, K. S. and Aulakh. B. S. 1990. Wheat yield and soil bulk-density response to some tillage systems on an Oxisol. *Soil Tillage Research* 18: 37–45.

Giller, K. E., Witter, E., Corbeels, M. and Tittonell. P. 2009. Conservation agriculture and smallholder farming in Africa: The heretics' view. *Field Crops Research* 114(1): 23–34. doi:10.1016/j.fcr.2009.06.017.

Gowing, J. W. and Palmer. M. 2008. Sustainable agricultural development in Sub-Saharan Africa: The case for a paradigm shift in land husbandry. *Soil Use and Management* 24(1): 92–99. doi:10.1111/j.1475-2743.2007.00137.x

Haggblade, S. and Tembo. G. 2003. Early Evidence on Conservation Farming in Zambia. International Food Policy Research Institute, Washington, DC.

Kanmegne, J. 2004. Slash and burn agriculture in the humid forest zone of Southern Cameroon: Soil quality dynamics, improved fallow management and farmers' perceptions. Tropenbos-Cameroon Series xii.

Lahmar, R. 2009. Adoption of conservation agriculture in Europe: Lessons of the KASSA project. *Land Use Policy* 27(1): 4–10. doi:10.1016/j.landusepol.2008.02.001.

Lal, R. 1979. Influence of 6 years of no-tillage and conventional plowing on fertilizer response of maize (*Zea mays* L.) on an Alfisol in the tropics. *Soil Science Society of America Journal* 43: 399–403.

Martin, C., Castella, J. C., Hoang Lan, A., Eguienta, Y. and Tran Trong. H. 2004. A participatory simulation to facilitate farmers' adoption of livestock feeding systems based on conservation agriculture in the uplands of northern Vietnam. *International Journal of Agricultural Sustainability* 2: 118–132.

Mortimore, M. and Harris. F. 2005. Do small farmers' achievements contradict the nutrient depletion scenarios for Africa?. *Land Use Policy* 22(1): 43–56. doi:10.1016/j.landusepol.2003.06.003.

Ngwira, A. R., Thierfelder, C. and Lambert. D. M. 2013. Conservation agriculture systems for Malawian smallholder farmers: Long-term effects on crop productivity, profitability and soil quality. *Renewable Agriculture and Food Systems* 8: 350–363.

Pittelkow, C. M., Liang, X., Linquist, B. A., Van Groenigen, K. J., Lee, J., Lundy, M. E., Van Gestel, N., Six, J., Venterea, R. T. and Van Kessel. C. 2015. Productivity limits and potentials of the principles of conservation agriculture. *Nature* 517(7534): 365–368.

Rockström, J. and Barron. J. 2007. Water productivity in rainfed systems: Overview of challenges and analysis of opportunities in water scarcity prone Savannahs. *Irrigation Science* 25: 299–311.

Sims, B. and Heney. J. 2017. Promoting smallholder adoption of conservation agriculture through mechanization services. *Agriculture* (64): 1–22. https://doi.org/10.3390/agriculture7080064.

Six, J., Feller, C., Denef, K., Ogle, S. M., Sa, J. C. D. M. and Albrecht. A. 2002. Soil organic matter, biota and aggregation in temperate and tropical soils – Effects of no-tillage. *Agronomie* 22(7–8): 755–775. doi:10.1051/agro:2002043.

Siziba, S. 2008. Assessing the adoption of conservation agriculture in Zimbabwe's smallholder sector. PhD Thesis. University of Hohenheim, Germany.

Thierfelder, C. and Wall, P.C. 2009. Effects of conservation agriculture techniques on infiltration and soil water content in Zambia and Zimbabwe. *Soil and Tillage Research* 105(2): 217–227. doi:10.1016/j.still.2009.07.007.

Thierfelder, C., Chivenge, P., Mupangwa, W., Rosentock, T. S., Lamanna, C. and Eyre. J. X. 2017. How climate-smart is conservation agriculture (CA)? – Its potential to deliver on adaptation, mitigation and productivity on smallholder farms in southern Africa. *Food Security* 9: 537–560.

Triplett, G. B. and Warren. A. D. 2008. No-tillage crop production: A revolution in agriculture!. *Agronomy Journal* 100: S-153–S-165.

Umar, B. B., Aune, B. J., Johnsen, H. F. and Lungu. I. O. 2011. Options for improving smallholder conservation agriculture in Zambia. *Journal of Agricultural Science* 3(3): 50–62. https://doi.org/10.5539/jas.v3n3p50.

Vogel, H. 1995. The need for integrated weed management-systems in smallholder conservation farming in Zimbabwe. *Der Tropenlandwirt* 96: 35–56.

Wall, P. 2007. Tailoring conservation agriculture to the needs of small farmers in developing countries: An analysis of issues. *Journal of Crop Improvement* 19: 137–155.

Wezel, A. and Rath. T. 2002. Resource conservation strategies in agro-ecosystems of semi-arid West Africa. *Journal of Arid Environments* 51: 383–400.

15

Technological and Social Innovations in the Context of Small Brazilian Rural Properties

Alexandre Marco da Silva
Department of Environmental Engineering, Institute of Science and Technology of Sorocaba – São Paulo State University (UNESP), Sorocaba, SP, Brazil

CONTENTS

15.1 Introduction

Land exploitation is the result of land scarcity. On one global context and always considering all continents, this scarcity has been considered relative. Sufficient fertile lands were accessible, but lands originally fertile and also producing naturally a set of useful plant and/or animal species for human nutrition – where then the activity was collecting and not planting food – became scarce since several thousand years ago in most continents (Vink, 1975). This scarcity is forcing the humans to choose two types of actions: (i) search for new lands that naturally still produce enough specific foods for human support without any local manipulation or (ii) stay in the same place and inject efforts to manipulate the land to get raw material – food and utensils (fiber, leather, wood).

The need to choose the second attitude has been often determined by local natural barriers and, even in the last centuries, comes being chosen to attend to the rapid growth of the human population in the world. The word agriculture means "art of farming." It is the set of techniques designed to cultivate the land aiming to obtain products from it, primarily food. However, with advances in techniques and technology, agriculture has increasingly served the provision of foodstuffs for the production of fibers, energy, raw materials for clothing, fuel, construction, medicines, tools, ornamentation, and countless other purposes. These products, as well as the agricultural methods used, might vary from one part of the world to another (Vink, 1975; Mishra et al., 2018).

People began to raise domesticated crops and livestock in several different regions of the world at different times (Tauger, 2020). However, markedly around 11,000 years ago, humans gradually learned to grow cereals and tubers and thus were able to settle in one place and establish a life based on agriculture. At the same time, livestock also initiated, with the gradual domestication and breeding of animals, which until then were wild. Agriculture permitted humanity to produce surplus food, which made it possible

DOI: 10.1201/9781003164968-17

to exchange goods for other types of food that were not produced by them. Surpluses also functioned as a source of food security in cases where cultivation was hampered by natural factors, such as prolonged droughts, frosts, or excessive rainfall events or periods (Sarkadi, 2019). Also, food surpluses have given people time to devote themselves to tasks unrelated to agriculture or obtaining food. From then on, the human being began to develop techniques to make his life easier and more comfortable, such as the construction of houses and objects that could facilitate his life and work, yet several important impacts on environmental and social have been generated since then.

15.2 The Brazilian Agricultural System

Historically, agriculture in Brazil is one of the main bases of Brazil's economy (da Silva e Souza et al., 2020). In 2017, we had in Brazil a total of 5,073,324 rural properties. All these properties occupied an area of 3,512,898 km^2, or 41.3% of the total Brazilian territory (IBGE (Brazilian Institute for Geography and Statistics), 2019). In Brazil, rural properties can vary from smaller than 1 ha to larger than 100,000 ha. Diversity constitutes a prominent feature of the contemporary Brazilian agricultural system (Figure 15.1). It is the result of the different relationships established with natural resources, distinct biomes, the remarkable presence of different social actors, different relations between rural and urban, different rural development projects and policies, development dynamics generated by the State and by the actors themselves and, more generally, of "global benchmarks" related to the rural and disseminated by several mediating organizations, usually in the form of good practices (Grisa et al., 2017). However, a substantial part of agricultural lands in Brazil is used for extensive livestock pastures (~45.0%) (Camargo et al., 2017; IBGE, 2017).

15.2.1 The Brazilian Rural Smallholdings

Brazilian agriculture is a specific case because of its regional heterogeneity (Buainain and Garcia, 2018). In Brazil, family farming is conceptualized by federal legislation (Brazilian Federal Government, 2012)

FIGURE 15.1 (a) Small rural property showing the low technological level (Bofete, SP, Brazil), the local predominance of pasture, and linear erosion arising (gullies); (b) small rural property also with a pasture (soil better conserved), furthermost we see a grange (raising chickens), depicting a multiple land use of some small rural properties (Laranjal Paulista, SP, Brazil); (c) a small rural property in Ibiuna (SP, Brazil) with the soil being tillage for horticulture, and (d) a big farming with an orchard (orange) illustrating a higher technological level of agricultural management (Tatuí, SP, Brazil).

according to a set of criteria, including the size of the property, depending mainly on your profits on the property and using more family labor than hired labor. The size criterion varies according to each municipality and is correlated to the achievement of a minimum agricultural income (Rada et al., 2019). Considering this size of rural property, in 2017, we had in Brazil a total of 3,897,408 properties sized as 50 ha or smaller (76.8% of the total of rural properties).

All these small farms occupied an approximated area of 810,000 km^2, representing 23.1% of the total ruralized Brazilian area or 9.5% of the total Brazilian territory (IBGE (Brazilian Institute for Geography and Statistics), 2019). This kind of rural property is normally seen as agriculture of subsistence (subsistence production is that one in which the food produced is geared toward meeting the vital needs of the landowner and his family) or production (and sale) of basic foods. However, of the approximate 81 million hectares of family farming, 48% were destined for pastures, while the area with forests or agroforestry systems occupied 31% of the areas, and the crops occupied 15.5%. This shows that the relationships among the farmers and the diverse segments of the society have changed, mainly because of the improvement of the network communication systems, as explained ahead.

So, what really is a small rural property? What are your features? Yet there is the concept in the Brazilian Federal legislation about familiar agriculture (Brazilian Federal Government, 2017), some additional features are normally added and discussed: (i) the income level of the production unit is low; (ii) the degree of integration of the producer with the market for inputs and products, in general, is reduced; (iii) the technological level is low, including the irrigation-related technologies (the water deficit is a frequent and persistent problem), and (iv) they are highly dependent on the broad sanity of the environment.

15.3 The Brazilian Rural Small Farms: The Processes of Modernization, the Technological, and the Social Advances

Historically, the Brazilian agricultural exploration has its conformation in colonial origins (markedly 18th and 19th centuries), based on three categories of land ownership (Rabaiolli et al., 2015): (i) the large monoculture export farm, which maintained the old and prolonged master/slave relationship that caused weaknesses and social dependencies; (ii) small property, originating from the colonization process that occurred around the 19th century, which gave rise to small producers; (iii) the average property that emerged with the cattle ranch next to the sugarcane plantation areas (in some Brazilian regions, as Northeast and Southeast).

The traditional process of modernization of Brazilian agriculture, at the same time that allowed strong growth in production, implied social exclusion and high environmental costs. The prediction of the early 1970s that Brazil would become a major agricultural power was confirmed. Despite its success, the traditional model of modernization – the main inspiration for agricultural policies since the 1970s – is the subject of several criticisms that reflect socio-environmental concerns which can be addressed in terms of crises (Sabourin et al., 2005): (i) the environmental crisis, with the increase in the quantity and diversity of pollutants from agricultural activity, which are systematically dumped in the rural environment and which also affect food security. Decreasing forested areas, increased soil erosion, and degradation of soils and water resources are also recurring problems, and (ii) the social crisis, which has been accentuated in recent decades, with serious repercussions on Brazilian society. The modernization of agriculture was – and is so far – a factor of social exclusion. The expansion of cities imposes, with increasing urgency, the relative, if not total, control of the rural exodus.

This situation led the federal government to develop, from the mid-1990s, a dual policy. On the one hand, the former Brazilian federal segment named Ministry of Agriculture, Livestock, and Supply tried to maintain, as a central goal, the competitiveness of the business sector, maximizing the opportunities of agribusiness. On the other, the Ministry of Rural Development officially dealt with agrarian reform and the development of family farming, recognizing the social, productive, and environmental importance of that segment for society in general.

In 1996, the Brazilian Government created one of the most significant programs of agricultural assistance for smallholders: the Pronaf – Brazilian National Program for Strengthening Family Agriculture (Vázquez and Niederle, 2018). Through the Pronaf, the familiar farmer might obtain financial support for individual or collective projects that generate income for family farmers and land reform settlers. The program has the lowest interest rates on rural financing, in addition to the lowest default rates among the country's credit systems. Pronaf aims to operate four major lines of action: (i) support for infrastructure and services in municipalities, (ii) training/professionalization of family farmers, (iii) public policy negotiations, and (iv) rural credit. Also, there are complimentary benefits and programs that aim, for instance, to secure the family farmers in case of loss of crops, due to drought (the family farm insurance, SEAF).

In January 2019, the Brazilian government created the Secretariat for Family Agriculture and Cooperatives and assigned this Secretariat to propose, standardize, plan, promote, guide, coordinate, supervise, and evaluate, within the scope of the current Ministry of Agriculture, activities related to agriculture family and agrarian reform settlements, rural cooperatives, and associations, urban and peri-urban agriculture, agroextractivism and infrastructure for rural areas within the scope of productive projects (Brazilian Federal Government, 2020).

However, the life of the Brazilian family farmers is far to be easy or relaxing. Medina et al. (2015) showed that 83.1% of the Brazilian family farms have unsatisfactory land to be plenty competitive, only 42.9% of household heads started attending primary school, and 62.3% of the households have a total remuneration of less than one Brazilian minimum salary. Farmers also face an unfavorable institutional context with only 12.8% benefiting from agricultural policies and 68.3% having access to electricity. Therefore, only 33.8% adopt elementary technologies such as soil fertilization, less than 25.0% consider reliably integrated into markets, and just 5.4% belong to cooperatives. Medina et al. (2015) considered a database until 2014 to elaborate the paper. However, currently (the year 2021), the situation changed little for the Brazilian family farmers.

Furthermore, and trying to be exemplifying a situation of a Brazilian region, Dal Moro and Brandli (2020), aiming to know about the potentialities and challenges of family agriculture in a region of South Brazil (a set of 21 municipalities sited in the northern region of the State of Rio Grande do Sul), they found a set of 27 challenges and 14 potentialities (the main are depicted in Figure 15.2).

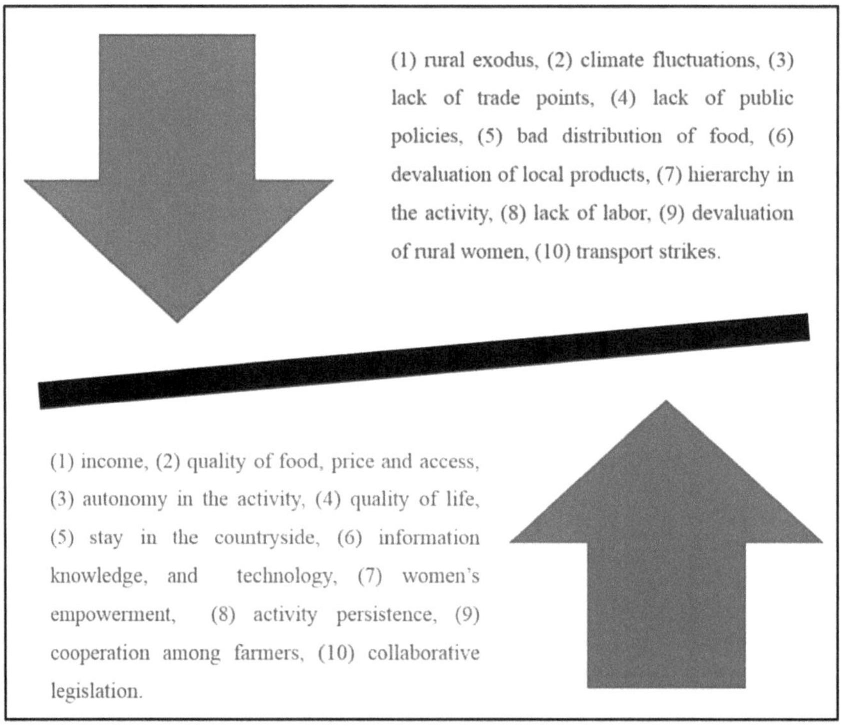

FIGURE 15.2 Representation of 10 from 27 challenges (top arrow pointing down) and 10 from 14 potentialities (bottom arrow pointing up). (**Figure elaborated with a database provided in Dal Moro and Brandli, 2020**).

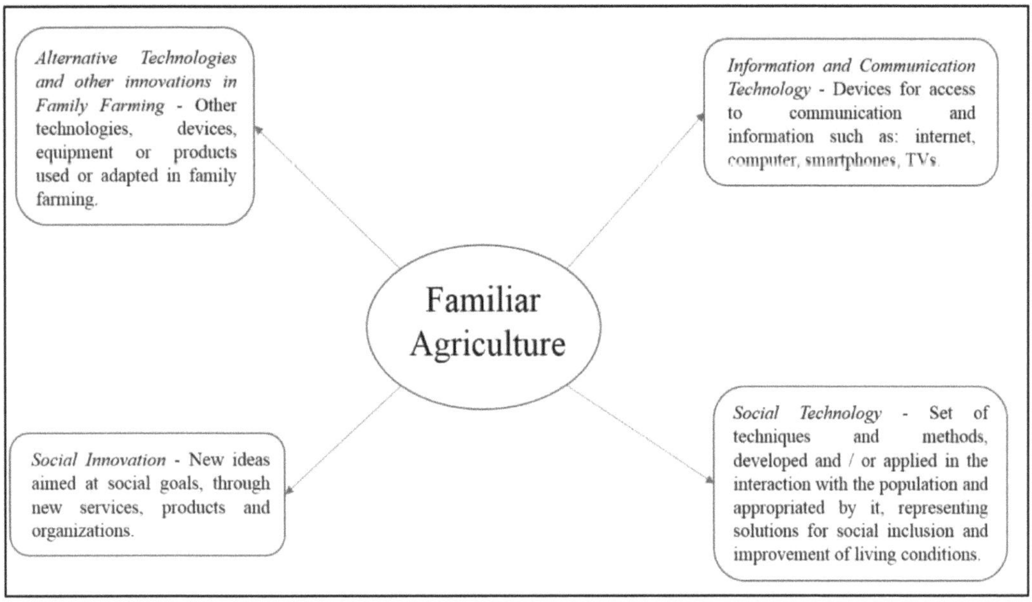

FIGURE 15.3 Some important matters that were identified by Carvalho and Lago (2019).

The authors mention that region has its local challenges and such challenges, if not well tackled, can cause inadequate sustenance in quantity and quality to serve families. On the other hand, at least in the studied region, income, and quality of food, price, and access were the main potentialities observed. In terms of technology, we have that the innovation in agriculture cuts transversely all dimensions and magnitudes of the production cycle and along the entire value chain – from the crop, forestry, fishery, or livestock production to the administration of inputs to market access (FAO (Food and Agriculture Organization of the United Nations), 2020). The technological development embraces two main goals: (i) inserting the familiar farmer in the digital world; (ii) provide conditions to initiate or improve the cooperative system of work. Figure 15.3 depicts four sets of technologies considered by CGI (Internet Management Committee in Brazil) (2012), Costa et al. (2014), and ITS (Brazilian Institute for Social Technology) (2004) (all of them cited by Carvalho and Lago (2019)) and embedded in the goals cited above.

According to Buainain and Garcia (2018), currently in Brazil, mainly because of wide geographical variation and regional contrasts, the technology and production systems range: (i) from a primitive cut and burn shifting agriculture used by smallholders in the rain forest regions (they largely use the "match stick technology") to diversified productions systems using up-to-date technology; (ii) from subsistence and maintenance farming to contract, commercial farming; (iii) from market isolated groups to smallholders producing high-quality food and nonfood products entirely integrated into the world markets (some of them organic). Following, we briefly describe each group of socio-digital technology related to familiar agriculture depicted in Figure 15.3.

15.3.1 Information and Communication Technologies (ICTs)

Digital technologies have the potential to revolutionize agriculture by helping farmers work more precisely, efficiently, and sustainably. Data-driven insights can improve decision-making and practices and help increase environmental performance while making the job more attractive to younger generations. Digital technologies also have the potential to offer consumers greater transparency as to how their food is produced (Rotz et al., 2019).

They offer opportunities to renew business models in value chains by connecting producers and consumers in innovative wa0ys. Beyond farming, digital technologies are key to make rural communities more attractive, smart, and sustainable, reducing problems related to remoteness and improving access to services. Research and Innovation (R&I) is vitally important to facilitate and accelerate digital transformation in agriculture and rural areas for the benefit of citizens and businesses (Carvalho and Lago, 2019).

However, especially in Brazil, the use of digital tools for commercialization, interest in digital solutions, usual bottlenecks in connectivity infrastructure, and lack of knowledge about available technologies. While more than 80% of landowners of large Brazilian farms make full use of digital technology (for purposes such as obtaining information and planning jobs, managing property, buying and selling inputs, and production), the vast majority of small farmers have zero or poor access to current technology (hardware, devices, drones, software and apps, databases, and/or networks).

Brazil's extensive territorial dimension, the lack of infrastructure in rural areas, and high investment constitute factors that make small- and medium-sized producers unable to access the internet and remain the main impediments to this expansion. In some rural properties, especially those that are furthest from urban centers, broadband internet access services are nonexistent. This fact is coupled with the financial investment necessary to acquire computers and software, training in the use of technology, and even cultural factors of resistance to innovations represent the challenges faced in the dissemination of these technologies in the rural scenario. On the other hand, when available, access to ICTs has a positive impact on producers and points to a trend in the sector. An important problem linked to digital technology remains in the fact that around 77% of Brazilian agricultural properties are headed by individuals with a very low level of schooling and instruction, which possibly results in low learning capability and with negative effects on the adoption of new technologies (Buainain and Garcia, 2018). This can make the process of disseminating a new technology more time-consuming and inefficient. Considering that digital technologies are "disruptive technologies" (i.e., a technological innovation, product, or service, with "disruptive" characteristics, which cause a break with traditional patterns, models, or technologies), the low level of educational instruction ends up being a factor that makes the farmer less flexible to agree to take this type of change in his/her life and his/her business. Further, the resistance of farmers and feelings of mistrust regarding the practice of making records, it is not experienced in their daily lives impede the use of instruments for decision-making and interfere negatively in the process of socializing the data. In this sense, the role of universities and research centers becomes fundamental, especially regarding ensuring that family farmers gain autonomy and can make their decisions safely (Deponti et al., 2017). Complementarily, educational, and technical training programs need to be continuously designed, offered, and developed in all parts of the country (Bolfe et al., 2020).

15.3.2 Alternative Technologies

When the goal is to develop solutions to solve problems and these solutions consider values such as minimum use of nonrenewable resources, minimum environmental interference, provide regional and subregional self-sufficiency, and collaboration in eliminating the exploitation and alienation of people, we are talking about alternative technologies. This kind of technology does not annulate the first group of technologies (digital technologies), but importantly complements the first group, however.

Alternative technologies are important in the regional context because they ponder the local characteristics (i.e., the technological development occurs in strict harmonization with the local potentialities and limitations). Alternative technologies constitute part of the local cultural knowledge and help local economic health. Some alternative technologies have the role to supply lacks or insufficiencies of any service inexistent in the rural property (for example, develop an alternative solution to pumping water from one local to above, without the use of electricity or fuel), or to develop new kind of products that might complement the economy of the rural property or rural community).

Exemplifying, as an alternative technology we have Minimal Processing Technology. Minimally processed vegetables (MPV) are generally defined as products that contain living tissues or those that have been slightly modified in their initial conditions, appearing fresh and maintaining their original qualities. Precut or MPV are those that undergo operations such as cleaning, washing with drinking water, sanitation, rinsing, peeling, cutting, packaging, and storage. They have the convenience of not requiring any significant preparation on the part of the consumer, in terms of selection, cleaning, washing, or cuts. Also, the added value of the product, through processing, increases the competitiveness of the productive sector and provides alternative means for commercialization. The success of this venture depends on the use of high-quality raw materials, handled and processed with good hygiene conditions (Pereira et al., 2015; Cruz, 2020).

In Brazil, an aspect that has contributed strongly to this growth is the expansion of fast-food services, hotels, lodges, inns, restaurants, hospitals, catering companies for airports, and also at the domestic level, which search for minimally preprocessed and high-quality products. In addition to adding value, another advantage of MPVs is the reduction in transport costs by up to 50%, since the parties do not market (Pereira et al., 2015). While the processing and processing of food generally take place in the kitchen of the farmers, the rural family agribusiness constitutes a new space and a new social and economic enterprise. The handmade food processing activity, especially for women farmers, also highlights the existence of cultural aspects associated with the eating habits of a given region (Cruz, 2020).

Another example: an alternative way, ecologically balanced, consistent with the characteristics of the region's soil and climate, comprises the cultivation, commercialization, and consumption of unconventional food plants (UFP). The UFP are plants that are often seen as invasive or harmful since they born spontaneously among other crops. However, these plants have great ecological importance and high economic potential. Although they are unknown to the majority of the population, many of these plants have a nutritional skill, in addition to serving as a relevant unexplored genetic resource. For this reason, the importance of plants is perceived in the composition of the menus of Brazilian families. These plants grow spontaneously and can be of different forms in human food, being a collaborator of food security in Brazil and South America, since many families are living under great fragility in this regard (Nascimento et al., 2018).

15.3.3 Social Technologies and Approaches

Conceptually, social technologies constitute a recently created approach (they are not a phenomenon isolated, and they have their historical heritage inherited from intermediate technology and appropriate technology, however). They are a technological modality idealized and implemented with the collaboration of the local population to promote social inclusion and the improvement of the quality of life of families, who normally have limited access to public services (Vázquez and Niederle, 2018). Originally, this kind of technology is not a ready, finished model. It is a moving and dynamic process, where people who need solutions are part of them, taking on the process of change and transformation. Communities take ownership of the technologies developed and assume the role of the protagonist of the process (Schwab and Freitas, 2016; Rover et al., 2017).

As an example of Social Technology, we have a system of teaching called "Alternation Pedagogy" which is a method that seeks interaction between the student who lives in the countryside and the reality he experiences in his/her daily life, to promote a constant exchange of knowledge between his/her life and work environment and the school environment. In the various tasks, jobs, and situations experienced by the student at school and in the family, it becomes very difficult to assess skill and coexistence separately, since we take the term "ability" more in the sense of expressing social, work, and study skills (Brazilian Federal Government, 2015; Valadão et al., 2017).

We consider skill/coexistence to be attitudes, relationships, ability to do, communicate, availability, openness, personal organization, leadership, responsibility and commitment, dialogue, respect, set of values, and persistence. The evaluation of the skills usually occurs every scholar semester, where through a script subsidized by a theme, the student makes a self-assessment, the family evaluates it and a classmate also makes the assessment of the student, and at school, the placement is made in common with the class doing the necessary reflections. An important part of the method is the experiential contents, which are all expressions of the knowledge extracted from the students' experiential reality, through a study plan. They are elements of reality that the adolescent/juvenile (or even the family) accurately perceived, added to the knowledge he already has of reality. In the educational process, it reaches the point of reflection making the relationship between the scientific and reality for its transformation (Brazilian Federal Government – Ministry of Education, 2015).

Another example is related to the concept or principle of "Solidarity Economy" (SE), which is a type of economy focused on human beings, instead of focusing on the incessant search for capital accumulation.

Despite this, it is consistent with the capitalist system, as the SE incorporates solidarity at the center of economic activity. An example of SE activity would be the opportunity of the people of a place commercializes their production through exchanges (barter), instead of national currency. Hence, the creation of a currency of social value and local use could be a promising path. This would comprise an unofficial currency used only by supporters of the SE system (Figueiredo et al., 2020).

15.3.4 Social Innovations – Establishing Cooperative Systems of Familiar Agriculture

The Brazilian family farming segment has achieved several significant advances, also accompanied by many limitations, however, in the bestowal of credit and technical support essential for the promotion of sustainable models of farming. This is because several of the actions adopted by the programs are associated with the conventional systems and organizations of production implemented during the period of modernization of Brazilian agriculture, which has socio-environmental consequences that sometimes contradict the goals of sustainable rural development (da Silva et al., 2018).

In addition to having access to new technologies – strongly associated with the development of social technologies to sustain the economic viability of establishments family members and to increase their ability to reproduce as a family social unit – small farmers should be focused on the way they make decisions and strategies productive process and its form of insertion in the markets. This might be facilitated by organizations as cooperatives (Bittencourt, 2020). Cooperatives can be defined as a nonprofit company that provides services to its members (Dal Moro and Brandli, 2020).

Associate rural landowner in cooperatives is a means for a group of individuals to achieve goals through a voluntary agreement for reciprocal collaboration. To this end, the cooperative operates in the market by developing credit, service provision, consumption, production, and marketing activities for its members. The exchange of experiences between members and the creation of an internal business chain are additional advantages of rural cooperatives. This means, for example, that a producer of sugarcane and corn can formalize a partnership with poultry farmers, swine farmers, and livestock farmers to act as a supplier directly (Jacto (enterprise), 2020).

Cooperatives are also a way of facilitating access to the up-to-date technology available and developing open-source models of digital information (acquisition, processing, storage, and exchange of data), which might help to guarantee that technology and data are maintained directly by the farmer, which can arm them with greater financial power and autonomy.

Although it is characterized as fundamental for the supply of food throughout the country, Brazilian family farming still faces many challenges for its survival. To support the sector and help to overcome any adversity, cooperative in agriculture proves to be an important possibility. Cooperative is a crucial attribute and it is credited to production and distribution, considering that family agriculture is a basis for the activity. What is desired when organizing cooperatives in agriculture is to solve problems, improve the economic situation, and satisfy the needs and common goals of a certain group of individuals (Dal Moro and Brandli, 2020).

15.4 Concluding Remarks and Future Perspectives

To continuously promote rural development for an ecologically sustainable, economically viable, and socially just family farming, it is necessary to provide innovative and multidimensional technologic alternatives for family farmers. The objective is to improve income generation, improve quality of life, make progress in the context of environmental sustainability, and improve the family succession process (Bittencourt, 2020). Altier (1989) followed by Borges et al. (2020) stated that the sustainability of family farmers should illustrate indicators that establish at least four criteria, regardless of the method used to assess this sustainability: (i) maintenance of the productive capacity of the agroecosystem; (ii) conservation of natural resources and biodiversity; (iii) strengthening the social organization and, as a consequence, reducing poverty; (iv) consolidation of local communities, preserving their traditions, their knowledge, and ensuring participation in the development process. Among the characteristics that motivate the farmers searching for the results for the indicators presented above, we have reduction costs of production, attendance of the requirements of the buyers, compliance with legislation, increasing the productivity (Natalli et al., 2020).

Food security and environmental conservation are a binomial to be adjusted and enhanced with the disciplined use of labor and technology and disciplined land use (da Silva e Souza et al., 2020). In the long term, sustainability has been the main objective in agricultural research (Teixeira and Ribeiro, 2020). This is being given priority to protect the environment. Integrating the notion and perception of sustainability into any technology and research design is essential to deal with the global challenges and tasks of environment safety and societal benefits (Mishra et al., 2018). In the context of the conservation of natural resources and biodiversity, biodiversity is an essential raw material for the future of the bioindustry, and Brazil has the largest biological diversity on the planet, with many assets of great interest to trade and the economy (Bittencourt, 2020).

Not less important, we also must address attention for problems as sanitation – including access to drinking water, and solid waste management – including the correct discard of pesticide packings (Matos, 2019; Medeiros et al., 2020). Considering that family farming in Brazil has been responsible for (i) approximately one-third of the cattle production, (ii) more than 80% of the annual production of cassava, (iii) more than half of the annual production of beans, and (iv) two-thirds of birds raised in the national territory, so we are aware of the socio-economic-environmental importance that this segment of society deserves to have in the country, especially from its leaders, in terms of generating income and the possibility of promoting environmental impacts (or conservation of this).

REFERENCES

Altier, M. A. 1989. *Agroecologia: As bases cientificas da agricultura alternativa.* Rio de Janeiro: AS-PTA.

Bittencourt, D. D. C. 2020. Strategies for family farming: vision of the future towards innovation (in Portuguese). Empresa Brasileira de Pesquisa Agropecuária, 15 p.

Bolfe, E., Barbedo, J. G. A., Massruhá, S. M. F. S., de Souza, K. X. S., and Assad, E. D. 2020. Challenges, trends and opportunities in digital agriculture in Brazil (in Portuguese). In Massruhá, S. M. F. S., Leite, M. A. de A., Oliveira, S. R. de M., Meira, C. A. A., Luchiari Junior, A., Bolfe, E. L. (Eds.). *Agricultura digital: pesquisa, desenvolvimento e inovação nas cadeias produtivas.* Empresa Brasileira de Pesquisa Agropecuária, Brasília.

Borges, I. M. S., Almeida, R. L. J., Fernandes, A. C. G., Silva, S. É., Silva, M. L. A., Barros, U. I. G., …, Freire, J. G. T. B. 2020. Family farming: analysis of sustainability through social economic and environmental indicators (in Portuguese). *Research, Society, and Development* 9: e54942832–e54942832.

Brazilian Federal Government. 2012. Lei 12.651 (of May 25th 2012) – Dispõe sobre a proteção da vegetação nativa. http://www.planalto.gov.br/ccivil_03/_ato2011-2014/2012/lei/l12651.htm (accessed December 2nd, 2020).

Brazilian Federal Government (Ministry of Education). 2015. Instituto Capixaba has 30 vacancies in alternation pedagogy. http://portal.mec.gov.br/busca-geral/212-noticias/educacao-superior-1690610854/21064-instituto-capixaba-tem-30-vagas-em-pedagogia-da-alternancia (accessed December 14th, 2020).

Brazilian Federal Government. 2017. Decreto 9.064 (of May 31st 2017) – Dispõe sobre a Unidade Familiar de Produção Agrária. https://www2.camara.leg.br/legin/fed/decret/2017/decreto-9064-31-maio-2017-785001-publicacaooriginal-152929-pe.html (accessed December 3rd, 2020).

Brazilian Federal Govrnment. 2020. Decreto 10.253 (of February 20th 2020 de 20 de fevereiro de 2020) - Aprova a Estrutura Regimental e o Quadro Demonstrativo dos Cargos em Comissão e das Funções de Confiança do Ministério da Agricultura, Pecuária e Abastecimento e remaneja e transforma cargos em comissão e funções de confiança. https://www.in.gov.br/web/dou/-/decreto-n-10.253-de-20-de-fevereiro-de-2020-244585023 (accessed December 8th 2020).

Buainain, A. M., and Garcia, J. R. 2018. Roles and challenges of Brazilian small holding agriculture. *Agroalimentaria* 24: 71–87.

Camargo, F. A. O., Silva, L. S., Merten, G. H., Carlos, F. S., Baveye, P. C., and Triplett, E. W. 2017. Brazilian agriculture in perspective: great expectations vs reality. *Advances in Agronomy* 141: 53–114.

Carvalho, E. S., and Lago, S. M. S. 2019. The appropriation of innovations in family farming: a systematic review of the literature (in Portuguese). *RECoDAF – Revista Eletrônica Competências Digitais para Agricultura Familiar* 5: 81–119.

CGI (Internet Management Committee in Brazil). 2012. Research on the use of information and communication technologies in Brazil: ICT households and companies (in Portuguese and English). https://www.cgi.br/media/docs/publicacoes/2/tic-domicilios-e-empresas-2012.pdf (accessed (again) December 12th, 2020).

Costa, J. S, Bastos, G. M. F., Lima, B. C. C., and Silva Filho, J. C. L. 2014. Social innovation, pleasure and suffering at work: the case of the Mandalla Project in Ceará, Brazil (in Portuguese). *Administração Pública e Gestão Social* 6: 11–18.

Cruz, F. T. 2020. Family farming, food processing, and advances and setbacks in the regulation of traditional and artisanal foods (in Portuguese). *Revista de Economia e Sociologia Rural* 58: 2–21.

da Silva, F. C., Sant'Ana, A. L. A., and Maia, A. H. 2018. Public policy on the family farming sector in Brazil: towards a model of sustainable agriculture. *African Journal of Agricultural Research* 13: 1719–1729.

da Silva e Souza, G. D. S., Gomes, E. G., de Andrade Alves, E. R., and Gasques, J. G. 2020. Technological progress in Brazilian agriculture. *Socio-Economic Planning Sciences* 72: 100879.

Dal Moro, L., and Brandli, L. L. 2020. Potentialities and challenges of family agriculture in a region of South Brazil. *International Journal of Sustainable Development & World Ecology* 27: 129–139.

Deponti, C. M., Kirst, R. B. B., and Machado, A. 2017. The interrelationships between Information and Communication Technologies (ICT) and family farming (in Portuguese). *RECoDAF – Revista Eletrônica Competências Digitais para Agricultura Familiar* 3: 4–23.

FAO (Food and Agriculture Organization of the United Nations). 2020. Innovation at FAO. http://www.fao.org/innovation/en/ (accessed November 25th, 2020).

Figueiredo, E., Montebello, A. E. S., and Norder, L. A. C. 2020. Solidary economy organization and practices with family farmers: the case of the Chão Institute (in Portuguese). *Brazilian Journal of Development* 6: 10348–10370.

Grisa, C., Kato, K. Y. M., and Zimmermann, S. A. 2017. The rural in public policies of contemporary Brazil (chapter III, in Portuguese). In *Tipologia regionalizada dos espaços rurais brasileiros: implicações no marco jurídico e nas políticas públicas.* IBGE, Brasília, p. 337–481.

IBGE (Brazilian Institute for Geography and Statistics). 2019. Agricultural census of 2017 – definitive database. IBGE. Rio de Janeiro, RJ, Brazil. https://censos.ibge.gov.br/agro/2017 (accessed December 2nd, 2020).

ITS (Brazilian Institute for Social Technology). 2004. Almanacs for debate: social technology in Brazil (in Portuguese). https://www.itsbrasil.org.br/tecnologia-social (accessed (again) December 12th, 2020).

Jacto (enterprise). 2020. Understand the importance of cooperation in family farming (in Portuguese). https://blog.jacto.com.br/cooperativismo-na-agricultura/ (accessed December 8th, 2020).

Matos, L. R. 2019. Pesticide use and solid waste management on family farms in Rondônia (in Portuguese). *Revista Presença Geográfica* 6: 35–47.

Medeiros, M., Cazella, A. A., Tecchio, A., and Marques, F. C. 2020. The constitution of an organizational novelty in Southern Brazil: advances and limits on the participation of family farming (in Portuguese). *Revista de Economia e Sociologia Rural* 58: e209211.

Medina, G., Almeida, C., Novaes, E., Godar, J., and Pokorny, B. 2015. Development conditions for family farming: lessons from Brazil. *World Development* 74: 386–396.

Mishra, S., Fraceto, L. F., Yang, X., and Singh, H. B. 2018. Rewinding the history of agriculture and the emergence of nanotechnology in agriculture. In Singh, H. B., Mishra, S., Fraceto, L. F., de Lima, R. (Eds.). *Emerging Trends in Agri-Nanotechnology: Fundamental and Applied Aspects.* CAB International, Boca Ratón, FL, p. 1–9.

Nascimento, S. G. S., Moraes, C. J., Hanke, D., Avila, M. R., and Nunes, O. M. 2018. Non-conventional food plants and family farming: limits and potential of commercialization in the municipality of Dom Pedrito, RS (in Portuguese). *Revista Agropampa* 3: 134–147.

Natalli, L. H., Munaretto, L. F., Bianchini, D. C., and Henkes, J. A. 2020. Environmental sustainability practices in Rural properties (In Portuguese). *Revista Gestão & Sustentabilidade Ambiental* 9: 351–374.

Pereira, E. M., Fidelis, V. R. L., Oliveira, M. I. V., Luíz, M. J. S., and Medeiros, A. R. H. 2015. Perception of family farmers of Rome-PB community on fruit and vegetables minimumprocessing (in Portuguese). *Revista Verde de Agroecologia e Desenvolvimento Sustentável* 10: 9–14.

Rabaiolli, J. A., Cunha, A. S., Oliveira, I. L, and Miorin, V. M. F. 2015. Evolution of land ownership categories in Brazil (in Portuguese). *Revista OKARA: Geografia em Debate* 9: 410–427.

Rada, N., Helfand, S., and Magalhães, M. 2019. Agricultural productivity growth in Brazil: large and small farms excel. *Food Policy* 84: 176–185.

Rotz, S., Duncan, E., Small, M., Botschner, J., Dara, R., Mosby, I., Reed, M., and Fraser, E. D. 2019. The politics of digital agricultural technologies: a preliminary review. *Sociologia Ruralis* 59: 203–229.

Rover, O. J., De Gennaro, B. C., and Roselli, L. 2017. Social innovation and sustainable rural development: The case of a Brazilian agroecology network. *Sustainability* 9: 3.

Sabourin, E., Le Coq, J. F., Fréguin-Gresh, S., Marzin, J., Bonin, M., Patrouilleau, M. M., Tonneau, J. P., De Aquino, J. R., and Teixeira, O. A. 2005. Modernization of family farming and social exclusion: the dilemma of agricultural policies (in Portuguese). *Cadernos de Ciência & Tecnologia* 22: 67–82.

Sarkadi, L. S. 2019. Impact of agriculture on food supply: a history. In *Chemistry's Role in Food Production and Sustainability: Past and Present.* American Chemical Society, Washington DC, USA. pp. 29–46.

Schwab, D., and Freitas, C. C. G. 2016. Social technology: implications and challenges of implementation (in Portuguese). *Revista Tecnologia e Sociedade* 12: 42–60.

Tauger, M. B. 2020. *Agriculture in World History.* Routledge Abingdon, Oxon.

Teixeira, M. L. C., and Ribeiro, S. M. C. 2020. Agriculture and sustainable landscapes: agricultural diversification in the state of Minas Gerais, Brazil. *Sustainability in Debate* 11: 17–28.

Valadão, J. A. D., Cordeiro Neto, J. R., and Andrade, J. A. 2017. Sociotechnical bases of social technology: translation of pedagogy of alternation in Rondônia (in Portuguese). *Organizações & Sociedade* 24: 89–114.

Vázquez, L. L., and Niederle, P. 2018. Public policies to support agroecology in Latin America and the Caribbean. *Perspective* 45: 1–4.

Vink, A. P. A. 1975. *Land use in Advancing Agriculture.* Springer-Verlag Berlin Heidelberg New York, Gießen, Germany, 394 P.

16

Nanofertilizers for Sustainable Crop Production: A Perspective in Small Farm Agriculture

Rubina Khanam[1], Pedda Ghouse Peera Sheikh Kulsum[2], Ruma Das[3*], and Sunanda Biswas[3]
[1]*ICAR-National Rice Research Institute, Cuttack, Odisha, India*
[2]*C V Raman Global University, Bhubaneswar, Odisha, India*
[3]*ICAR-Indian Agricultural Research Institute, Pusa Campus, New Delhi, India*

CONTENTS

16.1 Introduction

Growing at slower pace, the current world population will reach the mark of 9 billion by 2050, anticipating 60% more food (United Nations, Department of Economic and Social Affairs, Population Division 2019). The scale of the problem is being realized to be even greater with limited supply of resources and decline in arable land for sustainable crop production. Therefore, to address this problem, continuous innovation in the field of agriculture is very important and one such innovation is 'nanomaterials (NMs),' especially nanofertilizers (NFs), an efficient delivery system for plant nutrients (Chugh et al. 2021).

DOI: 10.1201/9781003164968-18

The nutrient use efficiency (NUE) of conventional chemical fertilizers is low and often does not coincide with the plant demand resulting in serious environmental consequences including eutrophication, air and ground water pollution, soil degradation, etc. The NFs, on the other hand, improve the use efficiency of plant nutrients either by improved delivery system or by reducing the losses of mobile nutrients. It helps to improve soil aggregation, carbon build up, and serve as growth promoters, such as titanium oxide (TiO_2) nanoparticles and carbon nanotubes (CNTs), in addition to nutrient delivery (Kah et al. 2019).

The NFs basically categorize into two categories: NMs which are made up of essential plant nutrient and referred to as 'NFs' and NMs, which act as carrier of nutrients and are referred to as 'nutrient-loaded NFs' or 'NMs enhanced fertilizers.' The high surface-to-volume ratio of NMs, high sorption and reaction capacity, and controlled or slow-release property make the NFs unique and more efficient than conventional one (Feregrino-Perez et al. 2018). The NFs are encapsulated or coated with NMs having control and slow-release property so that it satisfies the nutrient requirement of plants and coincide with their demand. However, type and concentration of NFs affect the uptake by different plants and also their germination, growth, and maturity. Thus, NFs play critical roles in maintaining soil quality and also crop productivity, which are the key ingredients for sustainable agriculture.

Small farms are vital for overall growth and development of agriculture in developing countries. In this regard, adoption of NFs in small farm system may be envisaged as safe, low cost, promising, and sustainable options. The purpose of this chapter is to discuss applications of different NMs in agriculture along with basic forms of NFs and their distribution. This chapter also presents the national and international importance and different ways for adoption of NFs in small farm holdings along with different regulation and legislation involved. The limitations of NFs are also critically discussed and conclusions are drawn by briefly stating its importance for sustainable development of the society.

16.2 Critical Applications of Nanomaterials in Agriculture

The progress of nanoscience (nano-devices and NMs) provided a way to novel applications, and it also permits broad advancement in agriculture research such as NFs, nano-pesticides, and other input management. The use of nanotechnology in agriculture can enhance crop growth and yield, accelerate seed germination, improve plant establishment, and enhance production. In rainfed areas, herbicide usage is far from satisfactory due to uncertainty in moisture availability. Encapsulation improved the efficacy of nano-emulsion and nano-capsule pesticides, resulting in less pollution of the environment for sustainable agriculture. The uptake of pesticide to target organisms is enhanced by nano-carriers through nano-encapsulation (Chugh et al. 2021).

The antifungal role of several types of nanoparticles (silver, copper, and silica) and their combinations have been demonstrated against different plant pathogenic fungi (Kah et al. 2019). Nano-fungicide is less harmful to environment as it is target-specific, thus help in reducing the damage to non-target plants and decrease the amount released into the surrounding environment. Cobalt and nickel ferrite nanoparticles ($CoFe_2O_4$ and $NiFe_2O_4$) were developed and found to show antimycotic activity against three plant-pathogenic fungi viz., *Fusarium oxysporum*, *Colletotrichum gloeosporioides*, and *Dematophora necatrix* and hence can be used for plant disease management. Therefore, the crystals of nano-herbicides, fungicides, and pesticides have a tremendous role in crop protection.

The NMs have been widely used to remove heavy metals from water/wastewater due to their large surface area and high reactivity (Mukhopadhyay et al. 2019). Metal oxide nanoparticles, including nano-sized ferric oxides, aluminum oxides, titanium oxides, magnesium oxides, carbon-based, and silica-based provide high surface area and specific affinity for heavy metal adsorption from aqueous systems. Besides, nano-devices, gadgets, and sensors are deployed to sense the stress of temperature, moisture, nutrient in plants and detect the hidden hunger of soil and plants (Figure 16.1). Similarly, nano biosensors had higher accuracy and sensitivity on residue detection. Under smart delivery systems (SDS), the nanoparticles, crystals, wires, tubes, and chips are effectively used for real time monitoring and are highly suitable for precision farming.

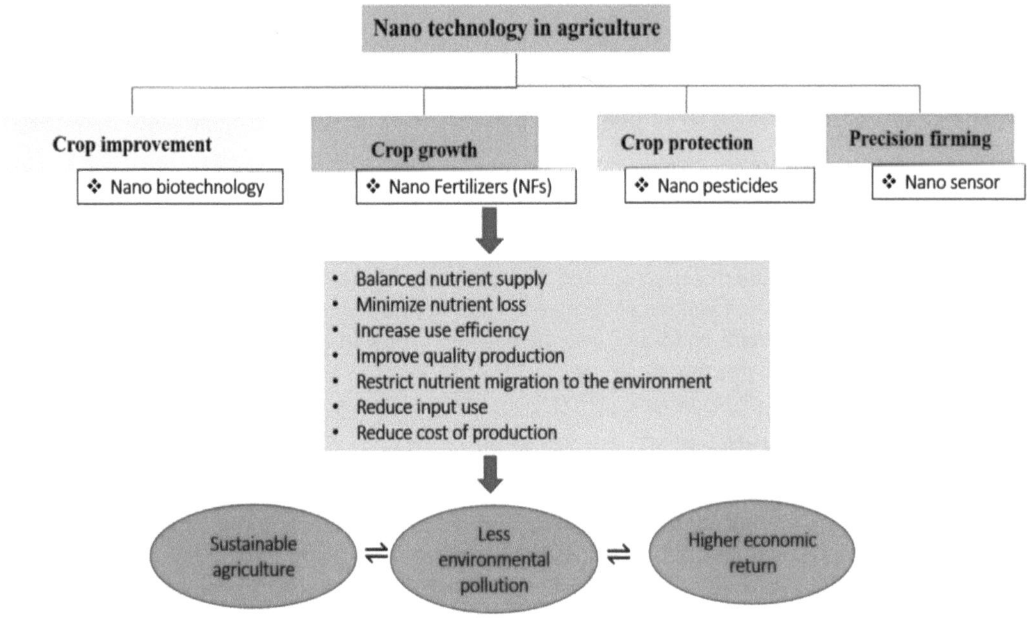

FIGURE 16.1 Application of nanofertilizers (NFs) for sustainable crop production.

16.3 The Basic Forms of Nanofertilizers and Their Distribution

Modern intensive farming systems utilize organic and mineral manures in order to supply essential plant nutrients, which eventually leads to serious deterioration of ecosystem and environment. A huge amount of fertilizer is being used to improve soil fertility and crop productivity. The International Fertilizer Industry Association (IFIA) reported that fertilizer consumption is increasing sharply and total consumption was reached about 192.8 Mt in the year 2016–2017 (Shang et al. 2019). It is reported that mineral fertilizers attributed one-third of the nutrient supply for crop productivity and the rest depends on other agricultural inputs. Nevertheless, the NUE of conventional sources of nutrients hardly surpass 30–40% (Shang et al. 2019). Consequently, gaseous loss of N as nitrous oxide and nitrates leaching has caused eutrophication and increased the impacts of global warming and environmental pollution. An appropriate nutrient management strategy apart from enhancing NUE, productivity, and profitability should also aim to enhance eco-efficiency and environmental sustainability. During past few years, tremendous progress has been made in application of nanostructure as the base materials for nutrients in order to satisfy the specific needs of the crops. The following base materials are most effectively exploited in commercial production of NFs. They include, zeolites, nanocomposites, super absorbent fertilizers (SAFs), and CNTs.

16.3.1 Zeolites-Based Carriers

These are versatile microporous, aluminosilicate minerals commonly used as commercial absorbents and catalysts. Almost 14% of the world consumption of these zeolites is as adsorbents. There are different types of mineral zeolites (clinoptilolite, stilbite, chabazite, analcime, natrolite), depending upon the ion exchange capacity and ability to retain or lose water. Its world production is highly decentralized. However, China accounts to 69% of its production followed by South Korea, Turkey, and Japan. Zeolites possess excellent properties such as ion exchange capacity, dehydration and rehydration abilities, and adsorption. They are molecular sieves that can selectively absorb ions that fit into the niches in their structures. The zeolites are very useful in agriculture due to their beneficial role on crop growth and yield, NUE (Mihok et al. 2020), as soil conditioner, and as fertilizer carrier in the form of nanoporous zeolite (Manikandan and Subramanian 2016).

16.3.2 Nanocomposites

These are hybrid solid materials with high surface area-to-volume and aspect ratio, incorporated in a continuous phase matrix (montmorillonite clay and some organic polymers). The NH_4^+-clinoptilolite with phosphorite acts as a carrier of nitrogen and phosphorus with slow-release property in soil (Kalia et al. 2020).

16.3.3 Super Absorbent Fertilizer (SAF)

The SAFs in agriculture are majorly prepared from acrylic acid and crosslinking agents (potassium by polymerization). High water retention capacity and slow-release properties are attributed to their potential use in improving nutrient and water use efficiency of crops. In arid regions, SAF have prominent application in agricultural practices.

16.3.4 Carbon Nanotubes (CNTs)

The CNTs or single-wall carbon nanotubes (SWCNTs) are one of the allotropes of carbon, weakly bound together by van der Waals interactions in a tree ring like structure. Water absorption and retention capacity of CNTs is exceptional that facilitate their extensive use in agricultural industry. The CNTs-based slow-release fertilizers demonstrated improved germination rate and seedling biomass. However, majority of studies reported positive effect of multi-walled carbon nanotube (MWCNT) under hydroponics or agar media, where soil culture has been used rarely. In several studies of CNTs phytotoxicity, critical dosage significantly exceeded the environmental safety limit. Besides, lower levels of CNTs (<100 mg L^{-1}) found beneficial or no adverse effect on plants. South Korea followed by United States, Belgium, Japan, and China are the major stakeholders in the production of CNTs. India is in fourth position under Asia-Pacific region.

However, fundamental questions for their adaptation in small farm holdings with regard to high performance and risk associated with toxicity need to be addressed to encourage active development and application of NFs. Regulation and legislation are also paramount to regulating the manufacturing, processing, application, as well as disposal of NFs.

16.4 Type and Beneficial Role of Nanofertilizers: National and International Perspective

Nano fertilizer has the potentiality to increase NUE owing to their larger surface area which ensures higher nutrient-surface interaction as well as higher nutrients uptake. Along with enhancing crops quality and sustainable yield, NFs may put a halt to environmental pollution resulted due to use of the conventional fertilizers. Researchers showed that, in order to enhance NUE and to overcome the chronic problem of eutrophication and macro- and micronutrient deficiency, NFs can be a best alternative for small- and large-scale agriculture (Shukla et al. 2019). Based on their actions, NFs are differentiated as control or slow-release fertilizers, nanocomposite fertilizers and controlled loss fertilizers (CLFs), etc. The NF as slow-release fertilizers restrict the huge losses of nutrients, which ultimately improves the efficiency of fertilizers and decrease the overall cost of production. It has been reported that the coating with porous NMs, viz., zeolites, clay, or chitosan drastically reduces the N losses through regulating the demand-based release (Shang et al. 2019). In addition, fertilizers coated with NMs significantly restrict the losses through leaching and denitrification. Moreover, controlled-release fertilizers coated with NMs have the possibilities of excellent source for supplying plant nutrients in near future. Ammonium-charged zeolites potentiality increase the solubility of phosphate fertilizers and, consequently, enhance phosphorus (P) availability by crops.

Furthermore, customized nano-formulation acts as a source for wide range of macro- and micronutrients in desirable proportions. Application of nano-calcite ($CaCO_3$ – 40%) along with nano-SiO_2 (4%), MgO (1%), and Fe_2O_3 (1%) can enhance the bioavailability of calcium (Ca), magnesium (Mg), and iron (Fe).

TABLE 16.1

Effect of Nano Fertilizers (NFs) and Nanoparticles (NPs) on Different Crops

NFs/NPs	Effects	References
Nano-coated slow-release N fertilizer	NO_3 – N leaching was decreased	Lawrencia et al. (2021)
Commercial NPK fertilizers with nanomaterials	Increases in grain yields of rice (10.29%), spring maize (10.93%), soybean (16.74%), winter wheat (28.81%), and vegetables (12.34–19.76%)	Subbaiah et al. (2016)
NPK nanofertilizers	Foliar application with NPK nanofertilizers improves the growth parameter, harvest index, and the economic yield in potato compared to commercial fertilizers	Abd El-Azeim et al. (2020)
N-NFs	Significant improved in plant growth, which subsequently resulted in higher essential oil yields in star flower	Mahmoodi et al. (2017)
Nano-sized hydroxyapatite (nHA)	Significant enhancement in the growth and antioxidant contents of *Adansonia digitata* plants treated with the foliar application of nHA	Patra et al. (2013)
Zn NFs	Increased Zn, germination, growth, yield in *Zea mays* (maize)	Subbaiah et al. (2016)
Zn NFs	Increased CO_2 assimilation and photosynthesis rate; reduced Na content in leaves in *Helianthus annuus* L. under salinity stress	Torabian et al. (2016)
Fe NFs	Increased Fe, growth, biomass, and yield in *Arachis hypogaea* (peanut)	Rui et al. (2016)
Nano-silica (SiO_2)	Improved growth parameters in *Z. mays* L.	Patra et al. (2013)
Nano-silica (SiO_2)	Increased yield in *Carthamus tinctorius*	Janmohammadi et al. (2016)
MnO NPs	Increased root length in *Lactuca sativa* L.	Liu et al. (2016)
Nano-silica (SiO_2)	Increased fresh and dry weights and chlorophyll and proline contents in *Ocimum basilicum* under salinity stress condition	Farhangi-Abriz and Torabian (2018)
Cu NPs	Improved the germination and root growth of soybean and chickpea (*Cicer arietinum* L.)	Rui et al. (2016)

It also remarkably boosts up the uptake of P with micronutrients zinc (Zn) and manganese (Mn). It was clearly observed that use of nano-formulated fertilizers significantly increased crop productivity. Subbaiah et al. (2016) reported higher crop yield in rice, wheat, maize, soybean, and vegetables in the tune of 10.29%, 28.81%, 10.93%, 16.74%, and 19.76%, respectively, with the application of NFs. Till date different NMs have been used as various sources of essential nutrients in agriculture for small-scale and large-scale farming. Some of the important sources are summarized in Table 16.1.

16.4.1 Nitrogen NFs

Nitrogen (N) fertilizers have an axial role in enhancing the food production. The consumption of urea as source of N has increased 29% after the green evolution era in India. Moreover, increased use of urea is responsible for 80% of the upsurge in atmospheric nitrous oxide (N_2O) level, which results in higher atmospheric temperature and pays a vital role in global warming. One of the main demerits of conventional sources of N is the higher rate of losses through volatilization and leaching, which starts immediately after its application in the field. To minimize those losses, N-based NFs is proven to be effective for slow release of N and effectively coincide with the demand of crops. Urea nanofertilizer (N-NF) with zeolites showed higher nutrient uptake, vigorous plant growth, and enhanced grain yield in maize compared to conventional N fertilizers (Manikandan and Subramanian 2016). Furthermore, use of nano-formulated N fertilizers to starflower (*Borago officinalis* L.) significantly improved the plant growth, which subsequently resulted higher yields of essential oil (Mahmoodi et al. 2017). Application of N-NF can reduce the cropping period up to 40 days in wheat with higher grain yield compared to the conventional fertilizers. Similarly, urea loaded zeolites were reported to improve the seed quality and yield of soybean (*Glycine max* L.) over conventional N fertilizers (Liu et al. 2016). Thus, nano-formulation of N sources is beneficial as it improves nutrient uptake and leads to better crop production, simultaneously, it reduces the volatilization and leaching losses and minimizes environmental pollution.

16.4.2 Phosphorus NFs

Diammonium phosphate (DAP) and single superphosphate (SSP) are the chemical sources of mineral P and used in agriculture to meet the crop demand. But most proportion of these fertilizers are lost as run-off or volatilized. It is estimated that 80–90% of the applied P fertilizers is lost to the environment and cannot be up taken by the crops which leads to environmental pollution and financial loss as well. Moreover, the use efficiency of P fertilizers is very less due to its immediate release and high fixation in soils. The recent approaches confirmed that nano-formulations can slowly release P for up to 40–50 days upon application during the pivotal crop need, whereas conventional P fertilizers deliver all the nutrients within 8–10 days of application (Liu et al. 2016). Thus, it has been recommended that the use of NFs or use of carrying materials like zeolites may have the potential to increase the use efficiency of P fertilizers. Use of nano-coated P fertilizers showed more stability decreases in the dissolution rate of the fertilizer and allow slow, sustained release of coated fertilizer to absorb more efficiently by the crops. Therefore, the use of nano-coated P fertilizers has become a recent approach to reduce fertilizer consumption and minimize environmental pollution. Several natural or synthetic polymers and Kaolin and polymeric biocompatible nanoparticles have been developed for this control release of P fertilizers. The given biodegradable polymeric chitosan nanoparticles are beneficial for slow release of phosphatic fertilizer.

16.4.3 Potassium NFs

Potassium fertilizers are applied in large quantities to enhance crop productivity but crops typically use 40–60% of the applied fertilizer. Kubavat et al. (2020) reported that release nano-formula of fertilizer reduces its losses in the soil while enhancing the long-term sustained supply of K to the crops. Therefore, application of nano-formulated K fertilizers can improve soil health and water quality by restricting K losses into soil and water; subsequently, it enhances quality and yield traits of crops.

16.4.4 Micronutrient and Beneficial Elements NFs

Farmers have been using high analysis commercial fertilizers extensively for many years, which contain the three main essential nutrients and absence of essential micronutrients needed for optimum plant growth of the plants. Although these micronutrients are required in a smaller amount, also vital for optimum plant productivity and quality, these essential elements also enhance the tolerance level of the crops against multiple stresses (Chugh et al. 2021). It was reported that more than 3 billion people across the world are suffering from micronutrient deficiencies (United Nations, Department of Economic and Social Affairs, Population Division 2019). Even when the plants are sufficiently supplied with macronutrients and water, micronutrient deficiency may inhibit plants from reaching their full potential. Hence, deficiency of micronutrients in crops reduces yield capacity and decreases the nutritional quality of crops. Among the different micronutrients, Zn, Fe, Mn, copper (Cu), boron (B), and molybdenum (Mo) are essential for plant growth, and each has its application in the agricultural sector. The main problem associated with the sources of these micronutrients is insolubility, low use efficiency, and less availability. This nano-sized micronutrient is more uniformly distributed in the soil, which decreases their interaction with soil particles and reduces their fixation and finally enhances the use efficiency. Exogenous application of ZnO NPs in soybean and *Brassica* showed improved Zn uptake with better growth and higher yield (Rahmani et al. 2016). It has been reported in several studies that nano-formulated Fe fertilizers improve germination and enhance the growth of the plants over commercial Fe fertilizers.

TABLE 16.2

Commercially Available NFs for Agriculture Use in Various Countries

Nanofertilizer	Commercial Names	Current Status	Nanomaterial Compositions	Manufacturer	Usage	Reference
Macronutrient based	NanoPro, NanoRise	Commercialized Commercialized	Not disclosed Not disclosed	Aqua-Yield® Operations, LLC	Fertilizer	Information from concern firm's website
	NanoPhos, NanoGro					
	NanoK, NanoPack	Commercialized	Unknown Unknown			
	Ready to use spray (concentrate)	Commercialized	Bio-humus in size range 100–700 nm	Green-Earth-Nano Plant, FL, USA	Fertilizer	
	NovaLand-Nano	Commercialized	Microelements as Mn, Cu, Fe, Zn, Mo, N NPs	Land Green & Tech Co., Ltd., Taiwan	Fertilizer	
Micronutrient based	NanoStress, NanoZn.	R&D	ZnO	Bio Nano Tech Ltd., Giza, Egypt	Fertilizer	Information from concern firm's website
	pH5saulodrip	R&D	Zn NPs		Fertilizer	
	Nano-Ag Answer			Urth Agriculture	Fertilizer	
	Nano Zn			IFFCO, India		
	Nano Zn			RFCL (RCF), Mumbai, India		

Rui et al. (2016) documented better root growth in peanut (*Arachis hypogaea* L.) plants treated with Fe NPs compared to nontreated plants under field conditions. Application of nano Mn fertilizers or Mn NPs on different crops has proved that nano-sized Mn can improve the root and shoot growth of mung bean by 52% and 38%, respectively, over commercially available manganese salt ($MnSO_4$).

In most of the countries, very few novel NMs containing products have been directly put into small farms. This attributed to the limited regulation due to complexity and case-by case legislating procedures involved in commercialization process. The key reasons commonly highlighted by most of the researchers for limited use of NFs in small farms are (1) long-term exposure to NFs and low toxic impact in agriculture produces, (2) a very little information about the bioavailability and biodistribution of NFs. The NFs available at present for commercial use and at the edge of research trials are furnished in Table 16.2.

16.5 Way out for Adoption of Nanofertilizer in Small Farm Holdings

The nature of farming in developing nations is majorly very diverse, small farm holding, and case-to-case basis. Low-cost and safe synthesizing technologies are imperative for introduction of NFs in small farm holdings in a viable way. The potential risks of traditional NFs, viz., plants and microorganisms are inconsistent and are under active research. In the subsequent section, possible solutions including green chemistry methods of NFs are discussed to consider it in small farm agriculture.

16.5.1 Biosynthesized NFs

Biosynthesis is one of the sustainable and eco-friendly green chemistry options for the design and development of NFs. Several biological systems, including microflora, yeasts, higher plant parts, and their extracts, have demonstrated capable suitability for the NFs biosynthesis, as listed in Table 16.3. The major advantages including: (1) biological system can act as reducing, stabilizing, and capping agents; (2) biosynthesis often occurs at neutral pH under ambient temperature and pressure, giving no scope

TABLE 16.3

Various Microorganisms and Their Biosynthesis of NPs

NPs	Name	Size (nm)
Zn	*Aspergillus* species	25
P	*Aspergillus niger*	15–20
	Alternaria alternata	27–79
	Bacillus methylotrophicus	10–30
Ag	*Bhargavaea indica*	111
	Brevibacterium frigoritolerans	97
	Weissella oryzae	20–50
	Ganoderma applanatum	20–25
P	*Penicillium fellutanum*	5–25
Ag, P	*Pseudomonas deceptionensis*	10–30
	Trichoderma harzianum	15–25

for usage of harmful reagents; and (3) less toxic and biocompatible. Low production and less controlled conditions are the two major setbacks in this method. These owe to the fact that the exact mechanisms and biological processes remain unclear.

16.5.2 Bioinspired Materials

It has been a difficult task to emulate the naturally occurring biological systems. Several bioinspired NFs are able to dynamically adjust their physicochemical properties in a controlled manner according to the environment. The new study of those materials in agriculture sectors is very limited. However, some bio-inspired devices have been found to be effective for pesticide delivery, disease detection, and environmental sensing, as listed in Table 16.4.

16.5.3 Regulation and Legislation

Regulation and legislation play a key role for the implementation of nanotechnology as well as commercialization of products. The legislation includes three major steps: (1) suggestions are made by academic bodies, viz., government organization, research institute, or even individuals, (2) sensible suggestions are made by a government organization and are proposed (by head of a state/country or a legislative body) as a

TABLE 16.4

Engineered Nanoparticles (ENPs) Used on Various Field Crops under Diverse Stress Conditions

Plant	Stress Type	Nanoparticles Used
Rice	High CO_2	$nTiO_2$
Wheat	Chilling	Biogenic Ag NPs
	Drought	Analcite NPs
	UV-B	Si NPs
	Heavy metals	ZnO NPs
Maize	Drought	Analcite NPs
	Salinity	ZnO NPs
Sorghum	High temperature	Se-NPs
Lentil	Drought	Silver NPs
Faba bean	Salinity	$nTiO_2$
Sesame	Drought	Iron oxide

directive, known as legislation. Legislation is then passed by a parliament of a country or other legislative arm of a government and becomes laws, (3) legislation (or law) usually covers broad and general aspect associated to an entity. Thus, regulations are generated, referring to the detailed and specific requirement.

16.5.4 Limitation and Urgent Need for Legislation

The NFs have a unique 'standard protocol' to follow. This includes both understanding and control of nanoscale properties/interactions/toxicities/functions, which further determine beneficial usage, consequences, transition of NFs, and their environmental fate and impact to ecosystem. Regulation and legislation have no basis to stand on because of prevailing uncertainty in scientific findings and evidences. To make it worse, each type of NFs has its unique properties, leading to a 'case-by-case' or 'product-by-product' evaluation regarding regulation and legislation (EU 2012). Given these constraints, current legislation is still at primitive stage of development. Currently, it covers an extremely broad and general aspect associated with NFs and NMs. It needs evaluation or reevaluation when it comes to a specific NM to put into the small farms.

16.6 Limitations of Nanofertilizers

Although NFs have enormous benefits over the conventional one, this should be carefully handled and due consideration must be given for its safe disposal before its promotion. The potential risk involved with the use of NFs must be fully evaluated before implementation of this technology. The application of NFs for long period may cause many unintentional and irreversible consequences (Kah et al. 2019) and showed harmful effect on food safety and security and on human health. Transportation, plant availability and toxicity of NFs as well as their impacts on environment limit their preference to adoption in sustainable agriculture sectors (Feregrino-Perez et al. 2018). Assessment of risk and hazards related to NMs and also life cycle of NFs in soil are very crucial for toxicological study. Therefore, in this section, some limitations of NFs have been discussed in brief.

16.6.1 Phytotoxicity

One of such harmful consequences is phytotoxicity of NMs when applied as NFs. The availability, translocation, and uptake of NM vary along with the physiology of plants and their wide range of uptake mechanisms. Different plants respond in different way to various NFs depending on the type, doses, and methods of application, and, therefore, phytotoxicity of NMs is also a great concern. The toxicity of NFs is usually associated with the micronutrients as their sufficiency and deficiency range is very narrow (Liu et al. 2016).

The application of different NMs such as oxides of titanium (Ti), Zn, cerium (Ce), and silver (Ag) results in their deposition on cell wall surfaces and causes oxidative stress to the cell (Buzea et al. 2007). Likewise, foliar spray of iron oxide in maize affects the photosynthesis and also causes oxidative stress in cucurbits. Application of MWCNT and Ag nanoparticle increases the reactive oxygen species production in crops. Application of Nano-Mg also increases the plasma membrane damage index (PMDI) which is very harmful and reduces membrane synthesis in crops. Foliar application of NF sometimes affects the crop quality of wheat by decreasing its protein content.

16.6.2 Human and Animal Exposure

NMs that are used as NFs are also very harmful for human and animal health. All the NMs are very reactive due to their minute size; therefore, the farm workers may also expose to xenobiotics during their application. During the respiratory exposure in human and animal, NMs easily make their way through the cell membrane into the biological system. Thus, the NMs enter into the blood circulation and flow to the liver, brain, and heart and cause serious health hazards. Prolonged exposure of NMs causes genotoxicity in human, which not only damages the DNA but also develops cancer and, further, adversely affects the future generations by affecting the reproductive cells.

Not only the humans but also many animals as well as the aquatic ecosystem are adversely affected by NMs and their contaminated water. The oxides of Ag, Zn, and Ti nanoparticle cause oxidative damage in rat spleen and thymus, and also zinc oxide (ZnO) nanoparticle causes sperm deformity in mouse (Guo et al. 2010). The titanium oxide (TiO_2) nanoparticle contaminated water adversely affects the activity of different enzymes and causes oxidative stress to the fish.

16.6.3 Microbial Community and Enzymatic Activity

Soil is the largest sink of NMs, where those are accumulated and cause a great risk in arable land by affecting their microscopic properties. The NFs affect the filtering and buffering capacity of soil, nutrient balance and, thus, can affect the soil fertility and plant growth. Introduction of NFs into the soil poses serious threat to many beneficial soil microbes. The impact of NMs on soil microbes can be determined through basal respiration rate and enzymatic activity ((Simonin and Richaume 2015). It was found that the arbuscular mycorrhizal fungi were greatly affected by TiO_2 nanoparticles even at low concentration. Both the TiO_2 and CuO nanoparticles induced perturbations on soil microbes in flooded paddy soil and disrupt the ecosystem of paddy by decreasing the soil microbial biomass C, total phospholipid fatty acids and activities phosphatase, urease and dehydrogenase enzyme activities. However, compared to TiO_2, the negative effect of CuO nanoparticles were more prominent on soil microbes (Xu et al. 2015). Similarly, ZnO, TiO_2, CeO_2, and Fe_3O_4 NPs also adversely affect the different enzymatic activities such as activity of invertase, urease, catalase, and phosphatase as well as bacterial communities in saline-alkali and black soils (You et al. 2018).

Toxicity of NMs also adversely affects the algae growth, chlorophyll content, and enzyme activity depending upon the type of NMs as well as physiological and biochemical characteristic of algae. Likewise, nano Ag with modified carbonate and nano TiO_2 also adversely affect the photosynthesis and growth dynamics (by causing oxidative stress) of *Chlamydomonas reinhardtii*, respectively. High concentrations of Fe_3O_4 and CuO nanoparticles decreased the bacterial community size in soil (Jiling et al. 2016). Also, ZnO and CeO_2 nanoparticles adversely affected the enzymatic activities and plate counts of *Azotobacter*, P-solubilizing, and K-solubilizing bacteria. Therefore, a proper and overall assessment of NFs efficacy mentioned in various studies must be considered before imposing this technology for sustainable agriculture.

16.7 Conclusions

The promising application of NFs in sustainable crop production cannot be overlooked. Their unique properties, i.e., high surface-area-to-volume and aspect ratio, ion exchange capacity, and adsorption triggered a great interest to improve production potential of different crops. Several researchers from different parts of the world also substantiated the multiple benefits of NFs, viz., improved crop yields and quality and also mitigating environmental issues of global warming and soil and water contamination that arises due leaching of nutrients from conventional fertilizers. More information needs to be made available related to case by case legislating procedures, bioavailability, and biodistribution of NFs. Furthermore, low-cost and safe synthesizing technologies including biosynthesized NMs, nano-biofertilizers are imperative for introduction of NFs in small farm holdings in a viable way.

REFERENCES

Abd El-Azeim, M. M., Sherif, M. A., Hussien, M. S., Tantawy, I. A. A. and Bashandy, S.O. 2020. Impacts of nano-and non-nanofertilizers on potato quality and productivity. *Acta Ecologica Sinica 40*(5): 388–397.

Buzea, C., Pacheco, I. I. and Robbie, K. 2007. Nanomaterials and nanoparticles: sources and toxicity. *Biointerphases 2*: 17–71.

Chugh, G., Siddique, K. H. and Solaiman, Z. M. 2021. Nanobiotechnology for agriculture: smart technology for combating nutrient deficiencies with nanotoxicity challenges. *Sustainability 13*(4): 1781.

EU. 2012. *Nanomaterials. Case by Case Safety Approach for Breakthrough Technology*. IP-12-1050EN. European Commission, Brussels.

Farhangi-Abriz, S. and Torabian, S. 2018. Nano-silicon alters antioxidant activities of soybean seedlings under salt toxicity. *Protoplasma 255*(3): 953–962.

Peregrino Perez, A. A., Magaña-López, E., Guzmán, C. and Esquivel, K. 2018. A general overview of the benefits and possible negative effects of the nanotechnology in horticulture. *Scientia Horticulturae 238*: 126–137.

Guo L. L., Liu X. H., GAO, L., Qin, D. X., Zhang, H. M., Liu, J. Y. and Cui, Y. G. 2010. Study on the effect of nanosized titanium dioxide for reproductive system of male mice. *Journal of Postgraduate Medicine 4*: 357–364.

Janmohammadi, M., Amanzadeh, T., Sabaghnia, N. and Ion, V. 2016. Effect of nano-silicon foliar application on safflower growth under organic and inorganic fertilizer regimes. *Botanica Lithuanica 22*(1): 53–64.

Jiling C., Youzhi, F., Xiangui, L. and Junhua, W. 2016. Arbuscular mycorrhizal fungi alleviate the negative effects of iron oxide nanoparticles on bacterial community in rhizospheric soils. *Frontiers in Environmental Science 4*: 10.

Kah, M., Tufenkji, N. and White, J. C. 2019. Nano-enabled strategies to enhance crop nutrition and protection. *Nature nanotechnology 14*(6): 532–540.

Kalia, A., Sharma, S. P., Kaur, H. and Kaur, H. 2020. Novel nanocomposite-based controlled-release fertilizer and pesticide formulations: prospects and challenges. In *Multifunctional Hybrid Nanomaterials for Sustainable Agri-Food and Ecosystems, Micro and Nano Technologies*, ed. K.A. Abd-Elsalam (pp. 99–134). Elsevier. Amsterdam, The Netherlands.

Kubavat, D., Trivedi, K., Vaghela, P., Prasad, K., Vijay, Anand, G. K., Trivedi, H., Patidar, R., Chaudhari, J., Andhariya, B. and Ghosh, A. 2020. Characterization of a chitosan-based sustained release nanofertilizer formulation used as a soil conditioner while simultaneously improving biomass production of *Zea mays* L. *Land Degradation & Development 31*: 2734–2746.

Lawrencia, D., Wong, S. K., Low, D. Y. S., Goh, B. H., Goh, J. K., Ruktanonchai, U. R., Soottitantawat, A., Lee, L. H. and Tang, S. Y. 2021. Controlled release fertilizers: a review on coating materials and mechanism of release. *Plants 10*(2): 238.

Liu, R., Zhang, H. and Lal, R. 2016. Effects of stabilized nanoparticles of copper, zinc, manganese, and iron oxides in low concentrations on lettuce (*Lactuca sativa*) seed germination: nanotoxicants or nanonutrients?. *Water, Air, and Soil Pollution 227*(1): 42.

Mahmoodi, P., Yarnia, M., Amirnia, R., Tarinejad, A. and Mahmoodi, H. 2017. Comparison of the effect of nano urea and nano iron fertilizers with common chemical fertilizers on some growth traits and essential oil production of *Borago officinalis* L. *Journal of Dairy and Veterinary Sciences 2*: 1–4.

Manikandan, A. and Subramanian, K. 2016. Evaluation of zeolite based nitrogen nano-fertilizers on maize growth, yield and quality on Inceptisols and Alfisols. *International Journal of Plant and Soil Science 9*: 1–9.

Mihok, F., Macko A., Andrej O, Renajta, O., Karol, K., Katarina, S. K., Ondrej, P. and Zuzana, K. 2020. Controlled nitrogen release fertilizer based on zeolite clinoptilolite: study of preparation process and release properties using molecular dynamics. *Current Research in Green and Sustainable Chemistry 3*: 100030, ISSN 2666-0865, https://doi.org/10.1016/j.crgsc.2020.100030.

Mukhopadhyay, R., Bhaduri, D., Sarkar, B., Rusmin, R., Hou, D., Khanam, R., Sarkar, S., Biswas, J. K., Vithanage, M., Bhatnagar, A. and Ok, Y. S. 2019. Clay–polymer nanocomposites: progress and challenges for use in sustainable water treatment. *Journal of Hazardous Materials 383*: 121125.

Patra, P., Choudhury, S. R., Mandal, S., Basu, A., Goswami, A., Gogoi, R., Srivastava, C., Kumar, R. and Gopal, M. 2013. Effect sulfur and ZnO nanoparticles on stress physiology and plant (*Vigna radiata*) nutrition. In *Advanced Nanomaterials and Nanotechnology* (pp. 301–309). Springer, Berlin, Heidelberg.

Rahmani, F., Peymani, A., Daneshv E. and Biparva, P. 2016. Impact of zinc oxide and copper oxide nano-particles on physiological and molecular processes in *Brassica napus*. *Indian Journal of Plant Physiology 21*(2): 122–128. doi: 10.1007/s40502-016-0212-9.

Rui, M., Ma, C., Hao, Y., Guo, J., Rui, Y., Tang, X., Zhao, Q., Fan, X., Zhang, Z., Hou, T. and Zhu, S. 2016. Iron oxide nanoparticles as a potential iron fertilizer for peanut (*Arachis hypogaea*). *Frontiers in Plant Science 7*: 815.

Shang, Y., Hasan, M., Ahammed, G. J., Li, M., Yin, H. and Zhou, J. 2019. Applications of nanotechnology in plant growth and crop protection: a review. *Molecules 24*(14): 2558. doi: 10.3390/molecules24142558.

Shukla, P., Chaurasia, P., Younis, K., Qadri, O. S., Faridi, S. A. and Srivastava, G. 2019. Nanotechnology in sustainable agriculture: studies from seed priming to post-harvest management. *Nanotechnology for Environmental Engineering 4*(1): 11.

Simonin, M. and Richaume, A. 2015. Impact of engineered nanoparticles on the activity, abundance, and diversity of soil microbial communities: a review. *Environmental Science and Pollution Research 22*(18): 13710–13723.

Subbaiah, L. V., Prasad, T. N. V. K. V., Krishna, T. G., Sudhakar, P., Reddy, B. R. and Pradeep, T. 2016. Novel effects of nanoparticulate delivery of zinc on growth, productivity, and zinc biofortification in maize (*Zea mays* L.). *Journal of Agricultural and Food Chemistry 64*(19): 3778–3788.

Torabian, S., Zahedi, M. and Khoshgoftar, A. H. 2016. Effects of foliar spray of two kinds of zinc oxide on the growth and ion concentration of sunflower cultivars under salt stress. *Journal of Plant Nutrition 39*(2): 172–180.

United Nations, Department of Economic and Social Affairs, Population Division. 2019. World Population Prospects 2019: Highlights. ST/ESA/SER.A/423.

Xu, C., Peng, C., Sun, L., Zhang, S., Huang, H., Chen, Y. and Shi, J. 2015. Distinctive effects of TiO_2 and CuO nanoparticles on soil microbes and their community structures in flooded paddy soil. *Soil Biology and Biochemistry 86*: 24–33.

You, T., Liu, D., Chen, J., Yang, Z., Dou, R., Gao, X. and Wang, L. 2018. Effects of metal oxide nanoparticles on soil enzyme activities and bacterial communities in two different soil types. *Journal of Soils and Sediments 18*(1): 211–221.

Section III

Innovative Agriculture-Multi-stakeholder Approach in Different Continents: Practice and Performance

17

Sustainable Intensification of Rice-Based Cropping Systems: Experiences from Eastern India

A. K. Srivastava[1], Malay K. Bhowmick[1,2], Kanwar Singh[3], Pardeep-Sagwal[1], S. Khandai[3], S. K. Dwivedi[4], Amit K. Srivastava[1], V. Kumar[3], Ashok Kumar[5], Sampad R. Patra[6], Virender Kumar[7], and Sudhanshu Singh[1]
[1]IRRI-South Asia Regional Centre (ISARC), Varanasi, Uttar Pradesh, India
[2]Directorate of Agriculture (Government of West Bengal), Kolkata, West Bengal, India
[3]International Rice Research Institute (IRRI), Guwahati, Assam, India
[4]Indira Gandhi Krishi Vishwavidyalaya (IGKV), Raipur, India
[5]International Rice Research Institute (IRRI), Bhubaneswar, Odisha, India
[6]Department of Agriculture (Government of West Bengal), Kolkata, West Bengal, India
[7]International Rice Research Institute (IRRI), Los Baños, Philippines

CONTENTS

17.1 Introduction

Rice is a preferred primary food around the globe. Asian countries produce and consume more than three-fourth of total rice production. More than 250 million farm households' livelihood depends on rice cultivation in Asia only. India contributes to about 21.5% of the world's total rice production. Indian agriculture is primarily dependent on predominant rice-based cropping systems (RBCS), covering around 43% of total rice-growing areas. In eastern India, major rice-based systems include rice–rice (R–R), rice–wheat (R–W), rice–rapeseed/mustard, rice–groundnut, rice–potato, and rice–pulses, based on agroecological conditions, market and domestic necessities and facilities available with farmers. The challenges in eastern India are intense due to the dependence on monsoon, small landholdings, lack of scale-appropriate mechanization, subsistence nature of farming, poor infrastructure for storage and marketing, etc. Furthermore, inefficient input use (fertilizer, water, labor), growing water scarcity,

DOI: 10.1201/9781003164968-20

increasing labor shortage, escalating cultivation cost due to high fuel price, and labor wages make the situation grim (Ladha et al., 2009; Kumar and Ladha, 2011).

Because of socioeconomic and other factors, many smallholder farmers mostly rely on agricultural production and suffer from malnourishment. However, these smallholder farmers have a huge potential for increasing productivity. Climatic and social factors along with the timely supply of quality agricultural inputs play an important role in productivity gains. Gaining economic advantages while maintaining the environment is the root of sustainability. However, environmental responsibility, economic viability, and social acceptability are the important crucial pillars for measuring sustainability in defined geography. Sustainable intensification (SI) has emerged as the best strategy to improve the productivity potential of geographical locations while reducing input use and minimizing environmental impacts (Gatto et al., 2020).

The prevailing RBCS, nowadays, is losing its potential ground due to a decrease in factor productivity. Hence, identifying suitable RBCS with higher resource-use efficiency fitting to the local agroecological situation is of prime importance. The SI of RBCS mainly aims to achieve maximum system productivity with minimum environmental impact by managing and reorienting crops in the systems to best utilize available resources (soil, air, sunlight, water, labor, equipment) and their beneficial interactions as well. Availability of climate-resilient varieties (inbreds and hybrids) of varying duration (short duration [SD] and medium duration [MD]), tailored management practices, and expanding irrigation facilities coupled with conservation practices provide an immense scope of cropping system intensification, higher water productivity, economic profitability, and long-term sustainability. Proper selection of component crops for diversification and intensification needs to be well maneuvered to harness the synergism among efficient resource utilization and improved crop varieties while mitigating the adverse environmental impacts. There is also a need to develop and manage site-specific and resource-efficient RBCS, particularly for the rainfed environments. To effectively address the trilemma of improving system productivity and profitability, ensuring environmental sustainability, and conserving natural resources in eastern India, it is imperative to adopt and disseminate the SI technologies in RBCS.

17.2 Opportunities for Sustainable Intensification in Eastern India

Eastern India comprises the plains of Assam, Bihar, Chhattisgarh, Jharkhand, Odisha, West Bengal, and eastern Uttar Pradesh. The region covers an area of 71.9 million hectares (m ha), representing about 21.9% of the country's area, and supports around 33.7% of the country's population. The majority of the farmers are smallholders, with an average earning of over half of the national per capita income (Ladha et al., 2009).

Rice cultivation in an area of 26.8 m ha accounts for around 63.3% of total rice acreage during the *kharif* season in eastern India (Kumar et al., 2019a). Although the prevailing climatic condition and natural resources suit for higher production, including livestock and fisheries, the production is restricted due to slow diffusion of information, fragmented landholdings, low seed replacement rate, lack of infrastructure facilities, intensifying incidences of biotic and abiotic stresses (like frequent floods, droughts, waterlogging), and social conflicts. Besides, small and fragmented landholdings along with inadequate infrastructure facilities for storage, processing, and marketing threaten agricultural development in the region. Moreover, there is a slow pace of procurement of farm produce, shortage of quality seeds, the problem of soil health, and less number of Farmer Producer Organizations in eastern states as compared to other states of the country.

The eastern states have a current record of about 150% cropping intensity. The region is rich in fertile soil with ample water resources. The average rainfall varies from 1,091 to 2,477 mm, with 18% of the country's utilizable water resources. However, farmers leave their fields fallow after rice harvest and do not opt for a second crop in the same cropping year. Efficient utilization of the fallow areas can have huge prospects and opportunities for improving system productivity and profitability of the RBCS. The total irrigated area is around 39% in this region. Thus, monocropped rice-based areas can be further intensified with second and/or third crops (Kumar et al., 2019a). The contribution of eastern states in rice production accounts for less than 50%, whereas it shares about 55% rice area. Thus, there exists a vast potential for improving productivity, profitability, and livelihood in the eastern states of the country.

Diversification and intensification of cropping systems with the proper use of residual soil moisture or available/limited irrigation facilities provide higher yields and better net returns under the prevailing agroclimatic conditions in the region. Due to soil and climatic conditions, rice cultivation cannot be replaced during the *kharif* season. Thus, there is a limited scope for crop diversification during the *kharif* season, whereas *rabi* and summer seasons offer great scope for crop diversification both under irrigated and rainfed conditions (Bhowmick et al., 2005, 2015, Biswas and Bhowmick, 2015). Conservation agriculture (CA) along with innovative crop establishment (CE) methods like direct seeding of rice (DSR), zero tillage (ZT), mechanical transplanting of rice (MTR), etc., can improve water use efficiency (WUE), soil health, and system productivity (Hellin et al., 2020). Remunerative and efficient crops like pulses, oilseeds, maize, and vegetables can be introduced during the *rabi* season to improve the system productivity and get a higher profit per unit investment. During the summer season, SD legume varieties can also be introduced. The addition of legumes ensures nutritional security, soil cover, and better soil health. Reorienting traditional cropping patterns provide the opportunity to improve WUE, increase soil organic content (SOC), water retention, and minimize nutrient losses (Kumar et al., 2008). Due to the imbalance in demand and production of pulses, oilseeds, and maize, the advanced and climate-resilient varieties should be included in nontraditional areas during nonconventional seasons with the utilization of residual moisture, and improved technologies.

17.3 Strategies for Sustainable Intensification of Rice-Based Cropping Systems

Eastern states offer enough potential for SI, but it should be economically viable, offering incentives to small and marginal farmers to intensify existing cropping systems. To successfully adopt SI options, there must be a trade-off between other income-generating activities and lower production constraints linked with storage facilities and unorganized sign-off procedures for farm produce (Gatto et al., 2020). Although there are various strategies for SI, farmers' adoption rate is relatively poor due to the knowledge gap and other factors.

Appropriate deployment of SI strategies can easily convert monocropped areas into double- or triple-cropped ones through better utilization and exploitation of rice fallows. Introduction of SD varieties (rice and non-rice crops) and scale-appropriate mechanization in the existing RBCS widens the possibilities of diversification and/or intensification along with improved input use efficiency and reduced turnaround time e.g., relay/*utera/paira* cropping of *rabi* pulses and oilseeds (Bhowmick et al., 2015; Biswas and Bhowmick, 2015; Kumar et al., 2019a).

Vegetables, pulses (legumes), oilseeds, and green manures can be introduced for diversification which may improve soil fertility as well as soil physical condition and reduce pest incidence. Integrated use of fertilizers, including organic-, inorganic- and bio-fertilizers, and the introduction of summer/pre-*kharif* crops can improve soil health and fertilizer use efficiency in RBCS (Dwivedi et al., 2008).

17.3.1 Selection of Crops and Their Suitable Varieties for Sustainable Intensification and Crop Diversification

Rice is the preferred crop during the rainy season (June–September/October) with more than 80–85% precipitation, whereas the lands remain fallow after rice harvest (November–May) in the absence of sufficient rainfall and inadequate irrigation facility. However, sufficient residual soil moisture after the rice harvest is available in the rice fallows of lowland areas, which can be exploited for raising a second crop during the post-rainy season. The traditional cropping sequences in double- or triple-cropped areas and the rice fallows in monocropped areas are nonproductive and nonremunerative. Migration of farm laborers to the nearby cities, inadequate farm mechanization, depleting natural resources, and environmental pollution are the major rice production constraints. Crop and varietal diversification and their timely introduction in the RBCS can substantially improve resource-use efficiency, productivity, and profitability, besides mitigating the risks of climate change. It has been found that SD rice cultivars followed by dry season crops give good yields and fetch profitable returns (Lal et al., 2017). Cropping system intensification can be made possible by including stress-tolerant rice varieties (STRVs) and hybrids during the *kharif* season, and remunerative and less input requiring crops like green manures, pulses, oilseeds, maize, and vegetables during the *rabi*-summer season. Adoption of early and mid-early rice varieties in the wet season will have a greater window for raising a good crop in the dry season, particularly the pulses and oilseeds having tolerance to low temperature. Green manures/grain legumes in the systems help to improve soil health with their good nodulation activity and ample nitrogen-fixing ability in a shorter window. These crops improve profitability, nutritional security, fertilizer use efficiency, and the economic conditions of farmers, and they can also be used as fodder. The introduction of SD pre-*kharif* crop preceding to *kharif* rice in the cropping systems would be feasible, just as catch crops, with uniform maturity, high yield, and disease resistance. Grain legumes play a crucial role in imparting nutritional security and soil health improvement as well, making them superior to green manure crops. For successful adoption of green manures/legumes/pulses in RBCS, there is a need to consider the rainfall, growing season, irrigation facilities, and production cost. The adoption of summer crops is very poor due to the nonavailability of quality seeds, higher labor wages, and increasing production costs. Hybrid maize, hybrid mustard, and high-value vegetable crops for their less water requirements can also be grown in irrigated *boro* rice areas as profitable alternative crops. Pulses and oilseeds can be grown in a wide range of cropping systems with minimum effort and also under rainfed environments to easily cope up with harsh environmental conditions. The deep root system of these crops favors the water and nutrient extraction from deeper soil zones, and better CE, requiring less irrigation. Improved varieties of pulses and oilseeds can also be introduced in nontraditional cultivation areas. It will support making the soil more fertile, improve nitrogen use efficiency and recycling nutrients, and increase SOC, water retention, promote mycorrhizal colonization, reduce pest and disease incidence (Singh et al., 2011). Thus, the introduction of pulses, oilseeds, and maize can enhance the production and productivity of these areas, thereby improving the resilience, productivity, and profitability of the diversified cropping systems.

17.3.2 Resource-Efficient Alternative Crop Establishment Methods

The CE method is the most critical for ensuring a good crop stand as well as productivity, particularly under rainfed situations. In RBCS, rice is commonly established by manual transplanting. A huge amount of water and labor requirements for transplanting reduces profit margins and makes rice farming less remunerative. Puddling makes the land unsuitable for subsequent crop production after rice. Thus, the R–R and R–W cropping systems are currently unsustainable. Switching from conventional puddled transplanting of rice (PTR) to alternative CA-based CE methods can reduce the production cost and increase productivity with minimum environmental impact. Since the conventional PTR method is highly input-intensive, precision dry-DSR (dry drill/mechanized DSR or improved broadcast-DSR) and MTR have emerged as alternative CE methods to reduce dependency on farm laborers. In heavy clay soils, where there is less chance of land preparation, ZT-DSR followed by relay cropping of pulses/oilseeds is suggested for higher system productivity. MTR brings higher yield, profit, labor productivity, WUE, and nutrient use efficiency (NUE) and is, thus, gaining popularity to the rural youths as well as farmers (Devkota et al., 2020). Besides producing more or less similar yields, DSR lowers the production costs and greenhouse gas (GHG) emissions as compared to the PTR (Kumar and Ladha, 2011; Devkota et al., 2020). Precision broadcast DSR with lower seed rate and cost-efficient weed management through herbicides may be helpful to displace the traditional method of *beushening/baisi*. However, the timely establishment of *kharif* rice in the RBCS helps to cope with the uncertain monsoon, leading to timely or even early harvest of rice, facilitating the utilization of residual soil moisture by the succeeding *rabi* crops at seeding and, thereby, minimizing the requirements of pre-sowing irrigation (Singh et al., 2020b).

17.3.3 Harnessing Rice Fallows

A total area of about 11.7 m ha area remains fallow after the rice harvest during *kharif* season in India (Gumma et al., 2016). Various reasons, such as variable and inadequate monsoon rains and low intensity of winter rains, lack of irrigation facility, low or very high soil moisture in the surface layer, water stagnation, cultivation of long duration (LD) rice varieties, lack of suitable winter crop varieties, soil compaction, and stray cattle, are identified for the existence of rice fallow areas (Kumar et al., 2019b).

However, the rice fallows, including medium lands and lowland, exist in the rainfed ecosystem, covering eastern, central, coastal, and northeastern hill regions. The majority of rice fallow areas are distributed in Bihar, Chhattisgarh, Jharkhand, Madhya Pradesh, Odisha, West Bengal, and the states of North Eastern Region (Singh et al., 2016). The southern states, Andhra Pradesh, Karnataka, and Tamil Nadu cover the remaining rice fallow areas (Subbarao et al., 2001). The prevailing practice of leaving land fallow after the rice harvest is no more economical and sustainable. The 11.7m ha rice fallow areas are almost equivalent to the combined net sown areas of Haryana, Punjab, and western Uttar Pradesh. If these fallow areas are brought under cultivation of pulses, oilseeds, and maize, which are generally deficient in the country, it may trigger another green revolution benefiting millions of farm families. Improved CE methods and matching agro-technologies during the *kharif* season under rainfed environments can easily accommodate a second crop in succession through proper moisture conservation practices (Singh et al., 2020a). Promotion of SD (3–4 months) and less input requiring crops (thriving best on residual soil moisture with adequate conservation practices) in the fallow areas, especially of eastern India, can intensify the RBCS with enhanced production of these deficit crops and augmentation of farmers' income.

There is a great challenge to efficiently and extensively utilize rice fallow areas in the country. The major bottleneck in growing the crops under rice fallows is the fast depletion of residual soil moisture, resulting in mid-season and terminal drought, coinciding with flowering and pod filling stages with the advancement of the winter season. Except for few regions and hilly terrains, another constraint for most of these rice fallow areas is the unsuitability of growing summer rice again during *rabi* season due to their high-water requirements (not adequately replenished by winter rains). In addition, the inherited soil physical and biological constraints affect seed germination, seedling emergence, and proper CE (Singh et al., 2017a). Lack of appropriate technologies and socioeconomic problems to bring the fallow areas under cultivation are major concerns for the sustainability of the rice ecosystem (Panda et al., 2018). Despite having tremendous potentials of rice fallows for raising a profitable crop with available modern agro-techniques and scientific crop husbandry using residual moisture, adequate attention has not been given in the past few decades. Consequently, a significant level of production in terms of food, feed, and fodder has not been realized over the years.

If the location-specific constraints are addressed, many unutilized areas can be converted into productive ones with appropriate crop planning (Kumar et al., 2019b). Based on the agroclimatic situation, several SD pulses (like chickpea, lathyrus, lentil, and pea) and oilseeds (like safflower, linseed, niger, and mustard), and also climate-resilient varieties of these crops, can be successfully grown in rice fallows with lifesaving supplemental irrigation. Rice-fallow intensification with the introduction of winter (*rabi*) season crops following improved tillage practices with rice residue retention further improves system productivity, benefit: cost ratio, energy use efficiency (EUE), and reduced GHG emissions (Kumar et al., 2020).

General strategies for exploitation of rice fallows through *rabi* pulses, oilseeds, and maize include the use of SD to MD rice varieties during *kharif*, selection of SD/water-efficient genotypes, adoption of location-specific management practices for growing *kharif* and *rabi* crops in the system, and special care for raising an assured second crop (seed treatment, seed priming, foliar nutrition, timely plant protection, lifesaving irrigation through sprinklers).

17.3.4 Efficient Nutrient Management

The productivity and profitability of a cropping system depend largely on nutrient management practices. Long-term fertilizer experiments indicate a decreasing productivity trend with the only use of chemical fertilizers in intensive RBCS.

Most of the farmers apply chemical fertilizers, especially nitrogen, in quantities higher than the recommended rates. Such indiscriminate nitrogen fertilizer uses cause loss of excessive N through leaching as nitrate and emission in the environment as nitrous oxide. Suboptimal or supraoptimal nitrogen application significantly affects both rice yield and quality, signifying the role of balanced crop nutrition in yield optimization of the system.

Farmers generally do not use organic manures although these manures help maintain soil fertility, improve biological and physical properties, increase water-holding capacity (WHC), enhance use efficiency of inorganic fertilizers, and thus make the cropping system more productive and sustainable.

Among the organic manures, farmyard manure is easily available, relatively cheap, and rich in nitrogen. Green manure crops improve the WHC of the soil, partially supplement the N need for rice crops, and improve the SOC. Retention of crop residues has also been found more beneficial than using inorganic fertilizers only.

An integrated nutrient management (INM) approach is very important in RBCS with greater scope for using different nutrient sources (inorganics, organics, and bio-fertilizers). Application of organic manures along with inorganic fertilizers can improve the SOC, retention of nutrients for prolonged supply to crop with improved NUE. It may curtail heavy usage of synthetic fertilizers, enhance soil health, and improve system productivity (Dwivedi et al., 2016). The INM strategy results in economically viable, agronomically feasible, and eco-friendly systems by improving carbon sequestration and reducing N losses and GHG emissions. Higher system productivity of rice–maize cropping system can be achieved with 50% N-P_2O_5-K_2O through fertilizers + 50% N through *Gliricidia* to rice and 75% recommended dose of fertilizers to succeeding maize crop. However, the uptake of these nutrients becomes more with 50% N substitution either through FYM or *Gliricidia* to *kharif* rice. In contrast, there is higher nutrient uptake by maize when 100% NPK is supplied through inorganic sources. Replacing summer rice with maize in R–R cropping systems is beneficial under limited irrigation facilities when INM can be followed using locally available organic nutrient sources like paddy straw/green manure crops (Kumar et al., 2017).

The site-specific integrated nutrient management (SSINM) strategy is ultimately needed to maintain or improve soil fertility and attain the desired levels of crop productivity along with improved SOC sequestration through optimization and integration of all available sources of plant nutrients (Singh et al., 2017b). Compared with the blanket recommendation, soil test-based recommendation, or farmers' fertilizer practice, the SSINM helps in improving NUE without depletion of soil fertility (Singh et al., 2015).

Deployment of nutrient-efficient crop varieties in RBCS is an important strategy for improving system productivity. The introduction of SD and MD rice varieties in RBCS makes it possible for multiple cropping and diversification through the introduction of oilseeds, pulses, or vegetables in subsequent crop seasons (*rabi*-summer/pre-*kharif*). It can improve the system productivity, profitability, and soil fertility compared with the cereal–cereal sequence. A SD and fast-growing green manure crops (*dhaincha*, sunnhemp, blackgram, cowpea, and greengram) can be introduced to cover the lean period (summer fallow for 80–90 days) that exists between the harvest of *rabi* crops and the next *kharif* season.

17.3.5 Relay (*Paira/Utera*) Cropping

In eastern and northeast regions, relay cropping is a general practice to utilize the residual moisture for sowing of succeeding winter (*rabi*) pulse crops like chickpea, lathyrus, lentil, and pea in the standing rice crop (Bhowmick et al., 2005, 2017; Gupta and Bhowmick, 2005, 2012; Singh et al., 2016, Singh et al., 2020a). It is also popularly known as *utera* or *paira* cropping. However, the susceptibility of chickpea to root rot, wilt, and pod borer (*Helicoverpa armigera*) constitutes the major constraints in these regions (Singh et al., 2017a). In the coastal agroecosystem of Odisha, mungbean and urdbean perform well under high humid conditions. In lower Assam, rice–lentil is the most productive among rice–*utera* (relay) crop sequences with higher production efficiency and net return despite lower land use efficiency than rice–chickpea or rice–lathyrus. In eastern India, rapeseed is a viable crop, suitable for system intensification and improving the family income of small and marginal farmers (Duary et al., 2016). Sometimes, toria crop with small seeds faces the problem of stand establishment, leading to a reduction in seed yield as well as net return.

In the post-rice *rabi* season, chickpea, lentil, mustard, lathyrus, and linseed have been evaluated under *paira* cropping (seed broadcasting of succeeding *rabi* crops at 15 days before rice harvest). Among them, lentil and lathyrus have been found to be potential crops under rice-fallow conditions. Mustard does not perform well under *utera* condition. ZT and ZT-mulching (ZTM) produce the highest WUE. Higher system productivity can be obtained from rice–*utera* lathyrus and rice–*utera* lentil.

Among different CE practices, *utera* cropping has the highest EUE, followed by ZT and ZTM, whereas among crop rotations, rice–*utera* lathyrus and rice–*utera* lentil exhibit higher EUE (Kumar et al., 2019a). *Utera* cropping also leads to less GHG emissions than ZT and ZTM. Nitrous oxide emission becomes lower in the rice–lathyrus, rice–lentil, and rice–chickpea systems than rice–mustard and rice–linseed rotations (Kumar et al., 2019a). Pea and lentil as relay crops give higher system yield and more net return, facilitate better moisture retention, and maintain beneficial microbial biomass than *toria* and niger in Assam. Relay cropping of pea at 15 days after flowering of *kharif* rice gives higher rice equivalent yield, better economics, and improved rain WUE (Sarma et al., 2017).

17.3.6 Efficient Water Management

Water availability is an important criterion for selecting component crops in a cropping system, particularly in the dry season due to the nonavailability of natural rainfall. Identification of appropriate crop combinations plays an important role in enhancing WUE. Agroecological intensification of RBCS largely depends upon water availability in the dry season. In the RBCS, water management in rice has a greater impact on the succeeding crop(s) due to changes in the water regime. Puddling, a time-consuming and capital- and water-intensive exercise, leads to subsurface compaction, which is not conducive to the succeeding crops. Innovative water-saving technologies such as alternate wetting and drying (AWD), saturation soil culture, raised-bed, and aerobic rice cultivation are found promising in enhancing the water productivity of rice cultivation in South Asia. Flooding to field capacity throughout the rice crop season considerably reduces the number of irrigations and gives higher WUE. Besides, the depth of submergence at 3–10 cm is sufficient for achieving optimum yield and control of most weeds. Singh et al. (2014) reported higher field WUE and yield of dry season crops after rice with two lifesaving irrigations in groundnut, lentil, rapeseed, and potato. Higher water productivity and lower water use can be achieved in rice by soil mulching in combination with the direct-seeded aerobic conditions.

17.3.7 Integrated Weed Management

Weeds are dynamic in nature, and weed infestation largely depends on the type of crop, cropping system, and cultural practices. Cropping system intensification in combination with new management practices greatly impacts the weed flora of a cropping system in the region. Cultural practices like crop rotation, crop geometry, mulching, tillage practices, etc., influence weed flora dominance and distribution, thus determining the weed control methods. Intensification of rice fallows at 300% cropping intensity considerably reduces the weed population and weed dry matter production as compared to 200% (R–W) and 100% (rice fallow) cropping systems (Prasad et al., 2013). Among the cropping systems (Prasad et al., 2013), rice–potato + wheat (1:1)-greengram produces the lowest weed population and weed dry matter (comparable with rice–potato–greengram). The weeds like *Phalaris minor* in wheat, and *Cichorium intybus*, and *Rumex* spp. and *Medicago denticulata* in Egyptian clover can be minimized by intensifying and diversifying them with other crops to obtain higher production efficiency. Infestation of few weeds is also associated with tillage practices. For example, the densities of *Echinochloa colona* and *Cyperus iria* increase in rice plots due to continuous ZT in the RBCS. ZT reduces the population of *Avena ludoviciana* and *Chenopodium album*, but increases the density of *Medicago hispida*, compared to conventional tillage systems. Research results from Modipuram (Uttar Pradesh) indicate a significant impact of stale seedbeds in reducing the weed incidence in the R–W systems as compared to the traditional seedbed preparation. Crop rotation using weed smothering and land covering crops such as cowpea, soybean, and blackgram reduces the weed population substantially.

Similarly, intercropping systems involving rice and non-rice crops like greengram, blackgram, pigeon pea, or groundnut can help suppress the weeds in upland. Intensification of the RBCS through the inclusion of summer greengram produces comparable grain yield to summer cowpea or *Sesbania* green manuring and reduces both the density and dry matter of grasses and sedges. Weed management by using herbicides is gaining popularity due to easiness and low-cost involvement, although the herbicides are highly crop-specific and have a residual effect on succeeding crops. Therefore the selection of herbicides for weed management in a cropping system, particularly in intercrops, should be made on consideration of their selectivity and residual effect on succeeding crops. However, an integrated weed management practice, employing all the available tactics and techniques would be the most effective approach for RBCS as a whole.

17.3.8 Integrated Farming System

Monocropped rice production systems may have a negative impact on the soil physicochemical and biochemical indicators, resulting in loss of biodiversity, and degradation of the natural resource base, thus making the farming system unsustainable in the long run. The integrated farming system (IFS) approach assumes greater importance for the sustainable livelihood of resource-poor small and marginal farmers. Because the IFS integrates different agricultural enterprises, such as cropping systems, livestock, aquaculture, agroforestry, agri-horticulture, and apiary in an optimum combination, resulting in multiple benefits. For example, the rice–fish cultivation system lowers the methane emission from the paddy field. The movement of the aquatic creatures disturbs the soil layers, thus breaking the chain of anaerobic soil reactions and lowering down the methane production. Recent estimates show that rice–fish culture may reduce methane emission than the mono-cultured rice field. It is also beneficial for restoring soil fertility, avoiding soil degradation, and imparting environmental sustainability through less fertilizer and pesticide application.

Integrated crop and livestock systems together can improve the efficiency and environmental sustainability of both the production systems. When livestock and crop production systems are integrated, the waste from one system can serve as a resource for the other system (Singh et al., 2017b). Bio-slurry (a common by-product from biogas plant) as a sole nutrient source for growing field crops at recommended dosage and rearing cows with own-produced green fodder can enhance profitability and sustainability of the farming system (Mukherjee et al., 2020).

In northeast India, monsoon precipitation forms diverse wet rice ecologies, which provide natural habitat, breeding environment, and feeding grounds for various aquatic lives like prawns, fishes, crabs, frogs, snails, tortoises, and other animals. The current farming practice can be broadly categorized into rice field capture fishery system, wild aquatic cropping system, mountain foothills and valley rice–fish farming system, and running-water terrace rice–fish farming system. However, the traditional rice–fish culture is less productive (in hilly areas) and with declining fishery harvest in flood plain areas (Das, 2018). The other major constraints are farmers' apathy, input shortage, security, poaching, lack of support services, and incidence of pests and diseases (Baruah et al., 2000).

17.4 Conclusions

Eastern India is bestowed with a bounty of natural resources such as water, fertile land, and rich biodiversity. Monocropping of rainfed rice, inefficient input use, unpredictable weather, problematic soil conditions, labor migration and scarcity, increasing production costs, low adoption of mechanization, and weak extension system are responsible for declining factor productivity in RBCS. It is imperative to develop and adopt improved technologies for SI as a whole. Converting monocropped rice areas into multiple (double or triple) cropping systems by utilizing rice fallows is an important strategy for diversifying and/or intensifying the existing RBCS. Efficient utilization of residual soil moisture by including SD and high-value crops, pulses, oilseeds, and vegetables, and their suitable resilient varieties with conforming management practices, including supplemental protective irrigation, help address the constraints in rice fallows. The geographic information system (GIS) and remote sensing (RS) tools can effectively help in mapping rice fallows and spatiotemporal soil moisture availability for effective targeting of the suitable technologies, including appropriate crops and their varieties. The negative impact of tillage-intensive puddling on soil quality can easily be addressed by introducing innovative and alternative CE methods such as DSR and MTR. CA practices in RBCS can improve crop WUE, soil health, and system productivity. ZT sown *rabi* crops with surface retained rice residue can offer many advantages, including soil moisture conservation, weed management, sustaining soil microbial activity, etc., besides facilitating early planting of *rabi* crops due to reduced turnaround time, thus intensifying the cropping systems and improving the system productivity. The infrastructure of irrigation, storage, product marketing, and value chain also needs to be developed for a comprehensive improvement in rice-based system productivity, profitability, and economic condition of the smallholder farmers.

REFERENCES

Baruah, U., Bhagawati, A. K. and Talukder, R. K. 2000. Prospect of fish farming in rice fields of Assam. *Indian Journal of Fisheries* **47**: 149–159.

Bhowmick, M. K., Aich, A., Aich, S. S. et al. 2005. Crop diversification through *paira* (*utera*) cropping with *rabi* pulses. *SATSA Mukhapatra – Annual Technical Issue* **9**: 43–60.

Bhowmick, M. K., Duary, B. and Biswas, P. K. 2015. Crop diversification through oilseeds in eastern India. In *Diversification of Agriculture in Eastern India, India Studies in Business and Economics*, ed. M. Ghosh, D. Sarkar, and B. C. Roy, 109–130. Springer India, New Delhi.

Bhowmick, M. K., Duary, B., Dana, I. et al. 2017. Low-cost technologies for improving productivity of lathyrus under rice-*utera* system in West Bengal. In *Contemporary Issues in Agriculture, Environment and Rural Development in India* Vol. 1, ed. A. Moktan, 113–118. New Delhi Publishers, Mohan Garden, New Delhi.

Biswas, P. K. and Bhowmick, M. K. 2015. Crop diversification through pulses in the Northeastern Plain Zone of India. In *Diversification of Agriculture in Eastern India, India Studies in Business and Economics*, ed. M. Ghosh, D. Sarkar, and B. C. Roy, 131–139. Springer India, New Delhi.

Das, D. N. 2018. Farming of fishes in rice fields of northeast India: A review. *Journal of Coldwater Fisheries* **1**: 27–41.

Devkota, K. P., Yadav, S., Khanda, C. M. et al. 2020. Assessing alternative crop establishment methods with a sustainability lens in rice production systems of Eastern India. *Journal of Cleaner Production* **244**, 18835. https://doi.org/10.1016/j.jclepro.2019.118835.

Duary, B., Teja, K. C., Bhowmick, M. K. and Chakraborty, N. R. 2016. Relay cropping of rapeseed: A viable agro-technique for system intensification and higher family income of small and marginal farmers in eastern India. In *Family Farming: Challenges and Opportunities*, ed. B. Mondal, D. Sarkar, S. D. Mukhopadhyay et al., 179–187. Renu Publishers, Mohan Garden, New Delhi.

Dwivedi, B. S., Shukla, A. K., Singh, V. K. and Yadav, R. L. 2008. Improving nitrogen and phosphorus use efficiencies through inclusion of forage cowpea in the rice-wheat system in the Indo-Gangetic plains of India. *Field Crops Research* **34**: 399–418.

Dwivedi, B. S., Singh, V. K., Meena, V. C. et al. 2016. Integrated nutrient management for enhancing nitrogen use efficiency. *Indian Journal of Fertilizers* **12**: 62–71.

Gatto, M., Petsakos, A. and Hareau, G. 2020. Sustainable intensification of rice-based systems with potato in eastern Indo-Gangetic Plains. *American Journal of Potato Research* **97**: 162–174.

Gumma, M. K., Prasad. S. T., Pardharsadhi, T. et al. 2016. Mapping rice fallow cropland areas for short-season grain legumes intensification in South Asia using MODIS 250 m time-series data, *International Journal of Digital Earth* **9**: 981–1003.

Gupta, S. and Bhowmick, M. K. 2005. Scope of growing lathyrus and lentil in relay cropping systems after rice in West Bengal, India. *Lathyrus and Lathyrism Newsletter* **4**: 28–33.

Gupta, S. and Bhowmick, M. K. 2012. Enhancing lentil productivity under rice-*utera* system in West Bengal, India. *Journal of Lentil Research* **5**: 23–32.

Hellin, J., Balie, J., Fisher, E. et al. 2020. Trans-disciplinary responses to climate change: Lessons from rice-based systems in Asia. *Climate* **8**: 35. ISSN 2225-1154, doi: https://doi.org/10.3390/cli8020035. Available at http://centaur.reading.ac.uk/89241/.

Kumar, A., Srivastava, L. K., Mishra, V. N., and Banwasi, R. 2017. Major and micro nutrient status of rice-chickpea grown in soils of Chhattisgarh plain region of India. *Indian Journal of Agricultural Research* **51**: 1–8.

Kumar, A., Tripathi, H. P., Yadav, R. A. and Yadav, D. S. 2008. Diversification of rice (*Oryza sativa*)-wheat (*Triticum aestivum*) cropping system for sustainable production in eastern Uttar Pradesh. *Indian Journal of Agronomy* **53**: 18–21.

Kumar, R., Mishra, J. S., Karnena, Rao K. K. et al. 2020. Crop rotation and tillage management options for sustainable intensification of rice-fallow agro-ecosystem in eastern India. *Scientific Reports* **10**: 11146. https://doi.org/10.1038/s41598-020-67973-9.

Kumar, R., Mishra, J. S., Rao, K. K. et al. 2019a. Sustainable intensification of rice fallows of Eastern India with suitable winter crop and appropriate crop establishment technique. *Environmental Science and Pollution Research* **26**: 29409–29423. https://doi.org/10.1007/s11356-019-06063-4.

Kumar, R., Mishra, J. S., Upadhyay, P. K. and Hans, H. 2019b. Rice fallows in the eastern India: Problems and prospects. *Indian Journal of Agricultural Sciences* **89**: 567–577.

Kumar, V., and Ladha, J. K. 2011. Direct seeding of rice: Recent developments and future research needs. *Advances in Agronomy* **111**: 297–413.

Ladha, J. K., Kumar, Virender, Chandna, P. et al. 2009. Integrating crop and resource management technologies for enhanced productivity, profitability, and sustainability of the rice-wheat system in South Asia. In *Integrated Crop and Resource Management in the Rice–Wheat System of South Asia*, ed. J. K. Ladha, Y. Singh, O. Erenstein, and B. Hardy, 69–108. International Rice Research Institute, Los Baños.

Lal, B., Gautam, P., Panda, B. B. et al. 2017. Crop and varietal diversification of rainfed rice based cropping systems for higher productivity and profitability in eastern India. *PLoS ONE* **12**: e0175709. https://doi.org/10.1371/journal.pone.017570.

Mukherjee, S. K., Bhattacharyya, P., Bhowmick, M. K. et al. 2020. Discovery of self-reliant regenerative sustainable agricultural model, an "On-Farm" research for higher profit of small and marginal farmers. *Discovery Agriculture* **6**: 193–198.

Panda, B. B., Satapathy, B. S., Nayak, A. K. et al. 2018. Agro ecological intensification of rice based cropping system. In *Rice Research for Enhancing Productivity, Profitability and Climate Resilience*. https://krishi.icar.gov.in/jspui/bitstream/123456789/19102/1/20190427164833%20B%20B%20PANDA.pdf (accessed on April 16, 2021).

Prasad, D., Yadava, M. S. and Kumar, A. 2013. Weed dynamics in transplanted rice after intensification of rice-fallow cropping system in Jharkhand. *Oryza* **50**: 146–150.

Sarma, D., Saikia, P., Sarma, P. K., et al. 2017. Crop intensification through relay cropping in rice fallows of Assam. *Indian Journal of Dryland Agricultural Research and Development* **32**: 75–77.

Singh, D., Bhaskar, B. P., Baruah, U. and Sarkar, D. 2014. Evaluation of rice (*Oryza sativa* L.) based cropping systems in major soil series of upper Brahmaputra Valley Assam. *Global Journal of Science Frontier Research: D Agriculture and Veterinary* **14**: 1–6.

Singh, N. P., Praharaj, C. S. and Sandhu, J. S. 2016. Utilizing untapped potential of rice fallow of east and north-east India through pulse production. *Indian Journal of Genetics and Plant Breeding* **76**: 388–398.

Singh, R. K., Bohra, J. S., Nath, T. et al. 2011. Integrated assessment of diversification of rice-wheat cropping system in Indo-Gangetic plain. *Archives of Agronomy and Soil Science* **57**: 489–506. 10.1080/03650341003641771.

Singh, S. S., Kumar, N., Praharaj, C. S. and Singh, N. P. 2017a. *Agro-technologies for Enhancing Pulses Production in Rice Fallows*. ICAR-Indian Institute of Pulses Research, Kanpur, Uttar Pradesh, 36 pp.

Singh, Sudhanshu, Srivastava, A. K., Dar, M. H. et al. 2017b. Climate resilient agronomy for sustainable rice production in rainfed environment of eastern India. *SATSA Mukhapatra-Annual Technical Issue* **21**: 1–20.

Singh, Sudhanshu, Srivastava, A. K., Bhowmick, M. K. and Patra, S. R. 2020a. Improving resilience and productivity of rice-based systems in drought-prone rainfed environment. *Extended Summaries. First Indian Rice Congress-2020 (Virtual) on "Rice Research and Development for Achieving Sustainable Development Goals"*, Dec. 8-9, 2020, ICAR-NRRI, Cuttack, Odisha, 371–372.

Singh, Sudhanshu, Kumar, Virender, Craufurd, P. et al. 2020b. Management of crop growth cycle as the frontier of yield growth in rice. In: *New Frontiers in Agricultural Extension* Vol. II, ed. A. K. Singh, P. Craufurd, Sudhanshu Singh et al., 165–184, International Maize and Wheat Improvement Center (CIMMYT), New Delhi.

Singh, V. K., Shukla, A. K., Dwivedi, B. S. et al. 2015. Site-specific nutrient management under rice-based cropping systems in Indo-Gangetic Plains: Yield, profit and apparent nutrient balance. *Agricultural Research* **4**: 365–377.

Subbarao, G. V., Kumar Rao, J. V. D. K., Kumar, J. et al. 2001. *Spatial Distribution and Quantification of Rice-Fallows in South Asia-Potential for Legumes*. International Crops Research Institute for the Semi-Arid Tropics, Patancheru, Andhra Pradesh, 316 pp.

18

Scale-Appropriate Mechanization for Improving Productivity, Profitability, and Sustainability of Rice-Based Cropping Systems in India

Pardeep-Sagwal[1], Suryakanta Khandai[2], Malay K. Bhowmick[1,3], Kanwar Singh[2], A. K. Srivastava[1], B. S. Dhillon[4], S. Rawal[5], V. Kumar[2], Ashok Kumar[6], Sudhanshu Singh[1], and Virender Kumar[7]
[1]*IRRI South Asia Regional Centre (ISARC), Varanasi, Uttar Pradesh, India*
[2]*International Rice Research Institute (IRRI), Guwahati, Assam, India*
[3]*Directorate of Agriculture (Government of West Bengal), Kolkata, West Bengal, India*
[4]*Punjab Agricultural University, Ludhiana, Punjab, India*
[5]*Chaudhary Charan Singh Haryana Agricultural University, Hisar, Haryana, India*
[6]*International Rice Research Institute (IRRI), Bhubaneswar, Odisha, India*
[7]*International Rice Research Institute (IRRI), Los Baños, Philippines*

CONTENTS

DOI: 10.1201/9781003164968-21

18.1 Introduction

Globally, around 80% of food is produced by ~5 billion smallholder farmers, and the onus to meet the increasing demand of food by 60% over the 2007 level lies squarely on this segment (FAO, 2014). But the productivity level is low commonly because of limited access to farm mechanization and farm inputs (FAO, 2013). Therefore, the densely populated global rice leaders (China, India, and Indonesia) need to produce more per unit area of land and per drop of water under diminishing availability of arable land and irrigation water. Farm mechanization played a crucial role in ushering benefits of the Green Revolution (GR) in India, although its impact was heterogeneous across regions as well as crops. The GR typically emphasized monocultures of improved crop varieties using heavy machinery and equipment, by design largely biased toward larger flat lands having higher potential crop yields. Currently, farm mechanization level in India varies from 40 to 45%, which is highly skewed toward the cradle of GR (Uttar Pradesh, Haryana, and Punjab), but with negligible access in northeastern states. The growing concerns about availability of labor, high labor wages, and diminishing income of farming business have necessitated the adoption of appropriate technologies that are energy-efficient for production and processing of focused cereal crops with less cost of cultivation, minimal postharvest losses, and value addition through location-specific management practices (Singh and Sahni, 2019). The proponents of mechanization uphold that prudent application of machinery can help save a farmer 15–20% of seed, 20–30% of fertilizer, 20–30% of time, 5–20% of laborers, and increase the cropping intensity by 10–15% and productivity by 15–20% (Singh, 2008; Shivananda et al., 2019). Propriety to small farms, lucidity of design and technology, adaptability to multiple farm operations, profitability, and most importantly, support of service provision are the essentials for wider adoption of mechanization.

The adoption of mechanization for crop establishment is only 5–10% in rice compared to 80–90% in wheat (AMMA, 2014). It indicates that still there is an ample scope for promoting the resource-efficient crop establishment methods, such as direct seeding of rice (DSR) and mechanical/machine transplanting of rice (MTR) to achieve long-term sustainability of rice-based cropping systems (RBCS). Zero tillage (ZT) with surface-maintained residues has a vast potential for increasing productivity and profitability of RBCS by increasing the gap between rice harvesting and sowing of *rabi* crops while eliminating the field preparation period and ensuring long-term sustainability by increasing cropping intensity, system productivity, input-use efficiency, and environmental safety. This has wider implications as an environmental savior for the protection of air quality in rice-wheat growing regions of northwest India grappled with intensive rice residue burning.

18.2 Current Perspective of Farm Mechanization

It is estimated in a World Bank study that nearly 50% population of India would be residing in urban areas by 2050. Further estimates show that the percentage of farmworkers in total workforce would be reduced to 49.9 in 2033 and 25.7 in 2050 from 54.6 in 2011. The share of agricultural workers was 16.3% in 1960–61 and is expected to be reduced up to 2.3% by 2032–33. Although the present growth of farm mechanization is at a faster rate among farmers in comparison to the past years, yet the overall farm mechanization level in the country is only 40–45%. To accomplish the farm operation in a timely manner and assuring quality as well, it is essential to get the farm power availability to a minimum of 2.5 kW/ha at national level. There is a significant regional disparity in farm mechanization. For example, the north Indian states like Punjab and Haryana are highly mechanized (70–80% overall, and 80–90% for rice and wheat) due to prevalence of highly productive and intensive cropping systems.

18.3 Challenges toward Adoption of Farm Mechanization

- The fragmented landholdings and small farm sizes in India restrict the economic use of big machineries. Such types of landholdings require efficient, but less costly agricultural tools and equipment.
- Farmlands are sloppy to variable extent in the hilly states like eastern part of the country, restricting the easy maneuvering and operation of big and heavy machinery. This necessitates the use of smaller and lighter machinery in sloppy lands.
- Though numerous farm implements are being developed by research institutes and private organizations, they do not readily offer to take them forward, which is one of the major constraints in the promotion and upscaling of mechanization.
- A few farm machineries are available only in urban areas of the region. However, the rural areas, especially the actual farming sites, still remain out of the sale and service networks.
- Transfer and adoption of new technologies are very slow in the northeastern region. Farmers are not aware about the improved and self-propelled kind of machinery and implements, and methods of using them.
- Till today, our technical know-how on mechanization is mainly restricted to the tractor, power tiller, rotavator, and power sprayer only.
- The majority of small farmers are poor who are not able to purchase costly machineries like tractors and combine harvesters Lack of access to farm power is an important challenge for slow adoption, limiting intensification and diversification of crops.

18.4 Need for Scale-Appropriate Mechanization

With the consistent decrease in landholding size of farmers in India, it is evident that the demand for small farm machineries or implements will increase due to diversity of crops, cropping patterns, and agricultural operations. Scale-appropriate mechanization is an alternative solution to address the increasing cost of cultivation, drudgery of farm operations and labor paucity. It can curtail the production cost, improve productivity as well as sustainability, and provide a viable solution to reduce environmental footprints by advancing the seeding and other farm operations in the RBCS. Thus, there are immense opportunities for sustainable intensification and crop diversification. Some of the important advantages for scale-appropriate mechanization are as follows:

- Farm mechanization helps in the timely completion of various farm operations and proper utilization of soil moisture for multiple cropping, leading to increased system productivity.
- Mechanized farming encourages farmers for crop diversification (to expand to new crops) since there is higher capital investment with the machinery and more time in hand.
- Custom hiring as a new livelihood option for unemployed youths will provide opportunities for production, repair and maintenance of machines, implements, and spare parts through skill up-gradation in various vocations, leading to a more self-reliant society.

18.5 Mechanization Options for Sustainable Intensification of Rice-Based Cropping Systems

18.5.1 Land Preparation

18.5.1.1 Tillage

Tillage can be defined as the mechanical manipulation of soil to create favorable conditions for raising a crop by plowing, ripping, or turning it. The main objective of tillage is to prepare better seedbeds, incorporate residues, fertilizers, and control weeds.

TABLE 18.1

Operational Charge and Cost of Different Machines for Land Preparation

Operation	Machine	Cost of Machine (USD)	Service Charge (USD)
Primary tillage	MB plow	350–700	25–27/ha
	Disc plow	650–1300	25–27.5/ha
	Cultivator	500–600	25–27.5/ha
Secondary tillage	Rotavator	1050–2100	35–37.5/hour
	Disc harrows	600–1300	25–27.5/ha
Leveling	Manual leveler	200–250	7–10/hour
	Laser land leveler	4050–6750 (full set)	12–16/hour

Primary tillage is the first operation done after harvesting of the previous crop for sowing of the succeeding crop when there is enough moisture in the soil, and the field is firm and strong enough to allow running of farm machinery with appropriate traction. Cultivator, mold board (MB) plow, disc plow, disc harrow, etc. are mainly used for primary tillage operation that is operated by a tractor (Table 18.1).

Secondary tillage involves lighter and finer operations after primary tillage mainly to reduce clod size, control weeds, prepare a fine seed bed, achieve a good tilth, incorporate fertilizers, and puddle and level soil surface. Various types of harrows, rollers, pulverizers, clod-crushers, levelers, rotary tillers, etc. are used for secondary tillage (Table 18.1).

18.5.1.2 Land Leveling

Water is the major limiting and crucial factor for crop production. Its precision management plays a critical role for future food security. Traditional methods of land leveling, being cumbersome and time-consuming, are not much accurate or precise, and hence, provide tremendous space for deployment of modern methods such as laser land leveler (LLL) (Table 18.1). LLL uses laser-equipped drag buckets for leveling (±2 cm) the land surface of agricultural field from its average elevation. Large horsepower tractors equipped with GPS and/or laser-guided instrument are used for leveling that help move soil to a desired cut or fill level. This is one of the key climate-smart interventions in the intensively cropped flood-irrigated systems in northwest India. LLL helps increase the cultivable area (~3–5%), improve water- and fertilizer-use efficiency, and weed control efficiency, leading to enhanced farm productivity (~5–20%), depending upon soil type and cropping system. The farm profitability may be increased by USD 150–200/ha in rice-wheat system.

18.5.2 Crop Establishment

18.5.2.1 Seed Sowing

18.5.2.1.1 Conservation Tillage

Tillage practice that retains crop residues on the soil surface to minimize the impact of water and wind erosion is called conservation tillage. With proper management, crop residues help in protecting soil resources, enhancing soil quality, restoring degraded ecosystems, improving nutrient cycling, increasing water conservation and availability, and suppressing crop pests. Zero till seed-cum-fertilizer drill, Happy Seeder (HS) and Turbo Happy Seeder (THS) are majorly used for seed sowing under conservation tillage condition (Table 18.2).

18.5.2.1.1.1 Zero Till Seed-cum-Fertilizer Drill This drill consists of inverted T-type furrow openers for minimal soil disturbance, a seed and fertilizer metering system (fluted roller/inclined plate/cup type), and drive wheel for transmission of power to metering system and depth control wheels. The multi-crop planter with inclined plate metering system can be used for sowing of other crops like paddy, wheat, pulses, soybean, and maize. ZT reduces tillage operations to merely a single pass for timely sowing of

TABLE 18.2

Operational Charge and Cost of Different Machines for Crop Establishment

Operation	Machine	Cost of Machine (USD)	Service Charge (USD/ha)
Seed sowing	Zero till seed-cum-fertilizer drill	1145–1750	20–22.5
	Happy seeder	2000–3500	35–37.5
	Super seeder	3000–3500	35–37.5
	Drum seeder	60–100	20–22.5
Transplanting	Mechanical transplanter	3250–13500	50–55

subsequent crops, which increases system productivity and alleviates costs by saving in terms of petrol/diesel or running costs for tractor, irrigation, fertilizer, and herbicides. It is an apt resource-efficient and cost-effective alternative to time-consuming and energy-intensive initiatives, which also reduces the risk of *Phalaris minor* in the case of wheat, besides enhancing yields and farm incomes (Singh et al., 2020a).

Compared with conventional tillage (CT), wheat sowing is best accomplished with the use of ZT-drill in the field having 3–4% more moisture level. ZT permits early sowing of wheat (October end to mid-November), leading to higher yield, owing to reduced weed infestation, more efficient nutrients-water interactions, and alleviated effect of terminal heat, compared to the late sown crop.

18.5.2.1.1.2 Happy/Turbo Seeder The HS is a tractor-mounted machine that cuts and lifts rice straw, drills wheat into the soil, and deposits the straw over the sown area as mulch. Looking at the issue in the combine harvested paddy field where farmers are forced to burn residue for *in situ* residue management, research institutes have developed HS or THS. In a paddy field harvested by Super Straw Management System (SSMS) fitted with the combined harvester, which chops and evenly spreads the stubble in the field, the HS helps farmers to directly sow wheat seeds to avoid burning of rice straw. The HS makes it possible to drill the crop seed without any soil disturbance and helps save time, energy, labor, water, and diesel. As the most profitable and scalable residue management practice, the use of HS offers 10–20% more profit than residue burning. Moreover, HS holds great promise in reducing air pollution and mitigating greenhouse gas (GHG) emissions, to a level around 78% higher than all burning options (CIMMYT, 2019). HS can also be used for seed sowing in DSR method of crop establishment.

18.5.2.1.2 Conventional Tillage

This is a system of cultivation mainly aiming to prepare seedbed and control weeds. This typically involves a sequential tillage to plow and harrow the field, to obtain a fine seedbed, and also to help remove most of the residues of the previous crop. Super seeder and drum seeder are used for seed sowing under CT condition (Table 18.2).

18.5.2.1.2.1 Super Seeder Super seeder is a combination of rotary tiller and seed planter with press wheel attached at the rear end which facilitates residue incorporation, seed drilling, and pressing of soil in a single pass. Using the machine saves time as it combines different operations, but there is a considerable increase in weight and tractor power to operate it, besides more fuel consumption (~20 l/ha). Poor yields due to nonsynchronous germination as well as crop maturity are other emerging concerns.

18.5.2.1.2.2 Drum Seeder For successful rice production, land preparation for nursery and main field requires good amount of water and involves labor for nursery raising and uprooting of seedlings and subsequent transplanting in the main field. Paucity of water at transplanting results in delayed transplanting with the use of overaged seedlings which curtails the tillering potential of rice crop, leading to reduced yield levels. To address these problems, drum seeding technology is a viable option for resource-poor farmers toward profitable rice production (Bhowmick et al., 2016). In drum seeding technique, the pre-germinated paddy seeds are put in drums made up of fiber material and evenly dispensed in lines ideally spaced at 20–22 cm in well prepared, leveled, and puddled fields. For drum seeding in 1 ha of land,

25- to 30-kg paddy seed is soaked overnight in water, followed by 10–15 hour of incubation to stimulate the sprouting. Sprouted seeds are put in the drum to fill to its three-fourth capacity. The drum is pulled manually across the field with a steady speed to attain uniform sowing and even crop stand.

18.5.2.2 Line Transplanting (Mechanical)

In conventional transplanting, a person puts 2–3 seedlings per hill in the puddled soil at a depth of 3–5 cm and draws backward in a bending posture with feet bogged into puddled soil. A worker, on an average, has to insert his/her fingers 350,000 times into puddled soil to transplant 1 ha of land. Manual transplanting process requires 250–300 person-hour/ha, which is roughly about 25% of the total labor requirement during the crop growing period (Samal et al., 2020). Timely transplanting confers stability to the RBCS and helps cope the uncertain monsoon with *kharif* rice crop, leading to early harvest of rice, facilitating the utilization of residual soil moisture by the succeeding *rabi* crops at seeding, and thereby minimizing the requirements of pre-sowing irrigation (Singh et al., 2020b).

In MTR, young rice seedlings (specifically grown in a raised mat) are transplanted with the use of a self-propelled rice transplanter. However, the use of self-propelled rice transplanter employs three persons to transplant 1.5–2.0 ha in a day (depending on size/make of the transplanter) with uniform spacing, optimum plant density (2–4 seedlings/hill), and added productivity (0.5–0.6 t/ha), compared to traditional methods (Table 18.2). The mat-type nursery is a prerequisite for machine transplanting of 14- to 18-day-old seedlings (18–20 cm in stature) with minimized seedling-root damage leading to minimal transplanting shock.

For MTR, well-puddled and leveled field is required with no much standing water on the surface to avoid floating hills. Before operating transplanter, the fields are puddled uniformly and left for 24 hours to settle down completely for better seedling establishment (Singh and Vatsa, 2006). Therefore, lowlands with standing water are not suitable for MTR. The MTR also provides better employment opportunities for rural youth by developing custom service business as an alternative job, and for women in the form of nursery enterprise.

18.5.3 Crop Management

18.5.3.1 Irrigation

To attain sustainable growth in agriculture sector, it is essential to potentially utilize the available water, being a scarce commodity. Concerted efforts should be made on the adoption of micro-irrigation technology for water-guzzling cereals, besides horticultural crops. Availability of liquid fertilizers opens up the scope, which can be exploited through proper awareness and mass campaigning among farmers on the benefits of fertigation. Micro-irrigation system needs to be energized by solar system that has the capacity to run majority of prevalent electrical pumping system. All the irrigation pumping systems (surface, subsurface, or deep bore wells) need to be connected with drip system to enhance the income from this setup. For instance, a solar water pump of 2000-Watt Peak (WP) can pump out water at the rate of around 80,000 l/day from up to 10 meters depth, and sufficient for irrigating ~1 ha of land. Efforts should be diverted to link tube well/river-lift irrigation projects with micro-irrigation systems to make the best use of energy both for lifting and pressurized irrigation.

Subsurface drip irrigation is an advanced technology to irrigate the rhizosphere with the help of emitters that are fitted on lateral tubes and inserted into the tubing as an emitting pipe. Different types of emitters are used, depending on specific requirements, crops and soil types. The success of an emitter system depends on judicious decision on crop water need, age of plants, plan-to-plant spacing, soil type, quality and availability of water, etc.

In sprinkler system of irrigation, water ejection in the air takes place under pressure via a string of nozzles attached to a web of high-density polyethylene (HDPE) or aluminum pipes, considering the available rainfall. The crops grown at a very high planting density (cereals, pulses, spices, and others) are most suited to this system, both in plains and undulating terrains. Introduction of solar-powered irrigation systems (SPIS) also seems to be important in the context of conservation agriculture (CA). The speed of the acceptability in using SPIS by the Indian farmers is far beyond its potential.

18.5.3.2 Weeding

Although hand-hoe is the most widely used manually operated weeding tool by smallholder farmers, yet it is highly tedious, laborious, and cost-prohibitive. In line sown and equi-spaced transplanted rice, cono-weeder is a useful tool to uproot and bury weeds under wetland condition. Singh et al. (2013) advocated using cono-weeder and other inter-cultivation implements for removal of weeds in between the rows of rice crop. It disturbs the topsoil, improves aeration, and helps in prolific root growth, thereby facilitating a suitable growing environment and better nutrient uptake, resulting in improved plant growth and productivity. While operating the weeder in between two equidistant rows of standing crop, the weeds are uprooted and buried under the soil by rollers. The cono-weeder can satisfactorily suppress the weeds in a single forward pass owing to its backward and forward movement in the 3-cm surface soil without a push-and-pull movement. A single operator can cover 0.05 ha/day, and the use of implement can reduce the cost by 40–45%. Even the implement is quite acceptable to women laborers for its higher efficiency. Contrarily, the use of manually operated weeder like cono-weeder or rotary weeder requires considerable labor (>20 man-days/ha). The motorized power weeder is an improved version, consisting of a 1.5 hp petrol engine, gearbox, mainframe, rotary wheel, float, handle, and controls. The paddy field should be line sown/transplanted with a spacing of 25 cm for easy movement of weeder to avoid any kind of root damage.

18.5.3.3 Fertilizer Application

Besides low factor productivity of fertilizers, imbalanced and inadequate use of nutrients as per need of high-yielding varieties is an issue against the philosophy of balanced nutrient application (Singh et al., 2017). Urea is the most widely used nitrogenous fertilizer in rice because of its high N content, low cost, and favorable physical properties. Upon broadcasting in the rice field, it is rapidly hydrolyzed into NH_4^- which is quickly transformed to NO_3^+ by the process of nitrification in the aerobic zone. Having a larger surface area, the use of prilled urea (PU) results in a loss of 55–70% applied N that percolates deep with the soil water or volatilizes in the air. Emission of N_2O from field soil leads to a loss of soil nitrogen, reducing the nitrogen-use efficiency (NUE). Deep placement of larger size urea (super granule/urea briquette) about 7–10 cm below the soil surface, either by hand, dibbler, or briquette applicator, not only enhances its use efficiency, but also delivers an environmental benefit by reducing runoff, N_2O emission, and volatilization losses of nitrogen. Improvement in NUE is also possible through the use of water-soluble fertilizers, crop- and stage-specific fertilizers, or very recently the nano-urea sprays that are able to cut the fertilizer need, thereby reducing costs, besides mitigating N_2O emissions. Farm machinery plays an important role in more precise application of nutrients at variable rates in intercrop as well as monocrop systems.

18.5.3.4 Plant Protection

Various plant protection equipment exist for different purposes such as application of insecticide, fungicide, and herbicide, preharvest sprays to defoliate or condition the crops for mechanical harvesting, hormones to increase fruit set or prevent early dropping of fruits, plant nutrients, and biological materials such as viruses and bacteria.

The success of pest control by pesticide use depends on quality of pesticides, timing of application, quality of application and coverage for proper dose to be applied, toxicant reaching the target with proper droplet size and proper density. Pesticide-application equipment is used differently, according to the energy required to operate the sprayer, the volume of spray handled, motive power used, and droplet spectra produced.

18.5.4 Harvesting

Harvesting cereal crops, especially wheat and rice, incurs a tremendous crop loss when untimely rain is experienced. Delayed harvesting causes grain shattering due to over maturity. During past decades,

TABLE 18.3

Operational Charge and Cost of Different Machines for Harvesting and Threshing

Operation	Machine	Cost of Machine (USD)	Service Charge (USD)
Harvesting	Reaper	2000–2050	8–10/hour
	Reaper-cum-binder	5300–7000	30–32.5/ha
	Combine harvester	9000–34000	50–70/ha
Threshing	Axial flow thresher	2150–3350	16–17/hour
	Open drum thresher	270–800	4–5/hour

the mechanization in rice harvesting has been rapidly evolved. The use of reapers or combine harvesters for mechanical harvesting (Table 18.3) is the best option for the farmers of northeast to improve the work and time efficiency but is not so common in use due to their limited availability and high costs.

18.5.4.1 Reaper

Reapers are used for harvesting crops mostly at ground level. The lifting and gathering unit of a reaper consists of row crop dividers, lugged conveyor belts, and star wheels. The row crop dividers are mounted on the reaper frame. They penetrate the standing crop, divide the crop into rows, and lift the lodged crop. A star wheel having curved teeth is mounted on the frame of each row crop divider with a ball bearing. The star wheels are driven by lugs attached to the upper lugged conveyor belt of a reaper. The velocity of the star wheel should be kept at 1.1–1.2 m/s for the better gathering of the crop toward the cutter bar. The ratio of cutter bar speed to forward speed of a reaper machine should ideally be between 1.3 and 1.4. A 60- to 70-cm high vertical platform is used for supporting the cut crop while it is being conveyed vertically to the right-hand side of the reaper by the lugged conveyor belts. In the end, the cut crop is laid in the form of a windrow. This facilitates manual collection and tying of crops in bundles. About 5–15% loss may occur, if proper care is not taken during harvesting, specifically during manual harvesting (Ojha and Nath, 1980). Different types of harvesting machines and techniques can save up to 85% labor and reduce 75% of harvest losses as well as the cost. In paddy, the harvesting cost is reduced by 55% (Khandai et al., 2021). Commercially available reapers in India include reaper attachment with power tiller, self-propelled vertical conveyor reaper, and tractor-front mounted vertical conveyor reaper.

18.5.4.2 Reaper-cum-Binder

It is suitable for harvesting and making bundles of wheat, paddy, and other crops like oilseeds and pulses. It is operated by 9-kW water-cooled diesel engine with riding-type, self-propelled vertical conveyor reaper windrower, single-cylinder, having a rated engine speed of 3000 rpm. The machine is provided with four pneumatic wheels – two driving wheels in the front with agricultural tread type tires and two rear steering wheels having automotive tires with clutch, brakes, steering, hydraulics, and power transmission, besides an operator seat. The system also comprises crop row dividers, star wheels, cutter bar with 76.2 mm pitch of knife section, vertical conveyor belts, and wire springs. The effective range of cutter bar is 1.2-m wide. While operating, firstly the crop row dividers make entry into the standing crop in field, and then the star wheels help guide the plants toward the cutter bar and assist in slightly lifting the cut crop, and turning it vertically, before it is conveyed by the lugged conveyor belt. The two flat belts having lugs send the cut crop toward the center of the machine and transfer back on a platform for making bundles weighing ~5 kg each, and finally, the crop is grounded in the rear side. The reaper-cum-binder works with an average capacity of 0.3–0.4 ha/hour. The machine weighs about 450 kg.

18.5.4.3 Combine Harvester

Combine harvester is mainly used for harvesting by integrating the function of cutting, threshing, and seed cleaning. Combine paddy harvester is especially suitable for both dry and wet paddy fields in plains, mountainous, and hilly areas. It harvests grain crops *viz.* cereals, oilseeds, and pulses, combining three separate operations like reaping, threshing, and winnowing into a simultaneous process. The straw is left behind on the field and can be used as mulch or bailed for feed and bedding for livestock. Among the rice harvesting machineries, rice combine harvester attains the leading position in the international arena as it satisfies various requirements of different users in terms of its wider adaptability, top security, high performance, and better efficiency. It is the most economically important laborsaving invention that is cheap, easy to operate, and time-saving. Different versions of combines are tractor-mounted, track type, and mini combine harvester with track.

18.5.5 Postharvest Management

18.5.5.1 Threshing

This operation provides employment opportunity for machine owner and field workers. Though the threshers are slowly giving way to combine harvesters, they still have a considerable standing in the postharvest procedure with their convertible nature. Guarded with a spinning thresh-drum with beaters and a fixed counter thresher, threshers are very efficacious and successful in bagging the basic harvest in a short span of time. Threshers (self-propelling or tractor-driven) usually straddle on rubber tire wheels for easy to-and-fro movement.

18.5.5.1.1 Open Drum Thresher

Open drum thresher (pedal operated) consists of mainly a well-balanced cylinder having a series of wire loops fixed on wooden slats in zigzag manner to transmit power with a gear drive mechanism (Table 18.3). After the cylinder attains a high-speed rotary-motion, the appropriate-sized paddy bundles are supplied to the teeth. Grains are generally separated out with the combining and hammering actions of threshing teeth. Threshing is the outcome of impact and rubbing action between threshing drawn loops and concave screen. The fan cleans the grains by blowing action when grains go down through the outlet at the bottom of the thresher. Open drum threshers are available in improved versions, with the electric motors or diesel engines having an additional winnowing fan. This type of machine is a women-friendly technology which can solve the problems of labor, time, and cost. The use of this machine for custom hiring can be an alternate source of earning for increasing their family income. This open drum thresher is an improved version, designed to have better safety features, with a threshing efficiency of about 5–6 q/hour (Khandai, 2018).

18.5.5.1.2 Axial Flow Thresher

It is a machine operated by prime movers (like tractor PTO/diesel engine/power tiller) to achieve a high level of grain separation through impact force with minimal damage to the kernels and reduced losses (Table 18.3). In threshing, the axial movement of grains and impact of rotating pegs mounted on the cylinder, over to the ear heads, forces the grains out from the sheath holding it. The axial flow of paddy crop is facilitated with the use of louvers provided on upper concave. Then the threshing unit throws out the straw by the paddles. Sieves, blower, or aspirators help in cleaning and separation of grains.

The paddy straws move axially several times in the axial flow thresher nearer to the threshing cylinder drum and louvers that separate out the grain cleanly from the stalk. The axial movement along with proper clearance between cylinder and concave provides a long threshing exposure with an adjustable speed. Grains are separated out from the straw and the chaff, and are further cleaned by two oscillating sieves. Moreover, axial flow thresher helps in vacating the field, facilitating sowing of the next crop (wheat) early by 8–10 days, leading to higher yield.

18.5.5.2 Drying

Harvesting of rice is normally done at the grain-moisture content (MC) between 24 and 26% (wet basis). Delayed, incomplete, or uneven drying leads to losses in quantity as well as quality of rice such as yellowing or discoloration caused by growth of mold and heat from respiration. Postharvest losses from harvesting to milling can reach about 14–40% as compared to 6–8% under the best management practice (Hellin et al., 2020). It is advocated that they should be stored at 13% moisture for grain and 14–16% for milling. This calls for immediate and appropriate drying grains of freshly harvested crop for maintaining high quality. Rice grain, called rough rice after drying, is amenable to processing. If dried properly, storage life of grains is extended, quality loss is prevented, and biological respiration is reduced that ensure optimum milling recovery. Sun-drying in open ground is traditionally followed for drying agricultural products due to some advantages, including ease of operation at no or low cost. However, sun-drying is prone to some problems such as extended drying period, contamination due to dust and microbes, and variation in the quality of the dried lot of grains. To circumvent these issues, a number of equipment for community drying have been developed, such as, recirculating batch dryer, solar bubble dryer, flatbed dryer, reversible flatbed drier, and LSU (Louisiana State University) dryer. Nowadays, drying techniques to dehydrate various farm and food products, several artificial methods such as convective hot air drying, microwave drying, vacuum drying, and infrared drying are in vogue.

18.5.5.3 Storage

Unlike traditional types of bulk storage systems, hermetic storage systems facilitate an effective control over grain MC as well as insect activity for grain storage in tropical regions. An airtight barrier is placed between the grain and the outside atmosphere. This type of storage system enhances grain quality, improves seed viability, reduces the attack of stored grain insects, and prevents entry of any rodents and birds to enter inside. The IRRI Super bags and cocoon are based on the principle of hermetic storage which can easily be used for safe storage of seeds or food grains in an extended period of time. Controlled grain-moisture conditions inside the bag help prevent the wetting and drying of grains. Discoloration of seed stored in super bag is also lesser as compared to that stored in jute bag (Kumar et al., 2020).

18.5.6 *In Situ* Residue Management

The sustainability of rice-wheat cropping system has been questioned. Residue burning has become an important issue, since it not only depletes essential nutrients due to burning, but also leads to large scale emission of harmful gases and pollutes the environment. Farmers prefer to burn residues as a cheap and timely alternative to vacate their field because of little window between harvesting of paddy and sowing of next season crop (wheat). Moreover, the paddy residues don't have an assured market due to high silica content, and as such are considered a waste, particularly by farmers of NW India, even though they realize its benefit as a source of carbon and other important nutrients. *In situ* management of crop residues is important not only for protecting the environment from air pollution, but also for the effective utilization of crop residues. Different machineries for management of crop residues (paddy straw) and seeding of wheat are available.

18.5.6.1 Straw Chopper

Paddy straw chopper is a straw chopping/shredding machine developed to chop and mix paddy stalks left behind after harvesting paddy by combine harvesters. This machine is extremely useful where SMS attachment in combine harvester machine is absent. With the use of a paddy straw chopper, paddy stalks are shredded, converted into small pieces, spread over the field, and can be easily mixed into the soil, which enriches the soil as organic fertilizer. Direct drilling of wheat seeds is also easier in chopped straw compared to lose straw left in the field. The height of the cut stubbles from the ground varies from 40 to 100 mm, whereas the length of the chopped straw varies from 10 to 180 mm.

18.5.6.2 Straw Reaper

Combine harvester leaves the wheat stalk in the form of loose and anchored stubbles that can be further chopped into fine pieces commonly known as *bhusa* through straw reaper. The recovered wheat straw after combine operation can be collected and used as cattle feed. The capacity of machine on an average is 0.4 ha/hour, and straw recovery is about 55–60%. The quality of straw (*bhusa*) is comparable with that made available for drum-type thresher.

18.5.6.3 Hay Rake

This machine is used for collection and cutting the hay or straw into windrows for collection after harvesting like a baler. It is also designed to fluff up the hay and turn it over so that it may get dry. It reduces the work of baler.

18.5.6.4 Baler

This machine is used to compress crop residues into compact bales for easy handling, transportation, and storage. The balers are of two types according to their baling shape, i.e. rectangular or cylindrical, of various sizes, bound with twine, strapping, and netting or wire. Crop residues are turned into bales which are further utilized for feeding of animal as well as biofuels. It creates an alternative business for farmers to sell bales to power plants and saves the environment from air pollution.

18.6 Conclusions

Farm mechanization optimally utilizes different resources such as labor, land, and water, saves time, and reduces the drudgery. The efficient and judicious utilization of time, labor, and resources helps sustainably intensify the RBCS through timely establishment of crops, resulting in an increased system productivity and farm income. The dominance of small and marginal holdings in India calls for the development and designing of scale-neutral machinery suited to different geographical terrains. Different machineries and their related services must be easily accessible to farmers on affordable rentals, in preference to owning them by the individuals at higher costs, and this can be achieved by promoting 'Custom Hiring Centers' (CHCs) at the village/block level with low duty machineries and 'Agriculture Machinery Banks' (AMBs) at the district/subdivision level with heavy-duty machinery (like combine harvester and LLL). To serve wider geographies, regional/state-level service centers may be promoted in the private sector to cater to a more specialized package of services. Availability of repair, service, and maintenance facilities must be ensured in the close proximity so as to address the operation and maintenance issues readily and promptly. Mobile service centers need to be promoted to cater to the minor repair demands and create enterprises. Creating awareness, skill development, and capacity building, need-based machinery refinement and policy support and reforms will help achieve mechanization at scale in RBCS.

REFERENCES

AMMA, 2014. Country presentation paper, Agricultural Machinery Manufacturers Association (AMMA) India, 2014. In: *Mechanization for Sustainable Agricultural Intensification in SAARC Region* (Gurung, T. R., Kabir, W. and Bokhtiar, S. M., eds.). SAARC Agriculture Centre, Dhaka, Bangladesh. p. 302.

Bhowmick, M. K., Duary, B., Bhattacharyya, P. and Dhara, M. C., 2016. Drum seeding technology: A viable option for resource-poor farmers toward profitable rice production in West Bengal. In: *Family Farming: Challenges and Opportunities* (Mondal, B., Sarkar, D., Mukhopadhyay, S. D., et al., eds.). Renu Publishers, Mohan Garden, New Delhi. pp. 125–130.

CIMMYT, 2019. CIMMYT News Release, August 08, 2019; https://www.cimmyt.org/news/happy-seeder-can-reduce-air-pollution-and-greenhouse-gas-emissions-while-making-profits-for-farmers/.

FAO, 2013. Cash Transfers: Promoting Sustainable Livelihoods in Sub Saharan Africa. Available online: http://www.fao.org/resources/infographics/infographics-details/en/c/178851.

FAO, 2014. *The State of Food and Agriculture. Innovation in Family Farming.* Food and Agriculture Organization of the United Nations, Rome. p. 139.

Hellin, J., Balie, J., Fisher, E., et al., 2020. Trans-disciplinary responses to climate change: Lessons from rice-based systems in Asia. *Climate* **8**(2): 35. ISSN 2225-1154, doi: https://doi.org/10.3390/cli8020035. Available at http://centaur.reading.ac.uk/89241/.

Khandai, S., 2018. Piloting of improved open drum threshers in Bihar. *International Journal of Current Microbiology and Applied Sciences* **7**(11): 1053–1056.

Khandai, S., Gupta, N., Baishya, S., Singh, K. and Kumar, V., 2021. Opportunity of mechanized harvesting methods of cereal Crops in India: A review. *International Journal of Current Microbiology and Applied Sciences* **10**(01): 3137–3145, doi: https://doi.org/10.20546/ijcmas.2021.1001.365.

Kumar, A., Khandai, S., Singh, M., et al., 2020. Hermetic sealed storage IRRI super bag reduces post-harvest losses in seed grains. *International Journal of Chemical Studies* **8**(4): 403–406.

Ojha, T. P. and Nath, S., 1980. Studies on harvesting of crop minimizing harvest & post-harvest losses. *Bulletin on Harvest and Post-harvest Technology (ICAR) Scheme*. IIT, Kharagpur. pp. 4–9.

Samal, S. K., Mishra, J. N., Pradhan, P. R., et al., 2020. Comparison of field performance of different paddy transplanter available in Odisha, India. *International Journal of Current Microbiology and Applied Sciences* **9**(3): 992–1000.

Shivananda, A. K., Yarazari, P., Devegowda, S. R. and Pavan, M. K., 2019. Custom hiring services of farm machinery in India. *Agrobios Newsletter* **XVIII** (4): 123–124.

Singh, S. 2008. Farm mechanization scenario in India. First machinery manufactures' meet, Karnataka. pp. 25–28.

Singh, R. S. and Sahni, R. K., 2019. Transformation of Indian agriculture through mechanization. *Economic Affairs* **64**(2): 297–303, doi: 10.30954/0424-2513.2.2019.4

Singh, S. and Vatsa, D. K., 2006. Performance evaluation of PAU manual paddy transplanter in hills of Himachal Pradesh. *Agricultural Engineering Today* **30**(3): 19–25.

Singh, K., Bhadraray, S. and Singh, U., 2017. Customised fertilisers: Methodology, guidelines and processes for manufacturing. *Training Manual on Enhancing Nutrient Use Efficiency Through Next Generation Fertilizers in Field Crops*. ICAR-Indian Institute of Pulses Research, Kanpur, Uttar Pradesh. pp. 17–25.

Singh, K., Kumar, V., Saharawat, Y. S., et al., 2013. Weedy rice: An emerging threat for direct-seeded rice production systems in India. *Journal of Rice Research* **1**: 106, doi: 10.4172/jrr.1000106.

Singh, M., Kumar, P., Kumar, Virender, et al., 2020a. Inter comparison of crop establishment methods for improving yield and profitability in the rice-wheat system of Eastern India. *Field Crops Research* **250**: 107776.

Singh, S., Kumar, V., Craufurd, P., et al., 2020b. Management of crop growth cycle as the frontier of yield growth in rice. In: *New Frontiers in Agricultural Extension*, Volume **II** (Singh, A. K., Craufurd, P., Singh, S., et al., eds.). International Maize and Wheat Improvement Center (CIMMYT, India). pp. 165–184.

19

Adoption Determinants of Improved Seeds and Technical Efficiency of Ugandan Small Farmers

S. Auci[1] and M. Coromaldi[2]
[1]Università degli Studi di Palermo, Palermo, Italy
[2]Università degli Studi Niccolò Cusano, Rome, Italy

CONTENTS

19.1 Introduction

The capacity of farmers to adopt smart agricultural practices, i.e. improved seeds, and enhance productivity becomes essential to reduce poverty and improve household food security in countries such as Uganda with a high percentage of the population dependent on agriculture (World Bank, 2008). The growing populations require increasing food needs and thus growth and development of the agricultural sector. This nourishment demand can be satisfied not only by expanding the area under cultivation but also by intensifying the agricultural production by adopting yield-enhancing technologies.

The dependence of sub-Saharan African countries' agricultural production on rainfall due to limited irrigation systems exposes farmers' crop income to high instability and, in some cases, with devastating socio-economic consequences (World Bank, 2010). Due to climate change, the number of African food crises per year has tripled in the last three decades (FAO, 2014). Production from rain-fed agriculture is reduced by up to 50% (IPCC, 2014). Climate change effects represent a "challenge" that sub-Saharan agriculture has to face in the immediate future, being subject to relevant risks generated by the new meteorological conditions. While such factors are likely to affect agricultural productivity, there is no doubt that the adoption of new technologies will play an important role in addressing climate change challenges.

Our study intends to merge two streams of literature; one related to the adoption of productivity-enhancing strategies such as hybrid seeds and the other one concerning farmers' technical efficiency. As regards the first branch, several studies consider different adaptation practices on crop productivity (Di Falco et al., 2011; Di Falco and Veronesi, 2013; Teklewold et al., 2019; Adego et al., 2019), while other studies focus only on the adoption of improved seeds. For Ethiopia, Cavatassi et al. (2011) found that the adoption of modern varieties of sorghum is a risk-reducing technology in prone drought areas.

DOI: 10.1201/9781003164968-22

There are few studies on the adoption of improved seeds in Uganda. Among these, Sserunkuuma (2005) pointed out that the adoption of improved maize varieties without fertilizer use and soil conservation practices accelerates land degradation, mining soil nutrients, and reducing yields.

The literature on farmers' technical efficiency in Asia and Latin America revealed that the human capital, access to formal credit, and extension services contribute positively to production efficiency (Battese and Coelli, 1995; Abdulai and Eberlin, 2001; Solís et al., 2009). Concerning Africa, Binam et al. (2004) showed that in Cameroon, technical efficiency increases when farmers are educated and located in fertile regions, have better access to credit, and are members of farmers' clubs or associations. On the contrary, the distance from the main access road and extension services affects negatively farmers' efficiency. In Swaziland, technical efficiency was found to be positively correlated with farmers' age, having an off-farm income, farmers' experience, intercropping, and use of hybrid seeds (Dlamini et al., 2012). Based on the Ghanaian maize farms, Abdallah and Abdul-Rahman (2017) identified policy variables such as credit access; education and extension access, and farm size as main determinants of technical efficiency. Moreover, Mwalupaso et al. (2019) found a positive relationship between agricultural informatization, i.e. the use of mobile phones, and technical efficiency in Zambia.

Finally, among the studies that looked at the adoption of modern varieties and agricultural productivity, Oyekale and Idjesa (2009) showed that in Nigeria, the adoption of improved maize seeds significantly increases production efficiency in a cross-sectional setting.

Our contribution to the existing literature on the adoption of modern varieties and technical efficiency is threefold. First, employing a nationally representative database, we estimate the determinants of improved seed adoption and technical efficiency by taking into account farmers' unobserved heterogeneity in a panel framework. Second, we provide information on farmers' adoption decisions related to hybrid seeds and what factors may affect farmers' efficiency by implementing a random-effects probit model and a stochastic frontier approach. Finally, we add empirical evidence to some country-specific studies on sub-Saharan countries. Therefore, we choose Uganda as a case study among African countries not deeply investigated. Besides, the Ugandan government has recently promoted several programs and organizations to incentivize the use of improved seeds and fertilizer. However, because of the limited widespread of these innovative technologies, farmers' agricultural productivity remains low.

19.2 Background: The Study Area

Specific to Uganda, more than 70% of people are employed in agriculture, mainly on a subsistence basis (UBOS (Ugandan Bureau of Statistics), 2015). The agricultural sector is dominated by smallholder farmers engaged in the production of food and cash crops, horticulture, fishing, and livestock farming, and contributing around 80% of the total agricultural output (NPA (National Planning Authority), 2015). Crop farming is rain-fed, with limited irrigation. The low crop productivity is tied to climate-related impacts (droughts, floods, rainfall variability), poor quality agro-inputs, diminishing soil fertility, poor land management and agronomic practices, disease, and pests, coupled with high harvest and postharvest losses (Uganda SPCR, 2017).

Food availability is a limiting factor, especially in regions such as Karamoja, East Central, and West Nile where frequent dry spells and lack of extension services affect production and productivity. Food insecurity represents a challenge for about 12% of the total population in the country (FAO, 2015). Uganda is among African countries greatly affected by current climate change and variability. According to the University of Notre Dame Global Adaptation Index, Uganda is ranked as the 27th most vulnerable and 25th least-ready country to address climate change. In the last 50 years, the average temperature has increased by 0.28°C per decade, and January and February experiencing a warming trend with an average temperature increase of 0.37°C per decade. The frequency of hot days has increased, and rainfall has decreased, becoming unevenly distributed and less reliable (Uganda SPCR, 2017). Several climate projections for Uganda foresee an increase in temperatures, frequency and severity of floods and droughts, and finally variability of rainfall. Specifically, the Uganda Climate Profile projects that the mean annual temperature across Uganda will increase between 1.2 and 2.3°C by 2050, and between 1.7 and 5.6°C by 2100. It also projects that the Heat Wave Duration Index will increase between 0.1 and 3.3 days per year by 2050, and between 0.8 and 71.1 days per year by 2100. Furthermore, most years will experience at least 30 consecutive dry days (CDD) with some years experiencing up to 60 CDD by 2050 and 80 on occasion by 2100 (USAID, 2013).

19.3 Data and Methodology Data Description

The data used for this study were collected by the Uganda Bureau of Statistics with financial and technical support from the Government of the Netherlands, and the World Bank Living Standards Measurement Study – Integrated Surveys on Agriculture (LSMS-ISA) project. Uganda National Panel Survey spanning from 2009 to 2014 is a nationally representative household survey consisting of an unbalanced panel composed of 7,445 households. Data include detailed information on household demographic characteristics such as education, household size, sex, and age of the household head and information on household shocks and assets. These data at the farmer and enumeration area level are combined with climatic variables, both temperature and rainfalls, collected by the Global Land Data Assimilation System (GLDAS) v2.1. GLDAS is a global gridded reanalysis dataset (Rodell et al., 2004) with a spatial resolution of 0.25°*0.25° and three-hourly temporal resolution. We try to capture droughts experienced by the farm household, including the CDD index that measures the maximum number of CDD per time period with a daily precipitation amount of less than 1 mm. As the length of CDDs increases, it may indicate enhanced dryness and a high risk for seasonal droughts. This trend may increase evaporation and stress water resources. Specifically, rain-fed agriculture is affected by the maximum number of CDD as it directly impacts soil moisture and crop growth. Generally, this indicator is projected to increase for the higher emissions scenarios (World Bank, 2021). As regards the dependent variables, we focus on the adoption of improved seeds and gross crop income. The first variable refers to the use of improved seeds in crop production by Ugandan farmers. Due to market constraints, the adoption of improved seeds as well as the use of chemical fertilizers is still low. Only 24 and 5% of the households adopt hybrid seeds and chemical fertilizers, respectively. The gross crop income is calculated as the sum of the annual value of consumed, stored, and sold Ugandan shillings (Table 19.1).

19.4 Methods of the Analysis

To analyze what socio-economic factors may affect farmers' adoption decision of using improved seeds and farmers' technical efficiency, which in turn has a direct effect on farmers' gross crop income, two different models based on unbalanced panel data are applied: the random-effects probit model and the panel stochastic frontier approach.

19.4.1 The Random-Effects Probit Model

The first model applied is a panel data binary choice model to estimate the probability of using improved seeds by a farmer. To account for the dichotomous nature of our dependent variable, we apply a random-effects probit model. The farmer's observable decision on the adoption of improved seeds A_{it} is represented by a dummy variable with values being 1 if adopted and 0 otherwise. The likelihood of adopting improved seeds may be estimated by the following model:

$$Pr(A_{it} = 1 | Z_{it}, c_i) = \phi(Z_{it}, c_i) = \phi - \beta_{\#} + \sum^{1\&} \beta_{\$} Z_{\$_t} + c_i 2 \quad (19.1)$$

where $\phi(.)$ is the normal distribution function, Z_{it} is the vector of independent variables that include farmer head and farm household characteristics, information sources and social capital, and the climate variable. c_i represents the heterogeneous unobserved effects based on the exogeneity assumption between observed covariates and unobserved heterogeneity (Wooldridge, 2010). The betas are the parameters to

TABLE 19.1

Summary Statistics

Activities/Components	Mean ± SD	Min	Max
Dependent variables			
Improved seeds (yes = 1)	0.24 ± 0.03	0	1
Gross crop income (log)	14.01 ± 0.40	13.39	15.32
Explanatory variables			
Farmer head and farm household characteristics			
Household head age	47.80 ± 15.06	14	102
Female household head (yes = 1)	0.30 ± 0.46	0	1
Married household head (yes = 1)	0.75 ± 0.43	0	1
Household size	6.47 ± 3.14	1	29
Household head highest educational level	7.53 ± 4.10	0	17
Intercropping (yes = 1)	0.83 ± 0.38	0	1
Dummy maize (yes = 1)	0.70 ± 0.46	0	1
Information sources and social capital			
Extension services (yes = 1)	0.25 ± 0.43	0	1
Household owns cellphone	0.54 ± 0.50	0	1
HH distance in km to nearest population center with +20,000	24.04 ± 17.92	0.21	102.03
HH distance in km to nearest market	32.75 ± 19.28	0.50	116.23
Climatic variable			
Consecutive dry days index	14.11 ± 8.04	0	37.47
Inputs of the production frontier			
Quantity of chemical fertilizer (log)	1.58 ± 0.44	1.50	8.96
Quantity of pesticide (log)	0.36 ± 0.53	0.21	7
Land cultivated (log)	1.08 ± 0.49	0.61	5.81
Family labor (log)	5.93 ± 0.37	5.31	8.19
Seed value (log)	10.85 ± 0.55	10.38	14.73
Slope (%)	7.28 ± 7.15	0.82	49.47
Low nutrient retention capacity (yes = 1)	0.01 ± 0.12	0	1
Observations	7,491		

be estimated. To successfully interpret probit coefficients, the marginal effects of Z_{it} on $Pr(A_{it})$ should be computed. The marginal effects show the change in the probability of A, assuming the value of 1 in comparison to the base category, given a unit change in one of the independent variables Z_{it}.

19.4.2 Panel Stochastic Frontier Approach

Applying the stochastic production frontier technique allows us to investigate the effects of improved seed determinants on technical efficiency and indirectly on farmers' gross crop income. Even if these factors may not be considered inputs to the production process, however, they exert a direct influence on the producer's efficiency and an indirect effect on the producer's performance. A farmer can be far from the optimal frontier due to randomness. Technical efficiency allows us to measure how close a farmer's production is to the farmer's optimal production based on the same bundle of inputs. In other words, a farmer could be considered technically efficient when it reaches the maximum level of output using the best combination of inputs within the existing technology. The estimated production frontier coincides with the optimal output level and represents the benchmark in respect to which the farmer-level efficiency is measured.

The advantage of using stochastic frontier analysis (SFA) to estimate productivity is threefold. First, it allows us to include several input dimensions in evaluating a synthetic measure of farmers' productivity.

The output is compared to all inputs like labor, physical, social, and human capital. Second, this methodology provides a relative productivity measure. Each farmer can be compared to the best practice in crop production. Finally, SFA allows disentangling the distance from the production frontier into an inefficiency term and a random error (Kumbhakar and Lovell, 2000). Following the Battese and Coelli (1995) model, we may estimate a stochastic production frontier for panel data in which the error term is composed of a systematic component expressed as an explicit function of a vector of exogenous variables (i.e. technical efficiency) and a random component. The technical inefficiency function is represented as a truncated normal random variable with the mean depending on exogenous variables. This approach allows us to estimate simultaneously the production input coefficients and inefficiency term parameters using a two-stage maximum-likelihood procedure. The production function model can be expressed as a Cobb-Douglas function:

$$Y_{it} = X_{it}\beta + (v_{it} - u_{it}) \tag{19.2}$$

where X_{it} represents the inputs of the production function of a Ugandan farmer such as labor, land, fertilizer, and pesticides, and the unobserved random noise ($v_{it} - u_{it}$) is divided into the two components. The first component, v_{it}, measures the standard statistical noise, such as errors in the measurement of variables, and follows the assumption of normally distributed error terms [independent and identically distributed (iid) (0, $\sigma^{\&}$)]. The second independent component, defined as u_{it}, represents the degree of inefficiency and measures the distance of farmers' production from the maximum output. It is distributed as a truncated normal random variable, (m_{it}, $\sigma^{\&}$).

The mean of this truncated normal distribution is a function of systematic variables that can influence the farmers' efficiency:

$$m_{it} = Z_{it}\gamma \tag{19.3}$$

where Z_{it} is a matrix of variables that include some of the determinants of adopting improved seeds and γ is the vector of parameters to be estimated.

19.5 Results and Discussion

To assess the household socio-economic and climatic factors that affect farmer's adoption decision of improved seeds, as well as to evaluate the impact of such determinants on farmers' efficiency, we first report the estimated coefficients of the random-effects probit model followed by the estimated coefficients of the panel stochastic frontier.

19.5.1 Factors Influencing Adoption of Improved Seeds

Table 19.2 shows the estimated parameters and the marginal effects for the adoption of improved seeds in Uganda. The Wald test confirms that the estimated model fits the data very well as shown in the last panel of the table. The probability of adopting improved seeds decreases by 1% when the household head is female and by around 3% as the distance to the nearest and highly populated city increases. This implies that female-headed households have to bear more difficulties in finding and adopting hybrid seeds compared to the male ones. The same occurs when farmers are located in isolated areas (Kassie et al., 2011). Policy interventions that make improved seeds more available to female farmers and facilitate access to populated cities for isolated farmers could promote adoption. On the other hand, farmers who have a higher educational level, access to extension services and live far from the market are more likely to adopt improved seeds (Oyekale and Idjesa, 2009; Di Falco et al., 2011). This is given to the fact that educated farmers have the adequate skills to decode new information and adopt new technologies. Moreover, as pointed out by Di Falco et al. (2011), our results confirm the crucial role of information provided by extension services in determining farmers' decision to adopt knowledge-intensive practices. Although we expected a negative relationship between the adoption probability and the distance from

TABLE 19.2

Estimation Results – Random Effects Probit Model

Variables	Coefficients	Marginal Effects
Household head age	–0.002	–0.023
Female household head (yes = 1)	(0.002)	(0.019)
	–0.212***	–0.015***
	(0.067)	(0.005)
Married household head (yes = 1)	0.061	0.011
	(0.073)	(0.013)
Household size	–0.002	–0.004
Household head highest educational level	(0.008)	(0.013)
	0.024*** (0.006)	0.043*** (0.011)
Intercropping (yes = 1)	–0.057	–0.011
Dummy maize (yes = 1) Extension services (yes = 1)	(0.057)	(0.011)
	0.384*** (0.052)	0.066*** (0.009)
	0.409*** (0.044)	0.025*** (0.003)
Household owns cellphone	0.065	0.008
HH distance in km to nearest population center with +20,000	(0.049)	(0.006)
	–0.005***	–0.028***
HH distance in km to nearest market consecutive dry days index	(0.001)	(0.008)
Constant	0.004*** (0.001)	0.035*** (0.010)
	0.000*** (0.000)	0.064*** (0.010)
	–1.638*** (0.148)	
Observations	7,599	
Log pseudolikelihood	–3,815.7258	
Wald chi^2 (12)	285.58	
Prob > chi^2	0.000	

Note: Margins calculated at means. Robust standard errors are shown in parentheses.
***$p < 0.01$, **$p < 0.05$, *$p < 0.1$.

the market, we find a positive and significant coefficient as in Adego et al. (2019). Since the hybrid seeds of maize are more available compared to other crop types, we control for maize production by introducing a dummy variable. Maize cultivation increases the probability of adopting innovative seeds by more than 6%. As expected, the parameter of climatic variable reveals that the probability of adoption significantly increases as the maximum number of CDD rises. This implies that farmers react to weather shocks by using modern varieties as adaptation strategies (Cavatassi et al., 2011; Di Falco and Veronesi 2013; Teklewold et al., 2019).

19.5.2 Production and Inefficiency Analysis

The production function and the inefficiency model coefficients are reported in Table 19.3. The estimated coefficients of the production frontier are reported in the upper panel of the table, while the lower panel presents the inefficiency coefficients. Four models are estimated by adding in the inefficiency model, three distinct variables such as the intercropping variable (Model 2), the distance to the nearest populated center (Model 3), and finally the CDD index (Model 4). The overall fitness of the specifications is guaranteed by the Wald test that is statistically significant at the 1% level. The significant effects of inefficiency in crop production are confirmed by the values of lambda that are statistically and significantly greater than zero.

As expected, in the estimations of the Cobb-Douglas production function, the input coefficients are stable across models with significant and positive signs (Battese and Coelli, 1995; Oyekale and Idjesa, 2009; Mwalupaso et al., 2019). Even the negative sign of low nutrient retention capacity of the soil is an

TABLE 19.3

Estimation Results – Stochastic Frontier – Battese and Coelli (1995) Model

Production Function	Model 1	Model 2	Model 3	Model 4
Quantity of chemical fertilizer (log)	0.013	0.010	0.010	0.009
	(0.013)	(0.012)	(0.012)	(0.012)
Quantity of pesticide (log)	0.063*** (0.010)	0.062*** (0.010)	0.063*** (0.010)	0.057*** (0.009)
Land cultivated (log)	0.075*** (0.011)	0.077*** (0.011)	0.076*** (0.011)	0.100*** (0.011)
Family labor (log)	0.378*** (0.013)	0.347*** (0.013)	0.347*** (0.013)	0.386*** (0.012)
Seed value (log)	0.028*** (0.009)	0.029*** (0.009)	0.028*** (0.009)	0.033*** (0.009)
Slope (%)	0.012*** (0.001)	0.011*** (0.001)	0.012*** (0.001)	0.002** (0.001)
Low nutrient retention	–0.079** (0.038)	–0.089** (0.040)	–0.084** (0.040)	–0.099*** (0.038)
capacity (yes = 1)	0.064*** (0.010)	0.066*** (0.010)	0.065*** (0.010)	0.074*** (0.010)
Extension services (yes = 1)	11.350*** (0.118)	11.537*** (0.117)	11.554*** (0.115)	11.450*** (0.122)
Constant				
Inefficiency Model				
Female household head (yes = 1)	0.041*** (0.015)	0.049*** (0.011)	0.049*** (0.011)	0.056*** (0.011)
Improved seeds (yes = 1)	0.023*	0.015	0.012	0.008
	(0.013)	(0.011)	(0.009)	(0.011)
Household head highest educational level	–0.024*** (0.004)	–0.020*** (0.001)	–0.021*** (0.001)	–0.015*** (0.001)
Intercropping (yes = 1)		–0.157***	–0.157***	–0.120***
		(0.010)	(0.010)	(0.011)
HH distance in km to nearest population center with +20,000			–0.001*	–0.001**
Consecutive dry days index			(0.000)	(0.000)
(log)	0.258***	0.384***		0.449*** (0.032)
Constant			0.400***	–3.074***
	(0.020)	(0.023)	(0.021)	(0.290)
Observations	7,491	7,491	7,491	7,491
Number of households	2,386	2,386	2,386	2,386
Wald chi^2 (8)	1,642***	1,482***	1,482***	1,585***
Sigma_u	0.0345***	5.30e–05	0.295e–03	0.0231*
Sigma_v	0.327***	0.324***	0.324***	0.307***
Lambda	0.105***	0.0001	0.00091	0.075***

Note: Robust standard errors are shown in parentheses.
***$p < 0.01$, **$p < 0.05$, *$p < 0.1$.

expected result. This index represents the low soil fertility and should be *de facto* interpreted as a production increasing factor because when the dummy is equal to zero, the plot has a good nutrient retention capacity.

Since the quantity of pesticide, family labor, land size, the value of total seeds sowed, soil characteristics, and the access to extension services are introduced as agricultural inputs, the estimations show that all the inputs are production enhancing for Ugandan farmers. Specifically, the main driver for crop income production is family labor input that shows elasticity equal to 0.386 in Model 4. This means that gross crop income increases with higher physical and social capital as well as labor and land inputs, conforming to *a priori* expectations. In the second panel of Table 19.3, the inefficiency model is estimated. In this model, a negative sign of the parameters implies a detrimental effect on technical inefficiency and consequently a positive effect on technical efficiency. The inefficiency components show mixed signs and different significance levels in the four specifications. As in the production function model, the estimated coefficients are robust across the four models. The signs and the significance of the parameters are stable in all the specifications. In line with Solís et al. (2009) and Mwalupaso et al. (2019), the positive and significant coefficient of the female household head variable suggests that female-headed households exhibit lower technical efficiency than male-headed households. This could be explained by the fact that women are mainly involved in unmeasured noneconomic activities (such as child care, cooking, and cleaning) performed in the household. In addition, they have less access to credit because of cultural prejudice and are less likely to attend agricultural extension training seminars. A noticeable finding is the positive and not statistically significant coefficients (except for Model 1 at the level of 10%) of the improved seed variable. The explanation of this result can be traced back to the impoverishing of soil nutrients when farmers adopt high-yielding varieties without using external inputs (fertilizer use and soil conservation practices) to replenish the loss of land fertility. This implies a worsening and decline of production efficiency and thus crop income (Sserunkuuma, 2005). The human capital variable shows a negative and significant effect on farmers' technical inefficiency. Farmers with more years of schooling are more efficient than their illiterate counterparts. As found in Abdulai and Eberlin (2001), Binam et al. (2004), Solís et al. (2009), and Abdallah and Abdul-Rahman (2017), education facilitates learning, adoption of new technologies, access to information and thus increasing productivity. The use of the intercropping practice has an expected positive and significant effect on technical efficiency. Farmers, who break away from monocropping and use appropriate rotation techniques, reduce soil exhaustion and depletion, improving land fertility benefits (Dlamini et al., 2012).

As regards the distance to the nearest city, we find that proximity to a highly populated center reduces farmers' technical efficiency. Households located in remote areas are projected to be more efficient in crop production than those living close to big cities. It is assumed that the distance to the city center, on the one hand, incentivizes farmers to mainly rely on their self-production and, on the other hand, limits the off-farm employment opportunities. This implies that family labor is primarily allocated to farm activities with the consequence of increasing returns from farming.

Finally, we introduce in the inefficiency model an indicator of weather shock, measured by the maximum number of CDD in the current year. The positive and statistically significant sign of this variable means that climatic extreme events negatively affect technical efficiency. In fact, soil quality and water resources, as well as plant growth, can be stressed by the length of drought periods in rain-fed land. When cultivated areas do not have enough irrigation facilities and dryness becomes unfavorable, i.e. more droughts during the growing season, the crop production would sharply decline.

19.6 Conclusions

This chapter presented the main drivers that affect both improved seed adoption and farmers' technical efficiency. Using data from a panel nationally representative survey conducted in Uganda, we analyzed the determinants of hybrid seeds adopted by farmers and we estimated the parameters of the crop production function and the inefficiency model. Based on the results of this study, we found that being a female-headed household would not confer any benefits for adopting improved seeds and for production efficiency. Households living close to city centers are more likely to adopt modern varieties; however, they are less efficient in crop production. This is because on the one hand farmers, more easily, have access to the input market and on the other hand they can rely on alternative income sources. As the highest educational level of the household head increases, the higher knowledge allows farmers to both decode new information and adopt new technologies and being more efficient.

Long periods of drought reduce farmers' efficiency; however, smallholders react to adverse weather events by using drought-tolerant seeds more resistant to climatic distress. The findings also revealed that maize cultivations and extension services are key determinants of improved seed adoption. Finally, intercropping practice improves land fertility by reducing soil exhaustion and depletion and, as a consequence, efficiency. Thus, policy interventions to increase agricultural productivity and enhance food security should alleviate obstacles to farmers' improved seed adoption. Policymakers should facilitate the distribution of inputs and disseminate information on agricultural practices able to improve land fertility and crop production. In addition, the government should incentivize children to complete their normal education instead of starting too earlier farming activity. This is because educated farmers have a higher probability of using improved seeds leading to higher efficiency and productivity.

REFERENCES

Abdallah, A. H., and Abdul-Rahman, A. 2017. Technical efficiency of maize farmers in Ghana: a stochastic frontier approach. International Journal of Innovation and Scientific Research, 29(2), 110–118.

Abdulai, A., and Eberlin, R. 2001. Technical efficiency during economic reform in Nicaragua: evidence from farm household survey data. Economic Systems, 25(2), 113–125.

Adego, T., Simane, B., and Woldie, G.A. 2019. The impact of adaptation practices on crop productivity in northwest Ethiopia: an endogenous switching estimation. Development Studies Research, 6(1), 129–141.

Battese, G. E., and Coelli, T. J. 1995. A model for technical inefficiency effects in a stochastic frontier production function for panel data. Empirical Economics, 20(2), 325–332.

Binam, J. N., Tonye, J., Nyambi, G., and Akoa, M. 2004. Factors affecting the technical efficiency among smallholder farmers in the slash and burn agriculture zone of Cameroon. Food Policy, 29, 531–545.

Cavatassi, R., Lipper, L., and Narloch, U. 2011. Modern variety adoption and risk management in drought prone areas: insights from the sorghum farmers of eastern Ethiopia. Agricultural Economics, 42(3), 279–292.

Di Falco, S., and Veronesi, M. 2013. How can African agriculture adapt to climate change? A counterfactual analysis from Ethiopia. Land Economics, 89(4), 743–766.

Di Falco, S., Veronesi, M., and Yesuf, M. 2011. Does adaptation to climate change provide food security? A micro-perspective from Ethiopia. American Journal of Agricultural Economics, 93(3), 829–846.

Dlamini, S. I., Masuku, M. B., and Rugambisa, J.I. 2012. Technical efficiency of maize production in Swaziland: a stochastic frontier approach. African Journal of Agricultural Research, 7(42), 5628–5636.

FAO, 2014. Food and Agriculture Organization of the United Nations, Rome, Italy. FAOSTAT. http://faostat.fao.org/default.aspx

FAO 2015. Responding to the Challenges of Poverty, Food Insecurity and Climate Change: Country Programming Framework 2015–2019, Kampala, Uganda.

IPCC 2014. Climate Change 2014: synthesis report. contribution of working Groups I, II and III to the fifth assessment report of the Intergovernmental Panel on Climate Change. R. Pachauri and L. Meyer (editors), Geneva, Switzerland, IPCC, 151 p. ISBN: 978-92-9169-143-2.

Kumbhakar, S. C., and Lovell, C. A. K. 2000. Stochastic frontier analysis. Cambridge University Press.

Mwalupaso, G. E., Wang, S., Rahman, S., Alavo, E. J. P., and Tian, X. 2019. Agricultural informatization and technical efficiency in maize production in Zambia. Sustainability, 11(8), 1–17.

NPA (National Planning Authority) 2015. Second National Development Plan II (NDP II) (2015/16–2019/20), Kampala, Uganda, http://npa.go.ug/wp-content/uploads/NDPII-Final.pdf (accessed March 11, 2021).

Oyekale, A. S. 2012. Vulnerability of peasant cocoa farmers to climate change in South-West Nigeria. Journal of Human Ecology, 40(1), 33–41.

Oyekale, A. S., and Idjesa, E. 2009. Adoption of improved maize seeds and production efficiency in Rivers State, Nigeria. Academic Journal of Plant Sciences, 2(1), 44–50.

Rodell, M., Houser, P. R., Jambor, U. E. A., Gottschalck, J., Mitchell, K., Meng, C. J., Arsenault, K., Cosgrove, B., Radakovich, J., Bosilovich, M., Entin J. K., Walker J. P., Lohmann D., and Toll, D., 2004. The global land data assimilation system. Bulletin of the American Meteorological Society, 85(3), 381–394.

Solís, D., Bravo-Ureta, B. E., and Quiroga, R. E., 2009. Technical efficiency among peasant farmers participating in natural resource management programmes in Central America. Journal of Agricultural Economics, 60, 202–219.

Sserunkuuma, D. 2005. The adoption and impact of improved maize and land management technologies in Uganda. electronic Journal of Agricultural and Development Economics, 2(1), 67–84.

Teklewold H., Mekonnen A., and Kohlin G., 2019. Climate change adaptation: a study of multiple climate-smart practices in the Nile Basin of Ethiopia. Climate and Development, 11(2), 180–192.

UBOS (Ugandan Bureau of Statistics) 2015. 2015 Statistical Abstract, Kampala, Uganda, http://library.health.go.ug/publications/statistics/uganda-bureau-statistics-2015-statistical-abstract (accessed March 11, 2021).

Uganda SPCR 2017. Strategic Program for Climate Resilience: Uganda pilot Program for Climate Resilience (PPCR), Kampala, Uganda, http://ccd.go.ug/wp-content/uploads/2019/10/Uganda-Strategic-Programme-for-Climate-Resilience_May_2017.pdf (accessed March 11, 2021).

USAID 2013. Uganda Climate Change Vulnerability Assessment Report. African and Latin American Resilience to Climate Change Project.

Wooldridge, J. M. 2010. Econometric analysis of cross section and panel data. MIT Press.

World Bank 2008. World development report 2008: agriculture for development. Washington, DC: World Bank.
World Bank 2010. Economics of adaptation to climate change – synthesis report. Washington, DC: World Bank. http://documents.worldbank.org/curated/en/646291468171244256/Economics-of-adaptation-to-climate-change-Synthesis-report (accessed March 8, 2021).
World Bank 2021.Climate Change Knowledge Portal, https://climateknowledgeportal.worldbank.org/country/uganda/climate-sector-agriculture (accessed March 11, 2021).

20

Model for Quality Assessment in Small Farms of Cape Verde

Margarida Saraiva[1], Álvaro Rosa[2], Elsa Simões[3], and António Ramos Pires[4]
[1]*Management Department, School of Social Sciences, Universidade de Évora and Researcher at BRU-Business Research Unit/Iscte-Instituto Universitário de Lisboa, Portugal*
[2]*ISCTE – University Institute of Lisbon and BRU-Business Research Unit, Lisboa, Portugal*
[3]*School of Agricultural and Environmental Sciences at the University of Cape Verde, Praia, Cabo Verde*
[4]*UNIDEMI – New University of Lisbon and Polytechnic Institute of Setúbal, Setúbal, Portugal*

CONTENTS

20.1 Introduction

Agriculture in Cape Verde is essentially of family based and of small dimension and it has been, since the discovery of the archipelago, a factor of support for families and development, despite numerous factors that limit it. The most threatening ones are the lack of water resources, characterized by erratic and unpredictable patterns of precipitation and lack of arable soil, insularity, and the orography of the islands (MAAP, 2005). To mitigate and minimize these limitations, the country has been investing for more than 40 years in public policies aimed at soil conservation and increasing the availability of water for agriculture, through various water harvesting techniques, among which, the construction of dams, in order to conserve water after the rainy season, so that they allow the practice of agriculture throughout the year. The costs of building infrastructures (e.g. dam, reservoirs, water distribution systems, drip irrigation systems) are high and they need to be adequately monetized for the sake of productivity in addition to the increase the farmers' incomes and the desired poverty reduction. Poverty still affects 27% of the population and its majority lives in rural areas. As a matter of fact, one of the essential government goals of Cape Verde is to fight poverty, and this is reflected in several governance programs that were implemented in the past 20 years. There is a close link between food insecurity and poverty: low income makes difficult for families to access essential food. With the country's commitment to the tourism sector, there is a new window of opportunity for growth of the agricultural sector, consequently, an opportunity for improvement of agricultural producers' income and in the latest stage, to the well-being of the population. However, in view of the scarcity of resources, whether natural or financial, it is believed that the bet on quality is a way by which the investments already made can be made profitable, improving their efficiency and effectiveness.

DOI: 10.1201/9781003164968-23

The certification of agricultural products produced in Cape Verde is a valuable issue once it will naturally facilitate the access to the tourism market; however, the certification requires implementation of quality improvement programs (Ryglová et al., 2017; Simões, 2013) in the entire value chain of local farming business The characteristics of the farms advises the choice of the methodology of the 5 Senses (5Ss), as a basis for the continuous improvement of quality. The implementation of 5S provides cost reduction, increase of economic efficiency, and the improvement of the quality of the products. It is also known that quality is strongly related to cultural habits, making the implementation of any quality program a unique work of experimentation and adjustment (Ho, 1999; Ablanedo-Rosas et al., 2010).

The growth of a sector, in this case, the agricultural sector, depends on not only the economic agents but also the institutional economic factors that are expressed in public policies and carried out by the governmental agencies. Assessing the degree of satisfaction from the farmers' side and assessing where public programs could be improved and adjusted can contribute not only to reduce economic agents' dissatisfaction but also to improve the management of available resources.

Thus, this chapter aims to present a diagnostic and monitoring model – SQual4Agri – that assesses the quality of management processes in the agricultural production unit and the satisfaction of farmers toward governmental services in the rural business cluster. The model is built in a participatory manner, through focus groups, with farmers from the irrigation perimeter of the Poilão dam, in Cape Verde, who participated in the discussion of the pertinence of the issues and their adequacy. This chapter is structured as follows: after this brief introduction, the characterization of the Cape Verde archipelago and the study area will be addressed; then there will be a theoretical framework and a description of the methodology was used and, finally, the obtained results will be presented.

20.2 The Archipelago of Cape Verde

The archipelago of Cape Verde located in the Atlantic Ocean, 450 km off the western coast of Senegal, consists of 10 islands, divided between the islands located in the North and those in the South. The islands of the Barlavento, in the North, are Santo Antão, São Vicente, Santa Luzia, São Nicolau, Sal and Boa Vista, and the Sotavento islands in the South are Maio, Santiago, Fogo and Brava. Only 10% of its area (4,033 km^2) is potentially arable. The essential family character of the Cape Verdean agriculture consists of 45,399 farms, of which 33,309 under rainfed and 8,580 under irrigated, the total area of the exploitation units is 36,456 ha, of which 31,692 ha are dedicated to rainfed and 3,913 ha dedicated to irrigation (RGA, 2015). In spite of the numerous factors that limit production, namely the cyclical droughts that lead to the drastic reduction of water resources, the lack of arable land, the insularity and the orography of the islands, agricultural production is still an important factor of support for families (MAAP, 2005). It is estimated that agriculture contributes 7.4% of GDP (MDR, 2004; INE, 2010). However, the country's economy is essentially service-oriented, despite that 38% of the population lives in rural areas, where the incidence of poverty is highest. The conclusions of a modeling exercise carried out in 2009, within the scope of the Comprehensive Africa Agriculture Development Program (CAADP)/Economic Community of West Africa States Agriculture Policies (ECOWAP), indicated that for every 1% of poverty reduction in Cape Verde, three-fourths results from the growth of the agricultural sector (DSERAN-MDR, 2015; MDR, 2012). This information has motivated public efforts to invest in agriculture, with emphasis on the construction of small and large surface water storage infrastructures, such as dams.

The increase in water availability for irrigation has allowed the passage from subsistence agriculture, practiced mainly in rainfed, in the production of corn and beans, to irrigated production, essentially in the production of horticultural products, roots and tubers, whose production is mainly channeled to domestic markets, therefore, to domestic food consumption (MAAP, 2005). Despite the advancement in the construction of these hydraulic systems in order to increase the water availability for agricultural production, there is still a strong need for investments in teaching agricultural and environmental sciences, in the training of technicians, farmers/producers, in investigation of the improvement of postharvest and packaging processes and the creation of logistic conditions such as the flow of production, from production areas to consumer markets (Ferreira, 2015).

Despite the growth of a country-wide market expressed in higher levels of demands in food, the increase in the supply, namely, the diversification of offer in fruit and vegetable has given a satisfactory response to the domestic demand. However, there are still challenges in the supply to hotels and restaurants due to the increase of tourists in the region (MTIE, 2013). The opportunity presented by the boom in tourism has to be addressed with strong commitment to quality, not only in the quantity of the product, but in total quality which will bring benefits throughout the value chain of agricultural production. The increase in the availability of water for production should result in an increase in production and productivity, with remarkable effects, not only on the standard of living (economic valuation), but also on the quality of life of the farmer/producer (MTIE, 2013).

The adoption of quality techniques and methodologies is very recent in Cape Verde and basically it falls into the reproduction of standardized systems and standards implemented in other geographical, social, and economic contexts different from those existing in Cape Verde. On the other hand, the quality vision has been fundamentally centered on the quality of the final product and not on the production process which would safeguard the benefits of those who buy and those who produce (Simões, 2013). The success of family farming business requires production, competitiveness, sustainability and market access ability, and management competence. The adoption of quality management systems would provide and reinforce all the above-mentioned factors (Lourenzani et al., 2008).

The Cape Verdean irrigated farms are essentially family owned and the output consists essentially of vegetables, fruit, roots, tubers, and sugar cane. The size of this kind of polyculture system ranges from 0.2 ha to 2 ha some authors consider this size of exploration "vegetable gardens" (Fonseca et al., 2014). Some previous studies (Nair, 2001; Reyes-García et al., 2012; Smith and Jehlicka, 2013) on tropical gardens have shown that these farming businesses accomplish excellent ecological, economic, and social functions and help explain the maintenance of local ecosystems (Reyes-García et al., 2012).

In spite of this, progress requires efforts in order to promote endogenous knowledge and training of the different actors so to enhance their sustainable exploitation (INIDA, 2016). For instance, the 5Ss methodology having been successfully applied in companies and industries of several branches and organizations and been proved to contribute to the sustainability of production through the optimization of resources may allow in the context of scarcity of resources, in the Cape-Verdean context, to contribute to the promotion of small agriculture, encouraging efficiency and effectiveness in the use of available resources, contributing to aspects of sustainability, as well as to increase farmer satisfaction.

20.3 Quality, the 5 Senses Program and Satisfactions

The term 5Ss was formalized, in 1980, by Takashi Osada (Lamprea et al., 2015). It is a methodology, which can be implemented in any organization or aspects of daily life (Delisle and Freiberg, 2014; Osada, 1991). Organizations use the 5Ss to lay the groundwork and the patterns for problem-solving and the foundation for a correct working environment for their teams (Ablanedo-Rosas et al., 2010; Hirano, 2009; Kobayashi et al., 2008). The original concept, developed by Osada, (1991), considers a valid practice for many environments and dimensions of life. Therefore, the 5S methodology must be suitable and adapted to each environment in which it is applied (Ablanedo-Rosas et al., 2010; Imai, 1986). Despite this, the result of several experiences in different countries has shown their relevance and their positive results, the most difficult part of which is to incorporate them into the daily practice of individuals (Jaca et al., 2014). The implementation of 5S requires training of people so that they can implement their own solutions (Davies, 2003; Ho, 1999). A balanced approach between the philosophical, methodological concept of Osada, (1991) and the technical or instrumental approach of Hirano (2009) can be the strategy to achieve the goals of the 5S, about achieving excellence in the workplace and in the processes (Lamprea et al., 2015; Kobayashi et al., 2008). On the other hand, quality can also be defined as the level that a group of essential characteristics able to satisfy the expectations and needs of the consumer (Zeithaml et al., 1990). In 1990, these same authors developed a quality tool, SERVQUAL, which uses a scale that assesses the gap between customer expectations and the provision of the service received to assess customer satisfaction with the service they receive.

TABLE 20.1
The Five Dimensions of the SERVPERF Model

Dimension	Meaning	Number of Questions
Tangibility	Physical facilities, equipment, appearance of personnel	4
Reliability	Ability to perform the promised service dependably and accurately	5
Responsiveness	Willingness to help customers and provide prompt service	4
Assurance	Knowledge of employees and their ability to inspire trust and confidence	4
Empathy	Caring individualized attention, the firm provides the customers	5
Total Statements		22

Following this logic, that under the proposal of Zeithaml et al. (1990), the quality of service is defined as the gap or discrepancy that exists between customers' expectations and their perceptions. This scale consists of two sections with 22 questions, grouped according to the attributes that can be evaluated by customers. According to Cronin and Taylor (1994), there is a fundamental difference between the quality of service and consumer satisfaction, while the first is a long-term attitude, the second is a transitional judgment made based on the provision of a specific service. Thus, they propose the most suitable way is to consider quality as the customer's perception and not so much in the technical specifications of their production process, hence the importance of listening and understanding the needs and expectations of the customer/consumer (Mota and Nascimento, 2011), and for this, the SERVPERF model is applied. This model is based on the study by Zeithaml et al. (1990) and measures the perceived quality, which is, therefore, the evaluation/judgment that the consumer has about the global superiority or not of a supplier. This concept differs from objective quality, as it is a form of relative attitude, but not equivalent to satisfaction, and results from the comparison between expectations and perception of performance. Table 20.1 summarizes the five dimensions of the SERVPERF model.

20.4 Methodological Procedures

Bearing in mind the nonexistence of quality programs implemented in small family agricultural production units in Cape Verde, we will perform for the present study the SERVPERF in order to diagnose the level of satisfaction of farmers in the perimeter of irrigation of the Poilão dam in relation to the public services of which they are beneficiaries; moreover, building on previous work to diagnose the level of implementation of the 5Ss Program in the islands of S. Nicolau and Santiago, Cape Verde (Simões and Saraiva, 2014; Simões, 2018), we will also diagnose the level of quality competence in the same sample of farmers. The aim is to develop a hybrid and multidimensional diagnostic tool (Awasthi et al., 2011; Mota and Nascimento, 2011; Yaya et al., 2012) for the assessment of quality management of agricultural production units and farmers' satisfaction. We believe this diagnostic procedure will play as an effective instrument for achieving improvements in the management of rural property and providing information adjustments in public policies and services aimed at these Cape Verdean farmers. The assessment model brings together three dimensions: production, quality, and satisfaction, in a perception questionnaire (SQual4Agri), resulting from the extension of an initial instrument, built from a diagnostic study, from the application of the 5Ss Program in five agricultural units on the island of S. Nicolau and one another applied to 34 agricultural units on the island of Santiago, both in Cape Verde (Simões and Saraiva, 2014; Simões, 2018). From this diagnostic questionnaire of the 5Ss, another component was added, the one referring to farmers' satisfaction, bearing in mind the public policies of which they are beneficiaries. The questionnaire was the output from a focus group of 15 farmers, all of them having agricultural production units in the Poilão dam. The focus group approach was essential in the construction of the questionnaire as it allowed obtaining information efficiently about what people think, feel, and understand how they act or react in relation to each individual item of quality (Boca, 2015; Oliveira and Freitas, 1998; Santos and Fogliatto, 2002). The SQual4Agri instrument consists of 79 items, divided into three sections: Production, Quality, and Satisfaction, are detailed in the dimensions of (1) EstProAgri, the (2) 5Ss tool,

TABLE 20.2

Organization of the Questionnaire, based on the Model SQual4Agri

	Organization	Variable Dimension	Item Number	Scale
Section II	Diagnosis of the Management Production Unit Quality	1st Sense of use (SEIRI)	5	Likert Scale of 3 levels and closed-ended questions
		2nd Sense of ordination (SEITON)	15	
		3rd Sense of cleanness (SEISO)	8	
		4th Sense of health (SEKETSU)	5	
		5th Sense of self-discipline, education, and commitment (SHITSUKE)	3	
Section III	Diagnosis of Farmers' Satisfaction	Tangibility	10	Likert Scale of 5 levels
		Reliability	5	
		Responsiveness	8	
		Assurance	14	
		Empathy	2	
		Global satisfaction	4	

and the (3) modified SERVPERF, which constitutes the conceptual model applied in this investigation where the "S" means satisfaction/SERVPERF/Service, "Qual" means Quality represented by 5Ss, "4" means For and "Agri" means Agriculture (see Simões et al., 2021).

This chapter addresses only two of these parameters: (1) Quality through the diagnosis of the 5Ss and (2) Evaluation of farmers' satisfaction, through SERVPERF. The questionnaire was applied in two moments, the first to a pilot group of 20 farmers that included the 15 who participated in the focus group and who actively collaborated in the construction of the questionnaire and a second group of 50 farmers to whom the new survey was applied with entered adjustments. Table 20.2 presents the organization, variables, type, and scales of the questions presented in the questionnaire, based on the SQual4Agri Model.

20.5 Results

20.5.1 5 Senses in the Management of the Production Unit

Regarding the quality assessment in the management of the production units and in order to assess the questions of each dimension of the 5Ss tool, the following scoring criterion was generally used: compliant (C) – 2 points; partially compliant (PC) – 1 point; nonconforming (NC) – 0 points, in order to classify and quantify each Sense. Also, in the multiple-choice questions, it was necessary to assign this same scoring criterion, in order to achieve a score for each Sense, using the formula:

Score = (sum of points × 5)/(no. of sense issues), to make easier the individual analysis of each variable/sense and to classify it, on a qualitative scale (Bad: 0–2; Insufficient: 2–4; Sufficient: 4–6; Good: 6–8; Very Good: 8–10).

In general, on a scale of 0–10, the obtained scores for each Sense, together with the 70 respondent farmers, in the Poilão dam, were low and ranged within the same sense and between the senses. Thus, as was observed in the diagnosis prepared by Simões and Saraiva (2014) and Simões (2018), there is a necessity to raise awareness and inform farmers about the advantages of improvements, regarding quality issues, in the studied agricultural production units.

The obtained results show that the 3rd Sense of Cleanliness and Zeal and the 4th Sense of Health/Standardization showed the lowest score (4 – INSUFFICIENT), the 1st Sense of Use and 5th Sense of Self-discipline, Education, and Commitment (6 – GOOD) were evaluated with the same score and the 2nd Sense of Ordination (7 – GOOD) was the best score.

Some situations found and that need improvement are (1) how to dispose of the property's waste and wastewater, which is mostly done in nature and which, due to its environmental and health implications, must be considered; (2) the existence of excess scrap materials also deserves clarification from the target audience; (3) aspects likely to cause accidents; (4) maintenance of work tools; (5) aspects of cleaning and hygiene on the farmer's property and personal hygiene, due to the impacts they may have on human, plant to animal health; (6) the supply of energy for lighting and the supply of water for domestic consumption. Despite these aspects that surely need significant improvements, the global results already show some attention from the farmer on them, which probably reflects his good motivation to implement a program like 5Ss. Table 20.3 presents the obtained results for each of the 5Ss questions, based on the SQual4Agri Model.

TABLE 20.3

Obtained Results for Each of the 5 Senses Questions, Based on the SQual4Agri Model

1st Sense of Use (SEIRI)	Score/Classification
The tools/equipment/agricultural inputs necessary for the work have their own storage place	PC
Easy access to the tools/equipment/agricultural inputs necessary for the job	PC
Conservation status of tools/equipment/agricultural inputs necessary for the job	PC
There are no excess or unnecessary materials and objects on the property (leftover materials, irrigation pipes, drums, unused machines, others)	PC
What is the main mode of evacuation of materials (garbage) in cleaning	PC
Partial score = (Sum of Points × 5)/(No. of Sense Issues)	**6**
Classification	***Good***
2nd Sense of Ordination (SEITON)	
The property is demarcated and/or identified	PC
The various subplots within the property are well demarcated and identified	PC
Existence of water tubes and/or other exposed wires/tubes preventing the passage or allowing an accident	PC
Existence of scrap material in the area	PC
Personal objects are kept in a specified location	PC
The circulation space allows the traffic of people and equipment without danger of incidents	PC
Existence of proper place for pesticide storage	PC
The agricultural producer checks the expiration dates of the pesticides	C
Existence of a suitable place for fuel storage	PC
The agricultural producer checks the expiration dates of the fertilizers	PC
The agricultural producer checks the expiration dates of the seeds	C
Defective equipment awaiting repair (hoes, cultivators, sprayers, motor pumps, etc.)	PC
Existence of motor pump and type of fuel	PC
Motor pump is protected	NC
If the pump is protected, what type of protection	PC
Partial score = (Sum of Points × 5)/(No. of Sense Issues)	**7**
Classification	***Good***
3rd Sense of Cleaning and Zeal (SEISO)	
Existence of tools and material available on the property to clean it.	PC
How often does the producer clean the property and its attachments	PC
How often does the producer maintain/clean work tools	C
Regarding hygiene has the property has a bathroom	NC
What are the main sources of dirt on the property	PC
What is the main source of energy that the owner uses for lighting	NC
What is the main mode of wastewater evacuation	NC
What is the main source of water supply	NC
Partial score = (Sum of Points × 5)/(No. of Sense Issues)	**4**
Classification	***Insufficient***

(Continued)

TABLE 20.3 ***(Continued)***

Obtained Results for Each of the 5 Senses Questions, Based on the SQual4Agri Model

	Score/Classification
4th Sense of Health/Standardization (SEIKETSU)	
The personal hygiene of the producer shows cleanliness	PC
There are unsafe conditions likely to cause an accident in the work area	PC
The resting hours from work and for meals are adhered to	NC
The producer protects himself with suitable clothing when in the workplace	PC
The producer protects himself with suitable footwear when at the workplace	PC
Partial score = (Sum of Points × 5)/(No. of Sense Issues)	**4**
Classification	***Insufficient***
4th Sense of Health/Standardization (SEIKETSU)	
In general, the producer considers that the items mentioned in the questionnaire are important for the proper functioning of the property	NC
The producer would be willing to take responsibility for some of the tasks mentioned in the questionnaire, if it meant improving his well-being and the profitability of his work	C
What is the motivation for implementing a quality management program?	C
Partial score = (Sum of Points × 5)/(No. of Sense Issues)	**6**
Classification	***Good***
5th Sense of Self-Discipline, Education, and Commitment (SHITSUKE)	
In general, the producer considers that the items mentioned in the questionnaire are important for the proper functioning of the property	NC
The producer would be willing to take responsibility for some of the tasks mentioned in the questionnaire, if it meant improving his well-being and the profitability of his work	C
What is the motivation for implementing a quality management program?	C
Partial score = (Sum of Points × 5)/(No. of Sense Issues)	**6**
Classification	***Good***

Compliant (C) – 2 points; Partially Compliant (PC) – 1 point; Nonconforming (NC) – 0 points/Classification: Bad: 0–2; Insufficient: 2–4; Sufficient: 4–6; Good: 6–8; Very Good: 8–10.

20.5.2 Farmers' Satisfaction

Regarding each dimension of the diagnostic study, carried out through the modified SERVPERF quality program, a 5-level Likert scale was used to assess satisfaction: 1 – Not Satisfied (NS), 2 – Little Satisfied (LS), 3 – Satisfied (S), 4 – Very Satisfied (VS), and 5 – Totally Satisfied (TS). The Satisfaction calculation followed the recommendations of Zeithaml et al. (1990). Thus, the average score of the Satisfaction assessment over each dimension was obtained through the following two steps: (1) for each respondent, the SERVPERF scores were added to the responses belonging to the dimension and then the sum was divided by the number of questions, which make up the dimensions; (2) the score obtained in step 1 was added to all 70 (N) respondents and the total divided by 70 (N). In general, on a scale of 0–5, the obtained scores for each of the five dimensions of satisfaction assessed, together with the 70 respondent farmers, in the Poilão dam, were low. However, within the same dimension, differences can be observed that show concrete and transversal aspects that require more urgent improvements (e.g. communication and technical assistance).

The obtained results show that the dimensions, Tangibility, Reliability, Responsiveness, Assurance, and Empathy were all evaluated with the same score (2 – little satisfied). Through the results, it is possible to verify the need for improvements:

- ***In the Tangibility dimension improving, namely:***
    1. the conditions for the reception and accessibility of farmers in the ministry delegation;
    2. equipping technicians with basic equipment for fieldwork;
    3. technical assistance in the field and regular visits by technicians and the delegate.

- ***In the Reliability dimension, it is observed that:***
  1. the issues of communication between the beneficiaries and the government need improvement, with regard to the better and greater involvement of farmers, in the development processes of the irrigation perimeter at the dam;
  2. trust (Reliability) in the institution to solve its problems and the deadline for resolving issues.
- ***When it comes to the Responsiveness dimension***, it is clear that the ministry delegation has little responsiveness when it comes to assisting the farmer and with regard to technical assistance for irrigation systems, pests, livestock, production planning, information about markets and training.
- ***The Trust/Security/Assurance dimension shows:***
  1. lack of confidence in technicians working in the delegation of the Ministry of Rural Development (MDR) of Cape Verde;
  2. they resent their little involvement in projects and programs developed by the ministry;
  3. they have little security in their own activity;
  4. difficulties in placing their products on the market, in the price of products and in the flow and delay in responding to their complaints.
- ***In the Empathy dimension***, it is perceived that the farmer expects greater and better attention in crisis situations and expects more collaboration to improve his productive activity. Although the results tend to indicate widespread dissatisfaction with the services provided by the Ministry of Agriculture and the Environment, there are differences in assessments from item to item within each of the five dimensions. This makes possible the identification of where dissatisfaction is greatest and can possibly been addressed in future policy planning. Despite little satisfaction demonstrated in the five dimensions, it is clear that the farmer is satisfied with his own activity. The general perception of satisfaction comes from the willingness to recommend this activity to their offspring or friends. Although some tiredness and discouragement demonstrated possibly due to inherent difficulties of the agriculture and unsatisfactory public policies, the farmers responded negatively when asked if he would change his profession if he could. Table 20.4 presents the results obtained for the six dimensions of the modified SERVPERF, based on the Model SQual4Agri.

TABLE 20.4

Obtained Results for the Six Dimensions of the Modified SERVPERF, Based on the Model SQual4Agri

Tangibility	Score/Classification
Reception conditions at the MDR Delegation	LS
Accessibility of the facilities of the MDR Delegation	LS
The technicians who assist us have materials and equipment	LS
The regularity of visits by MDR technicians	LS
Alternative technical assistance services to MDR	LS
The regularity of visits by the MDR delegate	LS
The UGBP water management service	S
Market availability of the varieties of seeds I grow	S
The increasing level in my production in the last 5 years	S
The diversification of the cultures I produce	S
Average	**2**
Classification	***Little Satisfied***

(Continued)

TABLE 20.4 ***(Continued)***

Obtained Results for the Six Dimensions of the Modified SERVPERF, Based on the Model SQual4Agri

	Score/Classification
Reliability	
Information I receive about the development goals of the Poilão Dam perimeter	LS
The collaboration of the delegation to solve my problems	LS
The services and support that the delegation has	LS
The deadline for problem-solving	LS
The price of water	LS
Average	**2**
Classification	***Little Satisfied***
Responsiveness	
The technical assistance provided by the delegation	LS
Technical assistance for the installation and operation of my irrigation systems	LS
Technical assistance for pests and diseases	LS
Technical assistance for livestock	LS
Information to schedule my production	LS
The trainings organized by the MDR are important for my activity	LS
The way I am waited in the delegation	LS
Management and distribution of irrigation water	S
Average	**2**
Classification	***Little Satisfied***
Assurance	
My participation in the projects and programs developed by MDR	LS
I feel respected by the technicians	LS
The technicians are friendly and polite	S
Trust in the technicians who work in the Delegation	LS
Confidence in the technical information they provide me	LS
I know my rights and duties	S
The security and profitability of my agricultural activity	LS
Placing my production on the market	LS
The price at which I sell my product	LS
Information given on changes to irrigation schedules	S
Information I get about collaboration possibilities	LS
Information on changes in production support projects and programs	LS
Respect for my opinion and suggestions	LS
Replies to my complaints	LS
Average	**2**
Classification	***Little Satisfied***
Empathy	
Technical visits in unforeseen situations	LS
The technicians talk to me about what I do well and what I need to improve	LS
Average	**2**
Classification	**Little Satisfied**
General Satisfaction	
In general, what is your degree of satisfaction	S
Would you pay for quality technical assistance?	YES
How much you would pay per month	500–1,500 CVE
Would you recommend farming to your friend or your child?	YES
If you could, would you change your professional activity?	YES

1 – Not Satisfied (NS), 2 – Little Satisfied (LS), 3 – Satisfied (S), 4 – Very Satisfied (VS), and 5 – Totally Satisfied (TS).

20.6 Final Considerations

The construction of the Poilão dam, aiming at the storage of water to use in irrigated agricultural production, presents itself from the point of view of this objective, as an appropriate measure of public policy, since the potential it holds promises to contribute positively in the poverty reduction and ensuring food security. However, for these objectives to be achieved, it is necessary that their beneficiaries/users have an active participation and that other public policy measures (e.g. research, rural extension, training) are adjusted to the needs of these users, in order to contribute to the achievement objectives as efficiently and effectively as possible. The implementation of quality methodologies and tools can contribute to this. The implementation of the 5Ss Methodology contributes significantly to success, which corroborates the arguments of Michalska and Szewieczek (2007), as it increases organizational performance, including cleanliness, health, and safety, as mentioned by Hirano (2009), as well as increasing capacity and productivity, which is also in line with Ab Rahman et al. (2010).

Assessing the satisfaction of farmers/users, through their perception of the quality of the benefits/services they receive is important, as their positive or negative perception plays an important role in the success or failure of the service or project, which confirms the conclusions of Fragoso and Espinoza (2017), Huang (2011), Rana et al. (2013), Reddy et al. (2013). From the analysis of the results of the study, the importance and relevance of the application of quality techniques and methodologies in agricultural production units is evident, and there is also a lack of greater awareness and information from farmers about their advantages, with regard to the issues of quality in the agricultural production units under study, according to what had already been observed in the studies by Simões and Saraiva (2014) and Simões (2018), respectively, on the island of S. Nicolau and on the island of Santiago.

The results tend to expose widespread dissatisfaction with the services provided by the Ministry of Agriculture and the Environment. Despite this dissatisfaction, it is clear that the farmer is satisfied with his own performance and business. This contradictory element may show that there is some tiredness and discouragement caused by the difficulties inherent to agriculture in Cape Verde and difficulties in the implementation of public policies in the area. We consider the application of these tools promotes: the most efficient use of available resources; increased productivity and more organized work environments; changing behavior and adopting more proactive attitudes; better use of physical spaces and reduction of accidents at work; greater discipline; an increase in cleanliness levels; greater collaboration between people and the improvement of human well-being; and an incentive, creativity and a commitment to continuous improvement (Simões and Saraiva, 2014). Scientific work, especially in the application of quality tools in the agricultural area, in Cape Verde, is scarce, so this work is of particular interest, both for what it represents for scientific development in general and for its practical application, which could be extremely useful for organizations in the sector. On the other hand, the smallness of the country is offset by the diversity of agricultural production systems, cultures and traditions in different regions. This scenario inhibits the generalization of results and requires a specific look at each reality.

In view of these scenarios, future research is considered essential, which allows a reevaluation: the motivation and interest of farmers; know and compare other areas of the country, due to the uniqueness that each one will have in the training process; the effectiveness of the implementation of quality programs, in particular in their contributions not only to the quality of the products, but also to the quality of life of the producers.

REFERENCES

Ab Rahman, M. N., Khamis, N. K., Zain, R. M., Deros, B. M., and Mahmood, W. H. W. 2010. Implementation of 5S practices in the manufacturing companies: A case study. American Journal of Applied Sciences, 7(8), 1182–1189.

Ablanedo-Rosas, J. H., Alidaee, B., Moreno, J. C., and Urbina, J. 2010. Quality improvement supported by the 5 S, an empirical case study of Mexican organizations. International Journal of Production Research, 48, 7063–7087.

Awasthi, A., Chauhan, S. S., Omrani, H., and Panahi, A. 2011. A hybrid approach based on SERVQUAL and fuzzy TOPSIS for evaluating transportation service quality. Computers & Industrial Engineering, 61, 637–646.

Boca, G. D. 2015. 5 S quality management. Oradea: University of Oradea.
Cronin, J. and Taylor, S. 1994. SERVPERF versus SERVQUAL: Reconciling performance-based and perceptions-minus-expectations measurement of service quality. Journal of Marketing, 58(1), 125–131.
Davies, E. C. 2003. Quality: Its historical context. Engineering Management, 13(2), 14–17.
Delisle, D. R., and Freiberg, V. 2014. Everything is 5 S: A simple yet powerful lean improvement approach applied in a preadmission testing center. Quality Management Journal, 21(4), 10–22.
DSERAN-MDR. 2015. Perímetro Irrigado da Barragem do Poilão: Estudo Socio-económico dos produtores/irrigantes. Sal – Cabo Verde: MDR.
Ferreira, V. S. 2015. Conflitos e participação no uso da água da Barragem de Poilão, (1a ed.). Praia: Imprensa Nacional.
Fonseca, C., Guiomar, N., and Pinto-Correira, T. 2014. Estimativa da Produção Hortícola Potencial no Concelho de Montemor-o-Novo. Évora: ICAAM-Universidade de Évora.
Fragoso, J. T., and Espinoza, L. 2017. Evaluation de la perception de la calidade de los servicios bancarios mediante el modelo SERVPERF. Contaduria y Administración, 62(4), 1270–1293.
Hirano, H. 2009. JIT Implementation Manual – The Complete Guide to Just-in-Time Manufacturing. The '5S' approach. Boca Raton, FL: CRC Press, 237–319.
Ho, S. K. 1999. 5-S practice: The first step towards total quality management. Total Quality Management, 10(3), 345–356.
Huang, L. C. 2011. The measurement of the service quality of rural wineries. Journal of Marketing Development and Competitiveness, 5(5), 29–45.
Imai, M. 1986. Kaizen. New York, NY: Random Business Division.
INE. 2010. Recenciamento Geral da População e Habitação. Obtido de Instituto Nacional de Estatísticas de Cabo Verde.
INIDA. 2016. Plan Strategique 2015-2022 du Systeme National de Research Agricole du Cape Verde. Praia – Cabo Verde: FAO.
Jaca, C., Viles, E., Paipa-Galeano, L., and Mateo, J. S. 2014. Learning 5 S principles from Japanese best practitioners: Case studies of five manufacturing companies. International Journal of Production Research, 52(15), 4574–4586.
Kobayashi, K., Fosher, R., and Gapp, R. 2008. Business improvement strategy or useful tool? Analysis of the application of the 5S concept in Japan, the UK and the US. Total Quality Management, 19(3), 245–262.
Lamprea, E. J., Carreño, Z. M., and Sánchez, P. M. 2015. Impact of 5 S on productivity, quality, organizational climate and industrial safety in Caucho Metal Ltda. Ingeniarc. Revista chilena de ingeniería, 23(1), 107–117.
Lourenzani, W. L., de Barros Pinto, L., Alves de Carvalho, E. C., and do Carmo, S. M. 2008. A qualificação em gestão da agricultura familiar: A experiência da alta paulista. Revista Ciência em Extensão, 4(1), 62.
MAAP. 2005. Estratégia de Desenvolvimento Agrícola no horizonte 2015. Praia – Cabo Verde: MAAP.
MDR. 2004. Recenceamento Geral Agrícola. Praia – Cabo Verde: Ministério do Desenvolvimento Rural.
MDR. 2012. Relatório de Elaboração DECRP-III/2012-2016. Praia – Cabo Verde: QIR-CV-OMC.
Michalska, J., and Szewieczek, D. 2007. The 5S methodology as a tool for improving the organization. Journal of Achievements in Materials and Manufacturing Engineering, 24(2), 211–214.
Mota, G. S., and Nascimento, D. R. 2011. Qualidade em Serviços de Atendimento ao Consumidor (SAC): Uma avaliação utilizando a escala SERQUAL e os critérios INMETRO. Gestão.org, 9(3), 565–584.
MTIE. 2013. Diagnostic trade integration study update. Praia - Cabo Verde: MTIE.
Nair, P. K. 2001. Do tropical homegardens elude science, or is it the other way around?. Agroforestry Systems, 53, 239–245.
Oliveira, M., and Freitas, H. 1998. Focus group, pesquisa qualitativa: resgatando a teoria, instrumentalizando o seu planejamento. RAUSP Management Journal, 33(3), 83–91.
Osada, T. 1991. The 5S's: Five keys to a total quality environment. Tokyo: Asian Productivity Organization.
Rana, A. S., Reddy, G. P., and Sontakki, S. 2013. Perceived service quality of agricultural organizations comparative analysis of public & private sector. International Journal of Advanced Research in Management and Social Sciences, 2(1), 286–295.
Reddy, G. P., Reddy, A. A., Sontakki, B. S., and Rana, A. S. 2013. Engendering agricultural research and extension: Public-private partnership options. Indian Journal of Agricultural Economics, 68(3), 420.
Reyes-García, V., Aceituno, L., Vila, S., Calvet-Mir, L., Garnatje, T., Jesch, A., … Pardo-De-Santayana, M. 2012. Home gardens in three mountain regions of the Iberian Peninsula: Description, motivation for gardening, and gross financial benefits. Journal of Sustainable Agriculture, 36(2), 249–270.

RGA. 2015. Recenciamento Geral Agrícola. Praia, Santiago: Ministério da Agricultura e Ambiente de Cabo Verde.

Ryglová, K., Rasovska, I., and Sácha, J. 2017. Rural tourism – Evaluating the quality of destination. European Countryside, 9(4), 769–788.

Santos, G. T., and Fogliatto, F. S. 2002. Grupos Focalizados: uma proposta de roteiro para identificação de atributos de referência. Anais. XXII ENEGEP – Encontro Nacional de Engenharia de Produção. Curitiba, PR.

Simões, E. 2013. Caracterização da qualidade física e química e sensorial da papaia Solo e Local da ilha de Santiago, Cabo Verde. Dissertação de Mestrado, Évora, Portugal.

Simões, E. B. 2018. Diagnóstico do Nível de Implementação dos 5S's em Propriedades Agrícolas, em três Municípios da Ilha de Santiago, em Cabo Verde. Revista TMQ – Techniques, Methodologies, and Quality, 9, 84–113.

Simões, E., and Saraiva, M. 2014. Cap. 8 – Diagnóstico do Nível de Implementação dos 5S's em Propriedades Agrícolas Cabo-Verdianas. Em M. J. Rosa, P. M. Sá, and C. S. Sarrico (Edits.), Qualidade em Acção: Casos de aplicação de ferramentas e metodologias de qualidade. Lisboa – Portugal: Edições Sílabo, 175–200.

Simões, E., Saraiva, M., Basch, G., Pires, A. R., Rosa, Á., and Chaleta, E. 2021. Hybrid conceptual model for assessing quality, production and satisfaction (SQual4Agri), in agricultural production units, in Cape Verde. Sustainability, 13(1), 37.

Smith, J., and Jehlicka, P. 2013. Quiet sustainability: Fertile lessons from Europe's productive. Journal of Rural Studies, 148–157.

Yaya, L. H., Marimon, F., and Fa, M. C. 2012. Assessing e-service quality: The current state of E-S-Qual. Total Quality Management & Business Excellence, 23(11–12), 1478–3363.

Zeithaml, V. A., Parasuraman, A., and Berry, L. L. 1990. Delivering quality service: Balancing customer perceptions and expectations. New York, NY: The Free Press.

21

Adoption of Improved Agronomic Practices by Small Farmers for Sustainable Crop Production

Case Study from Uttar Pradesh, India

P. K. Singh and O. P. Singh
Department of Agricultural Economics, Institute of Agricultural Sciences, Banaras Hindu University, Varanasi, India

CONTENTS

21.1 Introduction

Out of several operations of crop production process, tillage is one of the important operations that made possible to sow the crop seeds into the soil. Tillage of agricultural field before planting crops in the field is as old as human civilization, and its first evidence was found dated back 3000 BC in Mesopotamia (Singh et al. 2018). After introduction of machine-driven implements for agricultural operations, the farmers were steadily shifting from bullock power to tractor power for tillage operation. The tragic dust storm in the Midwestern United State in 1930 was a wake-up call to how human interventions in soil management and tillage led to unsustainable agricultural system (Singh et al. 2018). In 1930, Edward H. Faulkner

DOI: 10.1201/9781003164968-24

questioned the efficacy of intensive soil tillage (Faulkner 1943). After advent of seeding machine during 1940, it made sowing possible without soil tillage (Friedrich et al. 2012). Minimum soil disturbance along with organic soil cover can reduce soil erosion, soil physical property, and water losses (Serraj and Siddiqe 2012).

After the Green Revolution of Indian agriculture in 1968, farmers of the country were gradually shifting from resource extensive to resource intensive crop production system. Farmers have also shifted from bullock power to mechanical power, rainfed crop production to irrigated crop production, mono cropping to relay cropping, and organic farming to chemical fertilizer. The conventional method of crop production is characterized by intensive tillage for land preparation, crop residue burning which leads to soil erosion, carbon contain in soil, and deterioration physical property of soil. The continuous soil erosion from the agricultural field leads to reduce soil productivity. Uncontrolled use of inputs for crop production made possible to achieve the target of production, but at the same time, it leads to deterioration of natural resource (Singh et al. 2020). The researchers and policy makers are more worried about how to feed fast growing population of the country from depleting water and land resources.

Introduction of conservation agriculture tenders robust option for meeting future food-grain demand while contributing to sustainable agriculture and rural development. Conservation agriculture can augment the input use efficiency, farm income, and crop yield while preserving and regeneration of soil fertility, biodiversity, and the natural resource base. The conservation agriculture is promoted among the farming community as a notion of sustained and increased crop production to achieve acceptable profit from crop production while saving resources along with protecting environment (FAO 2008). Globally, total area under conservation agriculture was 105 million hectares in 2008. Out of this, more than 90 per cent area under conservation agriculture is cultivated in five countries viz., United States, Brazil, Argentina, Canada, and Australia. Conservation agriculture is a set of technologies that include minimum soil disturbance, permanent organic soil cover, improved crop rotation, and integrated weed management. The conservation agriculture helps farmers to reduce the negative consequences of conventional agriculture. The resource conservation technologies (RCTs) offer varieties of options. Out of these options, zero or reduce till-surface residue management system has benefited farming community of different regions of world including Indo-Gangatic plain by obtaining increased crop productivity, soil health, and narrowing in overall cost of cultivation by reducing use of inputs for wheat cultivation (Singh et al. 2018).

Zero-till drill-seeded wheat involves the sowing of crops without soil tillage by narrow slots opening and least width and furrow depth. This practice is popularly known as no-till/reduce tillage. In zero/reduce tillage, the soil surface covered by crop residues should be at least 32 per cent. Indo-Gangatic plain is dominated by paddy-wheat cropping system and farmers are adopting zero/reduce tillage for wheat crop. Zero-till drill-seeded wheat is the combination of farm management systems including better crop residue planning, sowing seeds into soil, weed and pest control, harvesting, and crop sequencing. By using zero or reduce tillage, a variety of crops can be sown, which includes wheat, maize, pulses, oilseeds, etc. There is huge opportunity for adoption of zero-till drill-seeded wheat in Indo-Gangatic region. All types of soils including moderately reclaimed soil are suitable for adoption of zero-till drill-seeded wheat. One of the important factors for acceptance of zero-till drill-seeded wheat among the farmers is timely sowing of wheat. There is positive relationship between delay in wheat sowing and reduction in wheat yield (Singh et al. 2020). Past researchers found in their study that after adoption of RCT, there was significant reduction in cost of wheat cultivation and escalation of net income, less use of fossil fuel which leads to lower emission of carbon dioxide, which is one of the greenhouse gases responsible for global warming (Holland 2004; Govaerts et al. 2009). Low cost of inputs and enhancement of wheat yield leads to overall progress of the socioeconomic conditions of farming community (Kaur et al. 2012; Singh et al. 2018).

Drivers of adoption of zero-till drill-seeded wheat technology are input cost reduction and higher earning, irrigation water saving, soil fertility enhancement by crop stubbles decomposition into the soil, and early sowing of wheat crop which leads to higher wheat production and productivity. Keeping in view the above aspects, present study was undertaken with the following objectives: (a) to evaluate the cost of wheat cultivation for zero-till drill-seeded adopters and non-adopters; (b) to estimate the impact of zero-till drill-seeded wheat on environment; (c) to analyze the factor influencing adoption of zero-till drill-seeded wheat; and (d) to find out reasons for adoption, non-adoption, and conditions for adoption of zero-till drill-seeded wheat in future.

21.2 Research Methodology

21.2.1 Profile of the Study Area

Present study was conducted in Chandauli district of Uttar Pradesh of India. The latitudinal and longitudinal extensions are from 25°16′ N to 25°27′ N latitude and 83°16′ E to 83°27′ E longitudes, respectively (Figure 21.1). The district is characterized with dual geological and physiological divisions covering 2541 km^2. The vast northern alluvial plain sections are belonging to middle Ganga plain and southern plateau is an extended part of Vindhyan Uplands. The district is popularly known as "*Dhaan Ka Katora*" in Uttar Pradesh because of extreme fertile lands of Indo-Gangetic Plain. From the administration point of view, district has been divided into three tehsils and nine development blocks. The major river flowing in the district is Ganga, Karmanasha, and Chandraprabha. The major canals are Chandraprabha, Mushakhad, and Narayanpur lift canal. The net and gross sown area of the district was 127,772 and 233,485 ha, respectively, during 2017–2018 (Government of Uttar Pradesh 2021). The cropping pattern of the Chandauli district is a rice-wheat cropping system. The soil types found in the district are loam, clay loam, clay, sandy loam, sandy, and rocky soils. The clay soil is best suited for adoption of RCTs, i.e., zero tillage.

21.2.2 Sampling Procedure and Data Collection

Uttar Pradesh state was selected purposively for the present study. Uttar Pradesh consists of 75 districts. Out of these districts, Chandauli district was selected purposively because farmers of the district are growing zero-till drill-seeded wheat from the last 13 years. There are nine development blocks in the Chandauli district. Out of total nine development blocks, Barhani development block was selected purposively for the present study. The criteria for selecting Barhani development block was based on large

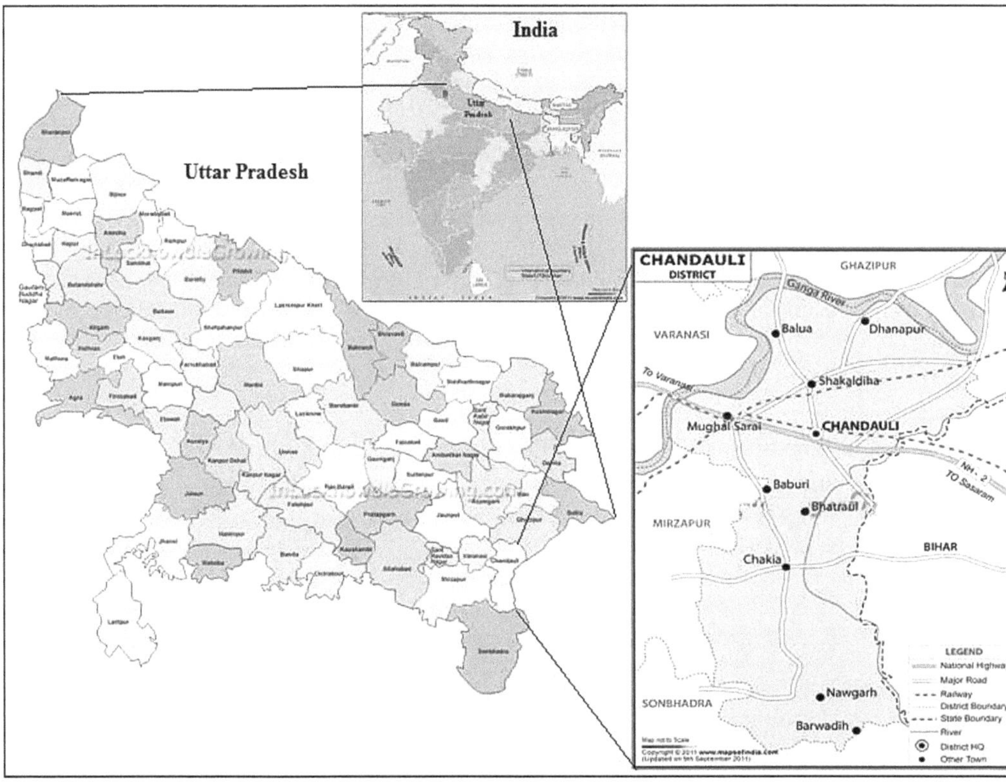

FIGURE 21.1 Location map of study area.

number of wheat growers who have adopted zero-till drill-seeded wheat cultivation. The Barhani development block consists of 196 villages. Out of these villages, Barhani village was selected purposively for the present study, because farmers of the village have adopted zero-till drill-seeded wheat cultivation. To conduct the study a list of zero-tillage adopters and non-adopters were prepared separately. A sample of 30 zero-till drill-seeded wheat adopters and 30 non-adopters was selected randomly using random table without replacement method. The primary data for the agricultural year 2018–19 was collected through personal interview using pretested schedule.

21.2.3 Analytical Procedure

21.2.3.1 Cost of Cultivation

The cost of cultivation of zero-till drill-seeded wheat adopters and non-adopters were calculated by using method suggested by the Commission on Agricultural Cost and Prices (Singh et al. 2018):

Cost A_1: All the input cost + depreciation on implements and farm buildings + land revenue, cesses and other taxes + interest on working capital + miscellaneous expenses
Cost B_1: Cost A_1 + interest on value of owned fixed capital assets (excluding land)
Cost C_1: Cost B_1 + rental value of owned land + rent paid for leased-in land
Cost C_2: Cost C_1 + imputed value of family labor
Cost C_3: Cost C_2 + 10 per cent of cost C_2 as account for managerial input of the farmer.

21.2.3.2 Economic and Environmental Benefit

The economic benefits of zero-till drill-seeded wheat were worked out using economic surplus model as suggested by Alston et al. (1995). The economic surplus model is given below:

$$\Delta CS = PQZ(1+0.5\ Zn)$$

$$\Delta PS = PQ(K-Z)(1+0.5\ Zn)$$

$$\Delta TS = \Delta CS + \Delta PS = PQK(1+0.5\ Zn)$$

where $Z = K\varepsilon/(\varepsilon+\eta)$; K is vertical shift in supply function as proportion of initial price; η is elasticity of demand (absolute) and ε is elasticity of supply.

The environmental benefits realized by adopters of zero-till drill-seeded wheat are reduction in carbon emission. For the estimation of reduction in carbon emission, first of all quantify the diesel saving on farm operations then multiplied by the 1 L equal to 2.6 kg of CO_2 (Jat et al. 2006) and 1 kg CO_2 is 0.27 kg of carbon (Paustian et al. 2006).

21.2.3.3 Factors Influencing Adoption of Zero-Till Drill-Seeded Wheat

Identification of relative importance of various factors influencing adoption of zero-till drill-seeded wheat in Chandauli district was worked out by using binary logit model (Mallada 1992). The algebraic form of model is given below:

$$p_i = \frac{1}{1+e^{-Z_i}}$$

where p_i is a probability of adoption of zero tillage for the ith farmer and ranges from 0 to 1. e^{-z} represents the base of natural logarithms and Z_i is the function of a vector of n explanatory variables and expressed as follows:

$$Z_i = \beta_0 + \sum \beta_i X_i$$

where β_0 is the intercept and β_i is a vector of the relationship between P_i and X_i, which is nonlinear, can be written as follows:

$$p_i = \frac{1}{1 + e^{-(\beta_0 + \beta_1 X_i + \cdots + \beta_n X_n)}}$$

Finally, the logit model is obtained by using the logarithm

$$L_i = L_n\left[\frac{p_i}{1 - p_i}\right] = z_i = \beta_0 + \beta_1 X_1 + \beta_2 X_2 + \cdots + \beta_n X_n$$

21.2.3.4 Ranking of Reasons for Adoption or Non-Adoption

The Garrett Ranking was used to rank the reasons for adoption and non-adoption of zero-till drill-seeded wheat. The percentage position of each rank thus obtained was converted into scores by referring to the table given by Henry Garrett. The score of all the factors arranged in order of their ranks (Singh et al. 2018).

$$\textit{Percent Position} = \frac{100\left(R_{ij} - 0.5\right)}{N_j}$$

where R_{ij} is the rank given for *i*th item *j*th individual and N_j is the number of items ranked by *j*th individuals.

21.3 Results and Discussion

21.3.1 Cost of Wheat Cultivation

It was observed from Table 21.1 that the average cost of cultivation (Cost C_3) for wheat was Rs 51,287.45 and Rs 55,188.78 for zero-till drill-seeded wheat adopters and non-adopters, respectively. In case of zero-till drill-seeded wheat adopters, the cost of wheat cultivation was lower (Rs 3901.33/ha) than the non-adopters. The responsible factors for declining cost of cultivation for zero-till drill-seeded wheat adopters were wages of hired labor, tractor charges, expenses on fertilizer, irrigation charges, and expenditure on plant protection measures.

In case of zero-till drill-seeded wheat adopters, out of average cost of wheat cultivation, highest share comes from rental value of own land (29.25 per cent), followed by expenditure on machine labor, cost of applying irrigation water, expenditure on chemical fertilizers, expenditure on human labor and remaining 11.47 per cent expenditure incurred by sample farmers on seed, plant protection chemicals, miscellaneous expenditure, and interest on working capital. And in case of zero-till drill-seeded wheat non-adopters, out of total cost of wheat cultivation maximum share comes from rental value of own land followed by expenditure on applying irrigation water, machine labor, cost of chemical fertilizer, expenditure on human labor and remaining 13.08 per cent expenditure incurred by sample farmers on seed, plant protection chemicals, miscellaneous expenditure, and interest on working capital.

21.3.2 Wheat Production and Gross Income

It was observed from Table 21.2 that the zero-till drill-seeded wheat adopters were getting higher yield of main product and by-product and ultimately higher gross income as compared to non-adopter farmers. The reason for obtaining higher yield under zero-till drill-seeded wheat was incorporation of crop residue into the soil that augment carbon contain in the soil. The wheat grain yield received by zero-till drill-seeded wheat adopters and non-adopters was 42.50 and 36.25 quintal per hectare, respectively. As for as wheat *bhusa* is concerned, it was 42.50 and 36.25 quintal per hectare for zero-till drill-seeded

TABLE 21.1

Average Cost of Cultivation of Wheat (Per ha)

	Zero-Till Drill-Seeded Wheat Adopters		Zero-Till Drill-Seeded Wheat Non-Adopters	
Particulars	**Physical Unit**	**Value (Rs)**	**Physical Unit**	**Value (Rs)**
1. Human labor				
a. Family labor	10 Man days	2500.00	10 Man days	2500.00
b. Hired labor	10 Man days	2500.00	12.5 Man days	3125.00
2. Machine labor	10 h	8000.00	18.75 h	8437.50
3. Seed	125 kg	2812.50	112.5 kg	2531.25
4. Chemical fertilizer				
a. Urea	135 kg	918.00	138.75 kg	943.50
b. DAP	125 kg	3125.00	125 kg	3125.00
c. Potash	50 kg	1200.00	55 kg	1320.00
5. Irrigation	37.5 h	7500.00	42.5 h	8500.00
6. Plant protective chemicals	–	1250.00	–	2500.00
7. Miscellaneous cost		750.00		1000.00
8. Interest on working capital		1069.45		1189.37
9. Rental value of owned land		15,000.00		15,000.00
Cost of cultivation				
a. Cost A_1		29,124.95		32,671.62
b. Cost B_1		29,124.95		32,671.62
c. Cost C_1		44,124.95		47,671.62
d. Cost C_2		46,624.95		50,171.62
e. Cost C_3		51,287.45		55,188.78

TABLE 21.2

Per Hectare Wheat Yield and Gross Income

	Zero-Till Drill-Seeded Wheat Adopters			Zero-Till Drill-Seeded Wheat Non-Adopters		
Particulars	**Yield (Qt)**	**Rate (Rs/Qt)**	**Value (Rs)**	**Yield (Qt)**	**Rate (Rs/Qt)**	**Value (Rs)**
a. Main product	42.50	1650.00	70,125.00	36.25	1650.00	59,812.50
b. By-product	42.50	500.00	21,250.00	36.25	500.00	18,125.00
c. Gross income			91,375.00			77,937.50

wheat adopters and non-adopters, respectively. The average farm harvest price of wheat grain and wheat *bhusa* was per quintal Rs 1650 and Rs 500, respectively. Per hectare gross income obtained by the respondents in the study area was Rs 91,375.0 and Rs 77,937.50 for zero-till drill-seeded wheat adopters and non-adopters, respectively. The incremental wheat yield (both main and by-product) received by the zero-till drill-seeded wheat adopters was 6.25 quintal per hectare over zero-till drill-seeded wheat non-adopters. The per hectare incremental gross income received by the zero-till drill-seeded wheat adopters over non-adopters were Rs 13,437.50.

21.3.3 Net Income and Benefit-Cost Ratio

Net income received by zero-till drill-seeded wheat adopters and non-adopters from wheat crop was Rs 40,087.56 and Rs 22,748.72/ha, respectively (Table 21.3). The increased net gain obtained by the zero-till drill-seeded wheat adopters was Rs 17,338.84/ha over non-adopter. The benefit-cost (B-C)

TABLE 21.3

Per Hectare Net Income and B-C ratio from Wheat Cultivation

Particulars	Zero-Till Drill-Seeded Wheat Adopters	Zero-Till Drill-Seeded Wheat Non-Adopters
1. Gross income (Rs/ha)	91,375.00	77,937.50
2. Cost of cultivation (Rs/ha)	51,287.45	55,188.78
3. Net Income over (Rs/ha):		
a. Cost A_1	62,250.05	45,265.88
b. Cost A_2	62,250.05	45,265.88
c. Cost C_1	47,250.05	30,265.88
d. Cost C_2	44,750.05	27,765.88
e. Cost C_3	40,087.56	22,748.72
4. B-C Ratio over:		
a. Cost A_1	3.14	2.39
b. Cost A_2	3.14	2.39
c. Cost C_1	2.07	1.63
d. Cost C_2	1.96	1.55
e. Cost C_3	1.78	1.41

ratio of wheat production for zero-till drill-seeded wheat adopters and non-adopters over cost of cultivation (cost C_3) was estimated to be 1.78 and 1.41, respectively (Table 21.3). From the above discussion, it was observed that adoption of zero-till drill-seeded wheat for wheat cultivation was more beneficial as compared to non-adopters.

21.3.4 Economic Benefit of Zero-Till Drill-Seeded Wheat Adoption

The prevailing average wage rate for human and machine labor was Rs 250/day and Rs 800/h, respectively, at the time of primary data collection. The average price of wheat seed was Rs 22.50/kg and average market price of wheat grain and wheat *Bhusa* was Rs 1650.00 and 500.00 per quintal, respectively. The economic benefit realized due to zero-till drill-seeded wheat adoption in the study area was found to be Rs 16,984.17/ha (Table 21.4). The economic gain from zero-till drill-seeded wheat adoption was due to low cost of human and machine labor, cost of fertilizers, irrigation charges, cost of plant protection chemicals, miscellaneous expenditure, interest on working capital, and incremental benefit of wheat yield (grain and *Bhusa*). High expenditure of money by the zero-till drill-seeded wheat adopters on wheat sced as compared to non-adopters in the study area.

TABLE 21.4

Economic Benefit of Zero-Till Drill-Seeded Wheat

Economic Benefit of Zero-Till Drill-Seeded Wheat	Amount (Rs)
Due to reduction in cost of hired human labor	625.00
Due to reduction in cost of machine labor	437.50
Due to reduction in cost of seed	–281.25
Due to reduction in cost of chemical fertilizers	145.50
Due to reduction in cost of irrigation charges	1000.00
Due to reduction in cost of plant protection chemicals	1250.00
Due to reduction in miscellaneous expenditure	250.00
Due to reduction in interest on working capital	119.92
Due to yield benefits (main and by-product)	13,437.50
Total	**16,984.17**

TABLE 21.5

Environmental Benefit of Zero-Till Drill-Seeded Wheat

Particulars	Zero-Till Drill-Seeded Wheat Adopters		Zero-Till Drill-Seeded Wheat Non-Adopters
Diesel consumption for land preparation and sowing of wheat crop (L/ha)	23.93		36.98
CO_2 emission (kg/ha)	62.21		96.14
Carbon emission (kg/ha)	16.80		25.96
Reduction in carbon emission (kg/ha)		9.16	

21.3.5 Environmental Benefit of Zero-Till Drill-Seeded Wheat

The average diesel required to burn for land preparation and sowing of wheat crop in the study area was found to be 23.93 and 36.98 L/ha for zero-till drill-seeded wheat adopters and non-adopters respectively. The CO_2 emission from zero-till drill-seeded wheat adopters' and non-adopters' field was 62.21 and 96.14 kg/ha, respectively, and per hectare carbon emission was 16.80 and 25.96 kg, respectively (Table 21.5). After adoption of zero-till drill-seeded wheat, carbon emission was decreased by 9.16 kg/ha.

21.3.6 Factors Influencing Adoption of Zero-Till Drill-Seeded Wheat

Out of several influencing factors, "B" value of farming experience (years) and land holding size with expected sign was found significant at five per cent level of significance. Age of respondents (years) was significant at 10 per cent probability level (Table 21.6).

Result shows that the farmers who come under the age group of less than 50 years have better possibility for zero-till drill-seeded wheat adoption. As age of respondents increases (above 50 years), there are fewer chances of adoption of zero-tillage in the study area. This is the fact that young farmers are more aware about the latest technology with risk taking ability. The year of farming experience coefficient was negative and significant at 5 per cent level of significance. It means as farming experience increased in the study area, there is chance for non-adoption of zero-till drill-seeded wheat. The land holding size coefficient was showing positive and significant, which suggests that as land holding size increases, there are probability of acceptance of zero-till drill-seeded wheat in the area under study.

TABLE 21.6

Binary Logit Model Estimates for Factors Affecting Adoption of Zero-Till Drill-Seeded Wheat

Variables	B	S.E.	Wald	df	Sig.	Exp(B)
a. Constant "a"	5.615**	2.982	3.546	1	0.060	274.488
b. Family size (No)	0.252	0.180	1.955	1	0.162	1.287
c. Farming experience (year)	–0.171*	0.066	6.636	1	0.010	0.843
d. Land holding size (ha)	0.108*	0.134	0.645	1	0.022	0.614
e. Age of respondents (year)	–2.342**	1.417	2.730	1	0.098	0.096
f. Adult male (No.)	–0.224	0.301	0.553	1	0.457	0.800
Model summery						
–2 Log likelihood	41.278					
Cox & Snell R Square	0.298					
Nagelkerke R Square	0.398					

*: Significant at 5.0 per cent level of significance.

**: Significant at 10.0 per cent level of significance.

TABLE 21.7

Reason for Adoption of Zero-Till Drill-Seeded Wheat

Reasons for Zero-Till Drill-Seeded Wheat Adoption	Garrett Score	Rank
Reduction in ploughing and sowing cost	73.25	I
Increase in wheat yield	67.80	II
Reduction in seed requirement	64.55	III
Reduction in cost of cultivation	58.15	IV
Reduction in irrigation water use	54.95	V
Early sowing of wheat	54.05	VI
Improvement of soil health	50.85	VII
Increase in net income	48.00	VIII
Reduction in fertilizer use	45.40	IX
Reduction in weed problem in field	29.80	X

21.3.7 Reasons for Adoption of Zero-Till Drill-Seeded Wheat

The results of the study showed that out of several influencing factors for adoption of zero-till drill-seeded wheat, utmost crucial factor for adoption of zero-till drill-seeded wheat was low ploughing and sowing cost of wheat, and it was having Garrett score of 73.25 (Table 21.7).

Second major factor was increase in wheat yield and the Garrett score was estimated to be 67.80. The Garrett score for third important factor viz., reduction in seed requirement was 64.55. The decline in cost of cultivation of wheat crop was given fourth rank. Reduction in irrigation water use for wheat crop and early sowing of wheat crop was the reason for adoption of zero-till drill-seeded wheat, and it was ranked fifth and sixth, respectively. Other reasons for adoption of zero-till drill-seeded wheat for wheat crop was improvement of soil health due to incorporation of crop residue into soil, increase in net income from wheat crop, decrease in fertilizer use, and weed population in wheat field and its rank was obtained seventh, eighth, nine, and tenth, respectively.

21.3.8 Reasons for Non-Adoption of Zero-Till Drill-Seeded Wheat

The major discouraging element for non-adoption of zero-till drill-seeded wheat in the study area was high weed population density and poor soil quality and its Garrett score was assessed to be 79.10 and 78.85, respectively, while its rank was first and second (Table 21.8).

TABLE 21.8

Reason for Non-Adoption of Zero-Till Drill-Seeded Wheat

Reasons for Non-Adoption Zero-Till Drill-Seeded Wheat	Garrett Score	Rank
High weed population density	79.10	I
Poor soil quality	78.85	II
Upland field	60.30	III
Uncertainty of irrigation water	58.70	IV
Not sure of profit	56.45	V
High cost of zero-till machine	53.15	VI
Less yield under zero-till drill-seeded wheat	49.50	VII
Does not own zero-till machine	49.05	VIII
Labor issues	48.85	IX
Not sure about zero-till drill-seeded wheat technology	47.30	X
Non-availability of zero-till machine on hire basis	46.50	XI
Non-availability of zero-till machine on time	44.90	XII
Custom hiring of zero-till machine is high	34.75	XIII

TABLE 21.9
Condition for Adoption of Zero-Till Drill-Seeded Wheat in Future

Condition for Adoption of Zero-Till Drill-Seeded Wheat	Garrett Score	Rank
If zero-till machine is available on subsidy	69.00	I
If custom hiring rate for zero-till machine is low	68.90	II
More observation of zero-till drill-seeded wheat technology on other field	58.80	III
Irrigation water availability	45.35	IV
If convinced of yield benefit	45.25	V
If better repair service of zero-till machine is available	38.70	VI
Non-availability of skilled labor	25.00	VII

Other discouraging elements for adoption of zero-till drill-seeded wheat in the study area were upland field and uncertainty of irrigation water availability, and they were ranked third and fourth, respectively. The rank attained for the factors discouraging farmers for zero-till drill-seeded wheat non-adoption were fifth, sixth, and seventh for not sure about profit, high cost of zero-till machine, and lower wheat yield under zero-till drill-seeded wheat, respectively. Other motive for non-adoption of zero-till drill-seeded wheat were farmers does not possess zero-till machine, labor problem, not sure about zero-till drill-seeded wheat technology, unavailability of zero-till machine on rent basis, unavailability of zero-till machine on time and high cost of custom hiring of zero-till machine, and it was ranked eighth, ninth, tenth, eleventh, twelfth, and thirteenth, respectively.

21.3.9 Conditions for Adoption of Zero-Till Drill-Seeded Wheat in Future

The most crucial factor for adoption of zero-till drill-seeded wheat by the farmers in future was found to be availability of zero-till machine on subsidy basis, and it was ranked first with Garrett score of 69.00 (Table 21.9). The second and third major conditions for adoption of zero-till drill-seeded wheat in wheat crop in future was lower rate of custom hiring for zero-till machine and more observation of zero-till drill-seeded wheat technology on other farmers' field. Similarly, fourth and fifth reason for using of zero-till drill-seeded wheat in the study area were irrigation water availability and if farmers would be convinced about yield benefit, then they can adopt zero-till drill-seeded wheat on their field. The other influencing factor for adoption of zero-till drill-seeded wheat was better easy services of repair for zero-till machine and availability of skilled labor for running zero-till machine.

21.4 Conclusions

In the case of zero-till drill-seeded wheat adopters, per hectare average cost of cultivation was lower than the non-adopters with Rs 3901.33. The incremental per hectare yield of wheat crop under zero-till drill-seeded wheat was 6.25 quintal. Per hectare average incremental gross income and net income over non-adopter was Rs 13,437.50 and Rs 17,338.84, respectively. The B-C ratio was estimated to be 1.78 and 1.41 for zero-till drill-seeded wheat adopters and non-adopters, respectively. The economic benefit due to adoption of zero-till drill-seeded wheat was found to be Rs 16,984.17/ha. After adoption of zero-till drill-seeded wheat, carbon emission can be reduced by 9.16 kg/ha over non-adopters. Age of respondents, farming experience and land holding size were the important factors that influenced the adoption of zero-till drill-seeded wheat. Major factors for adoption of zero-till drill-seeded wheat were reduction in cost of ploughing and sowing of wheat crop, increase in wheat yield, and decrease in the cost of cultivation. The high weed population in farmers' field was the major factor for non-adoption of zero-till drill-seeded wheat. The future conditions for zero-till drill-seeded wheat adoption were zero-till machine, which could be easily available on subsidy basis, lower rate of custom hiring of zero-till machine, and more observation on other field.

21.5 Future Prospective and Policy Implications

Government and manufacturers of zero-till machine should work together to popularize among the farming community. From manufacturer side, the efforts should be made to reduce the cost of machine. Manufacturers of zero-till machine should work in synergy with tractor manufacturer for improving complementarity, efficiency, and adoption of the technology on small and marginal farms. Mechanism for promotional incentives for zero-till machine should be linked with the purchase of tractor by the farmers. Government should frame policy to provide financial support to the farmers when they purchase zero-till machine. While designing subsidy/incentives, consumers' surplus and environmental benefits should be considered. There is a need to establish service providing centers/custom hiring centers by using IT tools (Apps) for zero-till machine. To achieve these purposes, government should provide necessary training and financial support to agricultural graduates for opening such service centers. These centers should provide zero-till machine on hiring basis to the marginal and small farmers of surrounding villages, because marginal and small farmers are resource poor, and they can't afford to purchase costly zero-till machine, and they are deprived from the benefits of zero-till drill-seeded wheat technology. Zero-till drill-seeded wheat sowing should be promoted in canal command areas as well as in the adjoining area of rivers having heavy soil/black soils with abundant moisture contained in the soil.

REFERENCES

Alston, J. M., Norton, G. W. and Pardey, P. G. 1995. *Science Under Scarcity: Principles and Practice for Agricultural Research Evaluation and Priority Setting.* Cornell University Press, Ithaca, NY.

Faulkner, E. H. 1943. *Plowman's Folly.* Michael Joseph Ltd, 26 Bloomsbury Street, W.C.I, London.

Food and Agriculture Organisation. 2008. What is conservation agriculture. In: FAO-Conservation Agriculture. http://www.fao.org/ag/ca/1a.html.

Friedrich, T., Derpsch, R. and Kassam, A. H. 2012. Global overview of the spread of conservation agriculture. *Field Action Science Report* 6: 1–7.

Govaerts, B., Verhulst, N., Castellanos-Navrrete, A., Sayre, K. D., Dixon, J. and Dendooven, L. 2009. Conservation agriculture and soil carbon sequestration: between myth and farmer reality. *Critical Review Plant Science* 28: 97–122.

Government of Uttar Pradesh. 2021. *Shyankhiki Patrika*, Department of Economics and Statistics, Ministry of Agriculture, Government of Uttar Pradesh, Lucknow. http://updes.up.nic.in/spideradmin/Hpage1.jsp

Holland, J. M. 2004. The environmental consequences of adoption of conservation tillage in Europe: reviewing the evidences. *Agriculture Ecosystem Environment* 103: 1–25.

Jat, M. L., Sharma, S. K. and Singh, K. K. 2006. Conservation agriculture for sustainable farming in India. Paper presented in winter school training at Department of Agronomy, Tamil Nadu Agricultural University, Coimbatore, January 21.

Kaur, B., Singh, S., Garg, B. R., Singh, J. M. and Singh, J. 2012. Enhancing water productivity through on-farm resource conservation technology in Punjab agriculture. *Agricultural Economics Research Review* 25(1): 79–85.

Mallada, G. S. 1992. *Introduction to Econometrics.* Second edition, Macmillan Publishing Company, New York, NY.

Paustian, K., Antle, M. Sheehan J. and Eldor, P. 2006. *Agriculture's Role in Greenhouse Gas Mitigation.* Pew Center on Global Climate Change, Washington, DC.

Serraj, R. and Siddiqe, K. H. M. 2012. Conservation agriculture in dry areas. *Field Crop Research* 132: 1–6.

Singh, O. P., Gautam, Y., Singh, P. K. and Singh, H. P. 2018. Enhancing input use efficiency through resource conservation technologies: empirical evidences from different agro-climatic zones of Uttar Pradesh. *Indian Journal of Agricultural Economics* 73(3): 249–268.

Singh, O. P., Singh, R. and Singh, P. K. 2020. *Economic and Environmental Impact of Resource Conservation Technologies in Different Agro-Climatic Zones of Uttar Pradesh*, Hanns Publication House, New Delhi & Varanasi.

22

Integrated Nutrient Management Technology for Sustainable Vegetable Production in Eastern India

Deblina Ghosh, Snigdha Chatterjee, Kaushik Batabyal, and Sidhu Murmu
Department of Agricultural Chemistry and Soil Science, Bidhan Chandra Krishi Viswavidyalaya, Nadia, West Bengal, India

CONTENTS

22.1 Introduction

Vegetables, being rich source of vitamins, minerals, dietary fibers, carbohydrate and various phytochemicals having nutraceutical properties, play a crucial role in nutritional security of the country. India is the second largest producer of vegetables next to China (Bera 2015). During 2017–18, the area under vegetables was 10.26 Mha with a production of 184.40 Mt in India. For this period, the total vegetable production was highest in case of Uttar Pradesh (283.16 Mt) followed by West Bengal (276.95 Mt) (NHB 2018). Vegetable production in India has grown remarkably since 1960s. The acreage harvested vegetables in India has been trebled from 2.9 Mha in 1961–63 to 8.6 Mha in 2017–18. The vegetables production in India has increased around 10 times, i.e., from 19.1 Mt in 1961–63 to 184.4 Mt in 2017–18. In spite of the spectacular progress made in vegetable production, the availability of vegetables in India is only about 245 g/day/person against the minimum dietary requirement of 300 g/day/person as recommended by the Indian Council of Medical Research and National Institute of Nutrition, Hyderabad. Among various factors responsible for the low productivity of vegetables in the country, inefficient nutrient management practice is of prime concern. Vegetables need further increase in the production and supply, which is possible only through further increasing in productivity with the help of various technological advancements.

Also, demographic pressure of a burgeoning population has kept researchers busy to find possible alternatives of raising productivity per unit land area and time. Achieving a balance between crop

DOI: 10.1201/9781003164968-25

nutrient requirements and nutrient reserves in the soil is essential for maintaining higher yield and soil fertility, besides safeguarding environmental degradation. Such an objective becomes further difficult to accomplish due to shrinking per capita land availability, more so in the developing world (Srivastava and Ngullie 2009). The present era of intensive high-yield agriculture is much dependent on chemical fertilizers. Agricultural intensification over the last few decades has strongly increased global food production, but at high environmental cost (Conway 1997; Tilman et al. 2001), and generally leads to imbalances in nutrient budgets and increased loss of potential pollutants to the atmosphere (Batabyal 2017). Thus, the main challenge in the forthcoming few years lies in the following question, i.e., can agriculture meet world's food needs which are expected to exceed 7.5 billion by the year 2020? (FAO 2012). Global food production needs to be increased at least by 70% over the current level by 2050 to meet the increased food demands (FAO 2009). To achieve this challenging target, agriculture should grow significantly, in consideration of the factors that contributes to increase the yield production, which are already reduced or tend to reduce, since they are placing unprecedented pressure on the natural resources (Selim 2020). Continuous use of synthetic fertilizers poses problems to soil by increasing soil acidity, which leads to hasty deterioration of soil health and productivity (Selim 2018). Organic inputs besides supplying nutrients to the first crop, also provides substantial residual amount of unutilized nutrients on the succeeding crop. At the same time, application of organic manures alone does not produce required yields due to their low nutrient status and low availability in bulk amount. Therefore, the nutrient needs can best be met through integrated nutrient management (INM). The concept of INM aims to increase the efficiency of all nutrient sources, be they soil resources, mineral fertilizers, organic manures, recyclable wastes, or bio fertilizers in an integrated manner. INM is an approach that seeks to both increase quality of production and protect the environment for posterity by addition of both organic and inorganic source of nutrients to the crops. It depends on nutrient application and conservation, new technologies to increase nutrient availability to plants, and the dissemination of knowledge between farmers and researchers (Palm et al. 2001). In the past, nutrient management had been driven by the need for increase in production in a sustainable way. But now, a sustainable nutrient management approach involves maximizing production, preventing on-site soil degradation, and minimizing off-site involvement of applied nutrients (Tagaliavini and Marangoni 2002). Moreover, INM reduces erosion, improves water infiltration, soil aeration, and plant root growth. Moreover, it minimizes the risk of downstream flooding (Gill et al. 2008). In this chapter, an attempt has been made to discuss the role of sustainable INM technologies toward augmenting vegetable production in eastern India.

22.2 Concepts of Sustainable Nutrient Management

There are three concepts of sustainable nutrient management:

i. IPNS (Integrated Plant Nutrition System) is a concept that aims to maintain or adjust supply to an optimum level for sustaining desired crop productivity by optimizing benefit from all possible sources of plant nutrients in an integrated manner;
ii. IPNSS (Integrated Plant Nutrient Supply System) is a method to achieve the objective of "IPNS." The supply of nutrients to the plants are from various sources, i.e., nutrient reserves in the soil and organic sources and;
iii. INM is the actual technical and managerial component of achieving the objective of IPNS under farm situations. It takes into account all the factors of soil and crop management including the management of all other inputs such as water, agrochemicals, etc. besides nutrients.

In the light of above three concepts, in order to address present or future nutritional problems, only two strategies – strategic and diagnostic approach are chosen. In strategic INM, potential problems on farms are characterized that fit into particular category, and recommendations are then accordingly made for each category. In the diagnostic approach, the problem of each soil type is diagnosed separately, and the remedies are then prepared as per soil type (Srivastava and Singh 2006).

22.3 Integrated Nutrient Management

INM refers to the maintenance of soil fertility and of plant nutrient supply at an optimum level for sustaining the desired productivity through optimization of benefits in an integrated manner from all the possible sources of organic, inorganic, and biological components. Several researchers discovered that various materials can be used as a constituent of INM combinations, viz., farmyard manures, natural and mineral fertilizers, soil amendments, crop residues and farm waste recycling, agroforestry, green manures, and compost (Zhang et al. 2012). Those materials are generally seen as a key practice in the eco-friendly system to ensure sustainable production and safety of the food production. The basic concept underlying INM is the maintenance or adjustment of soil fertility/productivity and of optimal plant nutrient supply for sustaining the desired level of crop productivity (FAO 1995). The objective is to accomplish this through optimization of the benefits from all possible sources of plant nutrients, including locally available ones, in an integrated manner while ensuring environmental quality. This provides a system of crop nutrition in which plant nutrient needs are fulfilled by a preplanned integrated use of: mineral fertilizers; organic manures/fertilizers (e.g., green manures, recyclable wastes, crop residues, and FYM); and bio-fertilizers. The appropriate combination of different sources of nutrients varies according to the system of land use and the ecological, social, and economic conditions at the local level (Roy et al. 2006).

22.4 Principles of INM

The points below are considered the main principles of INM (Figure 22.1) to be considered in planning the INM strategies (Selim 2020):

1. Adopting INM practices must be compatible with the local farming system, for example, field biological condition (weeds, insects, and diseases), soil texture, irrigation services, and equipment's availability and climatic conditions.
2. Using both sources of nutrients optimize fertilizer input, increases the use efficiency by crops and then lessens their nutrient requirements and reduces food contamination, environmental pollution, and total costs.
3. Corresponding soil nutrient supplies with crop demand spatially and temporally in order to conserve native soil contents.
4. INM practice reduces the fertilizer loss, improves the crop potential, and increases the profitability.

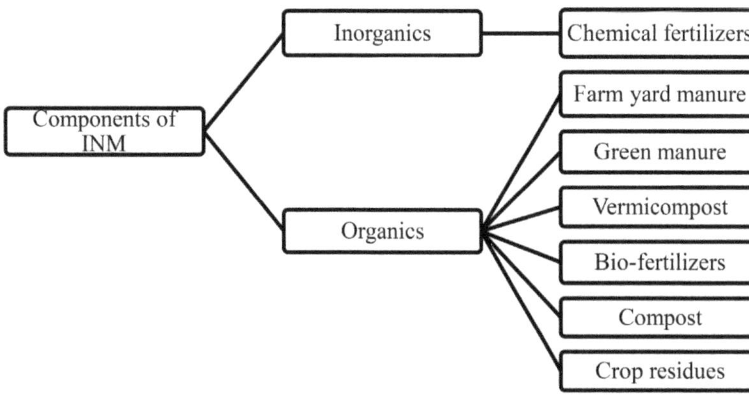

FIGURE 22.1 Different components of INM.

5. In long term, it improves the physiochemical properties of the soil (biological and hydrological properties).
6. Applying INM practice on root zone, where the most important process between plant and soil occur, is a "bottleneck" controlling nutrient conversions, solubility, availability, and release from soil to plant roots and absorption. Such an exercise can respond to maximize the biological potential through enhancing the soil microorganism's activity.

22.5 INM Impact on Vegetable Production in Eastern India

Targeting high yield is the most logical way to raise the total production of vegetables from the country's limited resources. The chemical fertilizers or the organic and biological sources alone cannot achieve production sustainability of vegetables. Even with the balanced use of chemical fertilizers, high yield level could not be maintained over the years because of deterioration in soil physical and biological environments due to low organic matter content in soils. In this context and as a further response to economic recession, and also to conserve and improve soil fertility, the concept of INM system has been adopted all over the world.

INM treatments remarkably affected growth characteristics and yield attributes of many vegetables. Growth characteristics like plant height, branches per plant, canopy diameter, root length, fruit weight, fruit length etc., all ultimately contributes to the yield of the vegetable. The integrated use of organics and inorganics resulted in increase in all the above growth characteristics as compared to sole application of inorganics in soils of hilly belt of Odisha (Table 22.1). Better growth and yield of plants receiving VC compared to other treatments may be due to higher nutrient content, i.e., available N, P, and K contents and higher microbial populations. Several enzymes and hormones present in VC promote plant growth and development (Venkateswarlu 2003). Favorable effects of CM and VC in the root zone could result in increased availability and uptake of nutrients by plants (Atiyeh et al. 2000; Baskar 2003; Roy and Singh 2006) that likely contributed to better growth and yield. Application of different levels and sources of nutrients significantly influenced the yield of cauliflower (Table 22.1). Organic, inorganic, and integrated sources of nutrients, on average, resulted in 20%, 217%, and 225% increase in total biomass

TABLE 22.1

Effect of Nutrient Management Technologies on Different Parameters of Yield and Crop Quality in Cauliflower (Batabyal et al. 2016b)

Treatments	Total Biomass Yield (Mg/ha)	Curd Yield (Mg/ha)	Curd Comp-action (g/cm^2)	Crude Protein (%)		Dietary Fiber (%)	Vitamin C (mg/100 g)	B:C ratio
				Curd	Stalk	Curd	Curd	
Control	9.99	2.63	0.96	10.3	8.9	41.7	48.6	1.0
NPK	30.55	7.75	1.30	17.2	19.9	44.0	54.2	2.7
VC10	11.68	2.93	0.79	15.8	12.9	48.0	60.5	0.9
FYM20	12.96	2.96	0.91	11.4	12.1	40.0	55.8	1.6
GM40	11.15	3.05	0.67	12.3	10.7	39.0	57.7	1.1
FYM10 + VC5	12.03	2.94	0.82	12.3	11.0	43.0	61.8	1.2
GM20 + VC5	12.37	2.88	0.73	11.2	9.1	41.0	64.9	1.1
GM20 + FYM10	11.68	2.68	0.69	13.7	11.8	42.0	57.8	1.3
VC3 + NPK	33.27	9.47	1.29	18.6	18.0	51.0	68.9	2.5
FYM5 + NPK	30.07	9.40	1.16	16.6	17.1	44.0	68.0	2.6
GM20 + NPK	30.85	8.32	1.12	20.3	20.2	48.0	70.2	2.3
LSD ($P \leq 0.05$)	3.69	1.42	0.47	1.1	1.3	2.2	9.46	0.2

NPK, N-P-K at 200-44-82 kg ha^{-1}; VC, Vermicompost; FYM, Farmyard manure; GM, Green manure; numbers followed by VC, FYM, and GM indicate dose in Mg ha^{-1}.

TABLE 22.2

Effect of INM on Growth, Yield Attributes, and root Characteristics of Winter Vegetables (Dass et al. 2008a)

	Plant Height (cm)		Canopy Diameter (cm)		Average Root Length (cm)		Root dry Matter (g/plant)		Root Volume (cm^3)		Branches/ Plant	Fruit Weight (g)
Treatments	**BP**	**C**	**BP**	**C**	**BP**	**C**	**BP**	**C**	**BP**	**C**	**BP**	**BP**
T_1-Control	40.0c	26.0c	26.0d	34.7e	23.5b	20.13c	5.7d	7.7d	12.9d	16.7d	3.6d	25.8f
T_2-RRF	46.7b	29.7b	37.7b	50.2a	29.3a	24.38ab	11.4b	13.3bc	26.7b	35.7bc	7.7b	83.5c
T_3-50% RRF+VC	49.2a	32.7a	40.0a	50.9a	29.4a	25.98a	12.6ab	15.6a	28.6ab	43.1a	8.7a	104.2a
T_4-50% RRF+BF	43.6c	27.6c	34.2c	40.4d	25.7b	22.65bc	8.4c	12.4c	18.0c	32.4c	6.5c	51.4e
T_5-50%RRF+CM	47.2b	30.6b	37.7b	45.6bc	30.1a	24.75ab	13.3a	14.6ab	29.5a	36.7bc	7.8b	83.4c
$T_6$50%RRF + VC + CM	48.0ab	31.2ab	38.5ab	47.6b	30.4a	26.92a	13.7a	16.1a	30.2a	44.6a	7.8b	88.8b
T_7-50% BF +CM	44.1c	29.5b	34.9c	43.6c	29.2a	25.85a	12.6ab	15.1ab	29.0a	37.8b	7.0c	71.5d

BP—Bell pepper, C—Cabbage, RRF—Recommended rate of synthetic fertilizer, VC—Vermicompost, BF—Phosphorus solubilizing bacteria (PSB) + Azotobacter, CM—Cow manure.

yield and 11%, 228%, and 240% increase in curd yield of cauliflower, respectively, over the control in soils of BCKV, West Bengal. The average curd yield of cauliflower with inorganic and integrated sources was two to three times higher than that with organics (Batabyal et al. 2016b).

22.6 INM Impact on Quality of Vegetable Production

Application of different levels and sources of nutrients significantly influenced the contents of value-added products, viz., crude protein, dietary fiber, and vitamin C in cauliflower. Combined application of organic and inorganic sources of nutrients produced curds with higher concentration of dietary fiber, crude protein, and vitamin C than their individual applications (Table 22.2), the relative magnitude being integrated > inorganic > organic (Batabyal et al. 2016b). In broccoli also, Tamang et al. (2017) observed that the INM technologies adopted for broccoli cultivation were found to influence its yield and quality. A balanced application of organic and inorganic nutrient sources to broccoli was productive, economically sound, and energy efficient.

22.7 INM Impacts on Economics of Vegetable Production

The highest gross return (benefit:cost ratio) can be obtained when integrated dose of inorganic fertilizer is applied with organic manure (Table 22.1). The combination of organics with inorganics resulted in significantly higher B:C ratio as compared to the sole application of organics (Batabyal et al. 2016b, Tamang et al. 2017). Three crops were taken for comparison, i.e., brinjal, broccoli, and cauliflower, and the common management practices were taken into consideration (Figure 22.2). It can be clearly seen that the management practices with integrated (inorganic + organic) source of nutrients resulted in higher B:C ratio than the sole application of inorganics.

22.8 INM Impacts on Soil Health

22.8.1 Physical and Chemical Health

Organic matter in the soil is crucial for soil productivity and better soil health. Addition of organic fertilizer results in increased soil organic carbon (SOC) levels in the soil, while chemical fertilizer result in decreased SOC and basic cation contents and lowering of soil pH. As a result, a positive effect on soil results in modification of soil structure thereby increases the yield in the long term. The integrated

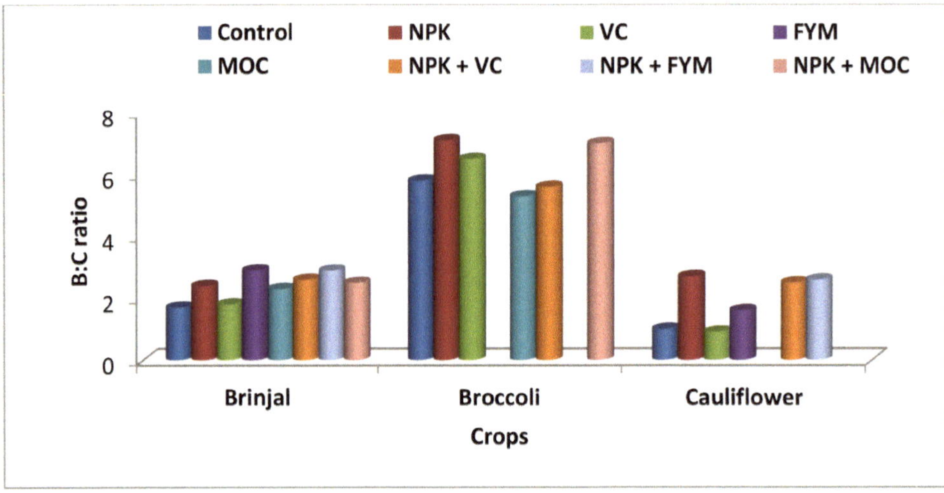

FIGURE 22.2 Benefit:cost ratio of some crops under INM.

(NPK—Recommended dose of fertilizers, VC—Vermicompost, FYM—Farmyard manure, MOC—Mustard oil cake) (Brinjal—Batabyal et al. 2017, Broccoli—Tamang et al. 2017 and Cauliflower—Batabyal et al. 2017).

application of organics with inorganics decreased the pH of the soil when compared to sole inorganic application (Table 22.3). Carbon supplementation in cauliflower through organics, on average, increased the TOC stock of soil by 26% over the initial soil. Only inorganics, however, did not show any significant change (+0.9%) in organic C content but, when integrated with organics, caused an increase (9%) (Table 22.3). The organic treatments, such as FYM20, FYM10 + VC5, and GM20 + FYM10 caused increase in TOC in soil as much as 45%, 34%, and 31%, respectively, over that to the initial soil (Batabyal et al. 2016b). This could directly attribute to direct addition of organic matter and increased root biomass and cycling.

TABLE 22.3

Changes in Soil Properties and Soil Quality Index under Different Nutrient Management Technologies after Six Years of Cauliflower Cultivation (Batabyal et al. 2016b)

Treatments	Soil pH (1:2.5)	BD (Mg m^{-3})	OC (g kg^{-1})	MBC (mg C kg^{-1})	SQI
Control	7.1	1.30	5.9	40.0	3.101
NPK	7.1	1.31	6.5	45.0	4.629
VC10	6.8	1.18	7.8	210.0	5.030
FYM20	6.7	1.23	9.9	330.0	5.804
GM40	6.8	1.22	8.0	199.0	5.352
FYM10+VC5	6.7	1.20	9.4	302.0	5.672
GM20+VC5	6.9	1.25	8.0	147.0	5.268
GM20+FYM10	6.5	1.25	8.8	192.0	5.646
VC3+NPK	6.9	1.23	7.0	218.0	5.464
FYM5+NPK	6.9	1.24	7.7	222.0	5.825
GM20+NPK	6.7	1.24	7.5	230.0	6.074
LSD ($P \leq 0.05$)	0.31	0.04	1.6	65.0	-

BD—Bulk density, OC—Organic carbon, MBC—Microbial biomass carbon, SQI—Soil quality index, NPK-N-P-K at 200-44-82 kg ha^{-1}, VC—Vermicompost, FYM—Farmyard manure, GM—Green manure, numbers followed by VC, FYM, and GM indicate dose in Mg ha^{-1}.

Along with the evaluation of soil chemical properties, soil physical properties are also enhanced by the use of INM practices, as it reduces soil erosion and increases cycling of the organic residues. Thus, it improves nutrient and water retention capacity of the soil. This management also improves the soil structure, water infiltration, and soil aeration (Das et al. 2014). It was reported in Eastern Ghats of Orissa that with application of FYM or VC, or both, the bulk density was relatively lower as compared to inorganics in tomato crop (Dass et al. 2008). This could be ascribed to better soil aggregation and aeration due to the organic amendments by adding various humus fractions and increase in microbial and enzymatic activity accelerated due to VC. Availability of all the elements (N, P, K, Ca, Mg, Fe, Mn, Zn, and Cu) in soil was found to be higher in organic and integrated treatments than in inorganic ones (Batabyal et al. 2016b). This is due to mineralization and subsequent release of these elements contained in the organics.

22.8.2 Biological Health

The microorganism plays a key role in nutrient cycling of the soils for sustainable productivity, because microbes are the source and sink for mineral nutrition and carry out various biochemical transformations. Moreover, an increase in soil microbial-biomass C and nitrogen (N) is obvious in soils receiving combined application of organic manures and chemical fertilizers compared to soils receiving chemical fertilizers only. The use of organic fertilizer together with chemical fertilizers, compared to the addition of organic fertilizers alone, has a higher positive effect on microbial biomass and hence soil health (Hati et al. 2008). Soil biomass is increased by INM as these amendments supply readily decomposable organic matter in addition to increasing root biomass and root exudates due to greater crop growth (Vineela et al. 2008). The application of organics alone or with inorganics provided a more favorable environment for rapid microbial growth, which caused a greater increase in MBC in the soils (Moscatelli et al. 2005). Management practices that include the incorporation of organic matter into soil, therefore, increase biological activity. Higher MBC content in soil associated with addition of FYM was also reported in alluvial soil of the hot humid subtropics of eastern India under rice based cropping systems (Majumder et al. 2008).A larger increase in MBC with FYM compared to VC or GM was observed by Batabyal et al. (2016b) in cauliflower (Table 22.3), which might be due to the presence of decomposition resistant fractions in the former (lignin 182 mg kg^{-1}; polyphenol 12 mg kg^{-1}) compared with the other two.

Soil quality is broadly defined as, "the capacity of a living soil to function, with in natural or managed ecosystem boundaries, to sustain plant and animal productivity, maintain or enhance water and air quality, and promote plant and animal health" (Doran 2002). Doran and Parkin (1996) proposed a minimum data set called indicators for characterizing and monitoring soil quality. The minimum data set includes soil attributes and properties such as, texture, soil and rooting depth, bulk density, infiltration, water retention characteristics, soil organic matter, electrical conductivity, extractable N, P and K, microbial biomass, and soil respiration.

Changes in SQI were higher with integrated treatments than with organic or inorganic fertilizer alone and the relative order of increase was as follows: integrated > organic > inorganic. Among the organic amendments, GM was the most efficient in improving soil quality followed by FYM, and VC. When effects of each organic in combination within organics were compared, GM was better than FYM and VC (Table 22.3). Quality of organics was more important than quantity in improving SQI. This was evident from variations in magnitude of improvement in SQI per unit organic C applied through organic sources (Batabyal et al. 2016b).

22.9 INM Recommendations for Vegetable Crops in Eastern India

A wide range of vegetables (Table 22.4) have shown a high magnitude of response through different combinations of INM in eastern India. These summarized results are more of interpretative than suggestive, affirming the practice leading to significant reduction in load on the use of inorganic chemical fertilizers under INM.

TABLE 22.4
INM Recommendations for Vegetables in Eastern India

Vegetable	Site	INM Recommendations	Yield Target (Mg ha^{-1})	Reference
Cauliflower	West Bengal	125% [N(200)-P(44)-K(82) kg ha^{-1}] + FYM (5 Mg ha^{-1}); and 125% [N(200)-P(44)-K(82) kg ha^{-1}] + VC (3 Mg ha^{-1})	32.31 and 34.57	Batabyal et al. (2016b)
Eggplant/Brinjal	West Bengal	150% [N(100)-P(22)-K(42) kg ha^{-1}] + VC (3 Mg ha^{-1})	22.31	Batabyal et al. (2016a)
Broccoli	West Bengal	125% [N(150)-P(33)-K(63) kg ha^{-1}] + CM (6.25 Mg ha^{-1})	41.1	Tamang et al. (2017)
Pointed gourd	West Bengal	NPK@220-125-125 kg N,P_2O_5 and K_2O ha^{-1} + FYM@ 2.4 t ha^{-1} + VC@ 1 t ha^{-1}	38.54	Tamang (2010)
Carrot	West Bengal	NPK@120-80-120 kg N,P_2O_5 and K_2O ha^{-1} + FYM@ 5 t ha^{-1}; and NPK@120-80-120 kg N,P_2O_5 and K_2O ha^{-1} + GM @ 20 t ha^{-1}	30.41 and 30.78	Murmu (2011)
Amaranthus	West Bengal	NPK@146-26-62 kg N,P_2O_5 and K_2O ha^{-1} + FYM@ 10 t ha^{-1}	30.0	Murmu (2011)
Radish	West Bengal	N(48)-P(4.8)-K(15.8) kg ha^{-1} + FYM (10 t ha^{-1})	45.0	Batabyal et al. (2015)
Cabbage	Odisha	50% [N(150)-P(22)-K(62) kg ha^{-1}] + VC (5 Mg ha^{-1})	53.94	Dass et al. (2008a)
Bell pepper	Odisha	50% [N(120)-P(26)-K(100) kg ha^{-1}] + VC (5 Mg ha^{-1})	8.51	Dass et al. (2008a)
Finger millet	Odisha	50% [N(40)-P_2O_5(20)-K_2O(20) kg ha^{-1}] + *Gliricidia* (2.5 t ha^{-1}) + *Azotobacter* and PSB (2.5 kg ha^{-1}) each	9.04	Dass et al. (2013)
Elephant's foot yam	West Bengal	NPK@200-100-150 kg N,P_2O_5 and K_2O ha^{-1} + MOC@ 1.75 t ha^{-1}	57.8	Das (2013)
Tomato	Odisha	50% [N(120)-P(75)-K(100) kg ha^{-1}] + VC (5 Mg ha^{-1}); and 50% [N(120)-P(75)-K(100) kg ha^{-1}] + VC (2.5 Mg ha^{-1}) + FYM (5 Mg ha^{-1})	20.75 and 19.70	Dass et al. (2008b)

FYM—Farm yard manure, VC—Vermicompost, CM—cow manure, PSB—Phosphorus solubilizing bacteria, MOC—Mustard oil cake.

22.10 Farmer's Success Stories on INM Technologies in West Bengal

West Bengal is the largest vegetable-producing state in eastern India. There are many farmers who have been highly successful and flourishing very well by adopting INM technologies despite having many other constraints. Through innovations and efforts, these farmers have not only transformed their own lives but also of others too. Documenting success stories of farmers can inspire millions of farmers. Followings are some of the examples of some promising farmers of West Bengal, who are very much successful in vegetable cultivation using INM practices in West Bengal:

Subal Sarkar lives in Balagarh village and Ajay Kath lives in Dhobapara village in Hooghly district. They cultivated onion, beans, cucumber, and bitter gourd with fertilizer dose of NPK (10:26:26): 60 kg ha^{-1}, urea: 75 kg ha^{-1}, NPK (10:26:26): 104 kg ha^{-1}, and NPK (10:26:26): 104 kg ha^{-1}, respectively, along with FYM 7.5 t ha^{-1} to each crop and achieved outstanding yield of 16–21 t ha^{-1} (onion), 5–6 t ha^{-1} (bean), 23 t ha^{-1} (cucumber), and 1.1 t ha^{-1} (bitter gourd).

FIGURE 22.3 Effect of INM technologies on (a) bitter gourd, (a) cucumber, and (c) onion in Hooghly district of West Bengal.

Ainul Haque and Parul Modak from Cooch Behar district cultivated tomato and cucumber with INM practice involving application of NPK (10:26:26) @ 75 and 100 kg ha^{-1}, respectively, along with application of VC @ 2 t ha^{-1} to each crop and achieved yield of 22 t ha^{-1} of tomato and 23.5 t ha^{-1} of cucumber.

Another farmer, Nazrul Mondal from Kasthodanga village in Nadia district achieved the yield of 40–45 t ha^{-1} of elephant foot yam, a high-value vegetable crop, and 12 t ha^{-1} of cabbage with application of urea: 75 kg ha^{-1}, MOP: 45 kg ha^{-1}, SSP: 45 kg ha^{-1}, and micronutrient: 8 kg ha^{-1}. While Rajkumar Ghosh (Baksha), with application of urea: 75 kg ha^{-1}, MOP: 45 kg ha^{-1}, SSP: 45 kg ha^{-1}, micronutrient: 6 kg ha^{-1} along with application of FYM @ 5 t ha^{-1} achieved yield of 16–21 t ha^{-1} of onion, 5–6 t ha^{-1} of bean, and 7.5 t ha^{-1} of okra. Rahidul Mondal from Rautari, Chakdah village in Nadia district achieved yield of 5–6 t ha^{-1} of bean and 4.2 t ha^{-1} of brinjal with application of urea, MOP and SSP at 60, 30, and 33 kg ha^{-1}, respectively, to each crop along with FYM @ 5 t ha^{-1}.

A field view of the different vegetable crops grown with INM technologies in West Bengal is shown below (Figures 22.3 and 22.4).

22.11 Constraints in INM

In our country, farmers gain more agricultural knowledge and experience from others than from the agricultural technology extension technicians. Lack of extension services also contributes to the low adoption of INM technology. The smallholding farms restrict nutrient management technologies. Small number of staff engaged in soil and fertilizer management in the extension system could not solve the

FIGURE 22.4 Effect of INM technologies on (a) cabbage, (b) chili, and (c) elephant foot yam in Nadia district of West Bengal.

technological problems of farmers. In the case of large-scale application of INM, the biggest problem lies in the judicious use of organic inputs with regards to differential nutrient release pattern and timely availability in bulk quantity. In order to add promising organic sources to INM practices, information is required on the crop-specific composition of the microbial community and the role of dominant microbial population for improving the yield or quality. Very limited studies have been carried out to separate the differences in yield potential arising out of genotype versus improved INM practices or genotype versus a plant's internal metabolic efficiency of nutrient use. The multi-prolonged effect of organic matter on soil quality changes can often be associated with a negative impact, and a decline in productivity warrants strict regulation of organic fertilizer quality and applied quantity to avoid any possible contamination of productive farmland.

22.12 Conclusion

INM is an important viable tool for sustainable vegetable production. Nutrient management technologies adopted for vegetable cultivation influenced not only the biomass yield of the crop but also its quality, the quality of soil, and the overall economics of the cultivation. Balanced application of organic and inorganic nutrient sources was proved to be productive, economically sound, and environment friendly practice. INM technologies helped to maintain soil quality by improving soil organic C stock, MBC, BD, and other available nutrients.

REFERENCES

Atiyeh, R. M., Domínguez, J., Subler, S. and Edwards, C. A. 2000. Changes in biochemical properties of cow manure during processing by earthworms (Eisenia andrei, Bouché) and the effects on seedling growth. *Pedobiologia* 44(6): 709–724.

Baskar, K. 2003. Effect of integrated use of inorganic fertilizers and farm-yard-manure or green-leaf manure on uptake and nutrient-use-efficiency of rice-rice system on an inceptisol. *Journal of Indian Society of Soil Science* 51(1): 47–51.

Batabyal, K. 2017. Nutrient management for improving crop, soil, and environmental quality. In *Essential Plant Nutrients* (pp. 445–464). Cham: Springer.

Batabyal, K., Mandal, B. and Hazra, G.C. 2016a. Nutrient management, energy input-output and economic analyses of eggplant production under subtropical conditions. *International Journal of Vegetable Science* 22: 409–419.

Batabyal, K., Mandal, B., Sarkar, D. and Murmu, S. 2017. Assessment of nutrient management technologies for eggplant production under subtropical conditions: a comprehensive approach. *Experimental Agriculture* 53(4): 588–608.

Batabyal, K., Mandal, B., Sarkar, D., Murmu, S., Tamang, A., Das, I., Hazra, G. C. and Chattopadhyay, P.S. 2016b. Comprehensive assessment of nutrient management technologies for cauliflower production under subtropical conditions. *European Journal of Agronomy* 7: 1–13.

Batabyal, K., Sarkar, D. and Mandal, B. 2015. Fertilizer-prescription equations for targeted yield in radish under integrated nutrient management system, *Journal of Horticultural Science* 10(1): 18–23.

Bera, S. 2015. Deficit rains spare horticulture, record production expected, Live mint, The Hindustan Times, January 2015.

Conway, G. 1997. *The doubly-green revolution: Food for all in the 21st century* (p. 335). London: Penguin.

Das, B., Chakraborty, D., Singh, V. K., Aggarwal, P., Singh, R., Dwivedi, B. S. and Mishra, R. P. 2014. Effect of integrated nutrient management practice on soil aggregate properties, its stability and aggregate-associated carbon content in an intensive rice–wheat system. *Soil and Tillage Research* 136: 9–18.

Das, I. 2013. Formulating nutrient management schedules for improving yield and quality of broccoli, elephant foot yam and onion. PhD Thesis, Bidhan Chandra Krishi Viswavidyalaya.

Dass, A., Lenka, N. K., Patnaik, U. S. and Sudhishri, S. 2008a. Integrated nutrient management for production, economics, and soil improvement in winter vegetables. *International Journal of Vegetable Science* 14(2): 104–120.

Dass, A., Lenka, N. K., Sudhishri, S. and Patnaik, U. S. 2008b. Influence of integrated nutrient management on production, economics and soil properties in tomato (Lycopersicon esculentum) under on-farm conditions in Eastern Ghats of Orissa. *Indian Journal of Agricultural Sciences* 78(1): 40.

Dass, A., Sudhishri, S. and Lenka, N. K. 2013. Integrated nutrient management to improve finger millet productivity and soil conditions in hilly region of Eastern India. *Journal of Crop Improvement* 27(5): 528–546.

Doran, J. W. 2002. Soil health and global sustainability: translating science into practice. *Agriculture, Ecosystem and Environment* 88: 119–127.

Doran, J. W. and Parkin, T. B. 1996. Quantitative indicators of soil quality: A minimum data set. In: *Methods for Assessment of Soil Quality* (JW Doran and AJ Jones, Eds). SSSA Special Publication No. 49, Soil Science Society of America, Madison, WI.

Food and Agriculture Organization. 1995. *Integrated plant nutrition system. FAO Fertilizer and Plant Nutrition Bulletin* No. 12. Rome. 426 pp.

Food and Agriculture Organization. 2009. Declaration of the World Summit on Food Security, Rome, Italy, 2009.

Food and Agriculture Organization. 2012. "FAOSTAT FAO database—statistics division food and agriculture organization of the United Nations, Rome," 2012.

Gill, M. S., Pal, S. S., and Ahlawat, I. P. S. 2008. Approaches for sustainability of rice (Oryza sativa)-wheat (Triticum aestivum) cropping system in Indo-Gangetic plains of India—A review. *Indian Journal of Agronomy* 53(2): 81–96.

Hati, K. M., Swarup, A., Mishra, B., Manna, M. C., Wanjari, R. H., Mandal, K. G. and Misra, A. K. 2008. Impact of long-term application of fertilizer, manure and lime under intensive cropping on physical properties and organic carbon content of an Alfisol. *Geoderma* 148(2): 173–179.

Majumder, B., Mandal, B., Bandyopadhyay, P. K., Gangopadhyay, A., Mani, P. K., Kundu, A. L. and Mazumdar, D. 2008. Organic amendments influence soil organic carbon pools and rice–wheat productivity. *Soil Science Society of America Journal* 72(3): 775–785.

Moscatelli, M. C., Lagomarsino, A., Marinari, S., De Angelis, P. and Grego, S. 2005. Soil microbial indices as bioindicators of environmental changes in a poplar plantation. *Ecological Indicators* 5(3): 171–179.

Murmu, S. 2011. Nutrient Management for improving yield and quality of some vegetable crops. PhD Thesis, Bidhan Chandra Krishi Viswavidyalaya.

National Horticultural Board. 2018. *Horticultural statistics at a glance 2018*. Retrieved from http://nhb.gov.in/statistics/Publication/Horticulture%20Statistics%20at%20a%20Glance-2018

Palm, C. A., Gachengo, C. N., Delve, R. J., Cadisch, G. and Giller, K. E. 2001. Organic inputs for soil fertility management in tropical agroecosystems: application of an organic resource database. *Agriculture, Ecosystems & Environment* 83(1–2): 27–42.

Roy, D. K. and Singh, B. P. 2006. Effect of level and time of nitrogen application with and without vermicompost on yield, yield attributes and quality of malt barley (Hordeum vulgare). *Indian Journal of Agronomy* 51(1): 40–42.

Roy, R. N., Finck, A., Blair, G. J. and Tandon, H. L. S. 2006. Plant nutrition for food security: A guide for integrated nutrient management. *FAO Fertilizer and Plant Nutrition Bulletin* 16: 348p. ISBN: 92-5-105490-8.

Selim, M. 2018. Potential role of cropping system and integrated nutrient management on nutrients uptake and utilization by maize grown in calcareous soil. *Egyptian Journal of Agronomy* 40(3): 297–312.

Selim, M. M. 2020. Introduction to the integrated nutrient management strategies and their contribution to yield and soil properties. *International Journal of Agronomy* 1–14: https://doi.org/10.1155/2020/2821678.

Srivastava, A. K. and Ngullie, E. 2009. Integrated nutrient management: Theory and practice. *Dynamic Soil, Dynamic Plant* 3(1): 1–30.

Srivastava, A. K. and Singh, S. 2006. *Nutrient Diagnostics and Management in Citrus.* In: *Technical Bulletin* No. 8, National Research Centre for Citrus, Nagpur, Maharashtra, India, pp 1–130.

Tagaliavini, M. and Marangoni, B. 2002. Major and nutritional issues in deciduous fruit orchards of north Italy. *Horticulture Technology* 12: 26–31.

Tamang, A. 2010. Integrated nutrient management for improvement of yield and quality of tomato *(Lycopersicon esculentum),* pointed gourd *(Trichosanthes dioica)* and broccoli *(Brassica oleracea).* PhD Thesis, Bidhan Chandra Krishi Viswavidyalaya.

Tamang, A., Das, I., Batabyal, K., Sarkar, D., Murmu, S., Mandal, B., Hazra, G. C. and Bhattacharyya, R. 2017. Assessment of nutrient management technologies for broccoli to improve productivity and quality and soil resources in the subtropics. *International Journal of Vegetable Science* 23(2): 102–124.

Tilman, D., Fargione, J., Wolff, B., D'Antonio, C., Dobson, A., Howarth, R., Schindler, D., Schlesinger, W. H., Simberloff, D. and Swackhamer, D. 2001. Forecasting agriculturally driven global environmental change. *Science* 292(5515): 281–284.

Venkateswarlu, J. 2003. *Rainfed agriculture in India: Research and development scenario.* New Delhi, India: Indian Council of Agricultural Research.

Vineela, C., Wani, S. P., Srinivasarao, C. H., Padmaja, B. and Vittal, K. P. R. 2008. Microbial properties of soils as affected by cropping and nutrient management practices in several long-term manurial experiments in the semi-arid tropics of India. *Applied Soil Ecology* 40(1): 165–173.

Zhang, F., Cui, Z., Chen, X., Ju, X., Shen, J., Chen, Q., Liu, X., Zhang, W., Mi, G., Fan, M. and Jiang, R. 2012. Integrated nutrient management for food security and environmental quality in China. *Advances in Agronomy* 116: 1–40.

23

Micropot Method of Nursery Establishment

An Innovative Approach for Greater Climatic Resilience and Higher Crop Productivity

Sampad R. Patra and Malay K. Bhowmick
Department of Agriculture (Govt. of W.B.), Kolkata, West Bengal, India

CONTENTS

23.1 Introduction

Despite the phenomenal increase in crop production with the use of improved technologies (including biotic and abiotic stress-tolerant varieties), Indian agriculture needs to address emerging and diverse challenges and constraints like natural resource degradation, cost escalation of inputs, and concerns of climate change to meet the demands of ever-growing human population as well as livestock population, including cattle and poultry. Sustainability of agricultural systems is thereby getting threatened due to a number of interrelated problems and issues.

Among the field crops in India, oilseeds are the second most important determinant of agricultural economy, next only to cereals. Rapeseed-mustard is the major group among oilseed crops. In general, the productivity of oilseeds in the State of West Bengal is very low that needs to be augmented through improved varieties/hybrids and technologies toward achieving self-sufficiency (Patra et al., 2020). Delayed sowing, short winter spell and marginalized/suboptimal management are the major bottlenecks in achieving higher productivity, profitability, and quality of mustard, and sunflower in West Bengal (Patra et al., 2019a, 2020). Besides, maize is one of the most versatile emerging crops having wider adaptability under varied agroclimatic conditions. Among the cereals, the maize crop has the highest genetic yield potential. Maize is also used as a prime feed grain in preparing poultry feed, cattle feed, and also fish feed as it is a rich source of energy and highly digestible for poultry. In addition, high-value vegetables are considered essential for well-balanced diets since they supply vitamins, minerals, dietary fiber, and phytochemicals. Introduction of these less water-requiring crops during *rabi* season is expected to provide huge opportunities for addressing the emerging challenges context due to sustainability threats of rice-based cropping systems, global climate change, building community immunity and nutritional security of common people under the perspectives of COVID-19 pandemic, and also ensuring market link and value chain of non-rice *rabi* crops, besides improving livelihood security and farmers' income (Patra et al., 2021). Micropot method involving raising of nursery seedlings and their subsequent establishment in the main field has been emerged as an innovative package of growing these crops under adverse climatic situation (Patra and Bhowmick, 2019; Patra et al., 2020).

DOI: 10.1201/9781003164968-26

23.2 Model Micropot Nursery

Establishment of a model micropot nursery unit for raising healthier crop seedlings can be made with the use of following guidelines:

- Select an open sunny area with the scope of getting profuse sunlight and without any chances of waterlogging.
- Make a raised bed (6.00 m length × 1.50 m width × 0.15 m height) with proper leveling.
- Spread a good quality polythene sheet (slightly more than the bed size i.e. 6.03 m × 1.53 m) over the bed.
- Arrange to place bricks (approx. 54 in no.) around the bed (just to make a compact brick boundary) so as to leave an inner space of 5.5 m × 1.0 m, looking like a box pattern.
- Excess polythene remaining outside the brick boundary to be folded upward (½″) with pegging (using 1½′ wooden/jute sticks) to make a bowl-like structure.
- Place a layer of sand (roughly 3 cm thick layer with a sand volume of 5 CFT) inside the box.
- Place 10,000 SMPs (Sampad Micropots with 25 mm top diameter, 23 mm bottom diameter, 30-mm pot height, and 12 cc inner volume) (see Figure 23.1) inside the box area in a triangular fashion (1,875 SMPs/m^2) so as to obtain a compact honey comb structure in a nursery unit.
- Prepare a structure over the brick-made model with the use of bamboo and wire for providing a shade.
- Provide a shade over the model nursery with the use of good quality transparent polythene sheet of at least 200 μ or 0.2 mm thickness as and when necessary.
- Prepare pot mixture (vermicompost + fine dust soil in 3:1 proportion) to be mixed with single super phosphate (SSP) and boric acid powder [10 g SSP + 55 mg boric acid + 1,000 cc (1 L) pot mixture for 75 SMPs]. The requirements of pot mixture (growing medium), SSP, and boric acid for 10,000 SMPs will be 5 CFT (3.75 CFT vermicompost + 1.25 CFT soil) or 200–250 kg (150 to 190 kg vermicompost + 50 to 60 kg soil), 1.4 kg and 7.5 g, respectively.
- Pour pot mixture loosely inside the pots and water thereafter.
- Place germinated/nutri-primed/treated seeds singly in each pot.

FIGURE 23.1 Sampad Micropot (SMP).

- If nutri-priming of seeds is not done, 1 g $ZnSO_4.7H_2O$ (21% Zn) per CFT of pot mixture (growing medium) may be added (optional).
- As an extra care (optional), use 4 to 5 g *Trichoderma viride*, 20 g phosphate-solubilizing bacteria (PSB), and/or 20 g *Azotobacter* per kg of growing medium.
- Pour pot mixture loosely over the seeds, if needed anymore.
- Irrigate in the sand layer as and when required, and water will move to the SMPs through capillary action. Never apply water directly over the pots.
- Seedlings will be ready within 12–15 days after sowing (4-leaf stage). Spray the solution [1 g $ZnSO_4.7H_2O$ + 10 g urea + 1 g boric acid to be dissolved in 1 L water] on foliage at 2 days before uprooting of seedlings.

Table 23.1 shows an outline estimate (INR 10,000/- as fixed cost along with variable recurring cost) for the establishment of model micropot nursery. (see Figure 23.2) Such a simple unit of specialized

TABLE 23.1

An Outline Estimate for the Establishment of Model Micropot Nursery

Items/Materials	Cost (INR)	Purpose
Fixed cost		
Raised bed of 6.0 m (length) × 1.5 m (width) × 6″ (height)	–	Only labor, no input cost
Good-quality polythene sheet (6.0 m × 1.5 m)	700	To spread over the bed
Bricks (approx. 54 nos.)	600	Formation of a tray-like structure by placing bricks around the bed, keeping an inner space of 5.5 m × 1.0 m
Sand (5 CFT)	100	Sand filling (1″) within the tray
Micropots (10,000 nos.) per unit	4000	To place 10,000 SMPs within the tray, making a compact honeycomb structure
Bamboo and wire	700	To prepare a structure for providing shade over the brick-made model
Good-quality transparent polythene sheet of at least 200 μ or 0.2 mm thickness (6.0 m × 1.5 m = 9.0 sqm.)	1400	To provide a shade over the model nursery
Soil hole maker (5 nos.)	2500	Each model unit should have five nos. of soil hole maker on rental basis for transplanting of ready seedlings.
Total	10,000	–
Recurring cost[a]	*Quantity*[b]	
Vermicompost/organic manure	3.75 CFT or 105 L (approx.) (150–190 kg)	Growing medium (200–250 kg)
Soil	1.25 CFT or 35 L (approx.) (50–60 kg)	
Single super phosphate	1.4 kg	Nutrition
Boric acid	7.5 g	
$ZnSO_4{\cdot}7H_2O$	1 g/CFT medium	Nutri-priming of seed
Trichoderma viride	4–5 g/kg of medium	Optional
PSB	20 g/kg of medium	Optional
Azotobacter	20 g/kg of medium	Optional
Spray	Nominal	As per guideline

One unit = 10,000 SMPs; SMP: Sampad Micropot.

[a] Recurring cost for each cycle: Variable (duration of each cycle: 12–15 days i.e. two cycles in a month).

[b] Pot medium/mixture for one unit of 10,000 SMPs.

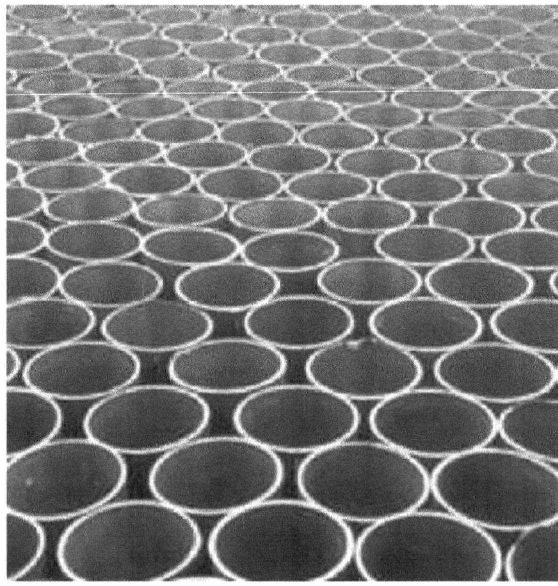

FIGURE 23.2 Model Micropot Nursery.

micropot nursery can be envisaged at a time to raise about 10,000 healthy seedlings or quality planting materials of different crops like oilseeds (mustard, sunflower), maize, high-value vegetables, etc., which may be subsequently transplanted in the main field under normal or zero tillage condition with the use of soil hole maker during *rabi* season. The said nursery may be used in a cyclic fashion with nominal recurring/maintenance cost (Patra et al., 2021).

23.3 Crop Establishment in Main Field

Seedlings can be transplanted in the main field under normal tillage or zero tillage condition. Tillage involves land preparation, basal fertilization (P_2O_5, K_2O, and B), planting of seedlings at recommended spacing, application of 0.5% urea solution just after transplanting, top dressing of urea at 5–7 and 20 days after transplanting (DAT), foliar spray of Zn and B along with 1% urea solution and need-based management practices. Under zero tillage condition, it is to dig out the holes only where the seedlings are planted at recommended spacing. Pot mixture [preparing growing medium with one- to two-part (s) compost and one part top soil, mixing the same with 850 g SSP and 2 g boric powder per CFT of growing medium] is placed in each hole before transplanting whilst 0.5% urea solution needs to be added just after transplanting, followed by top dressing of N and K_2O at 15 and 20–40 DAT (Patra et al., 2021). No other special cultural practices are needed (Patra et al., 2020, 2021).

23.4 Crop Performance in On-Farm Locations

A number of experiments have been conducted on micropot method of crop cultivation at different research stations viz. Water Management Research Station, Ranaghat (Nadia); Pulses and Oilseeds Research Station, Berhampore (Murshidabad); Dry Land Research Station, Bankura; and Zonal Adaptive Research Station, Nalhati (Birbhum)/Mohitnagar (Jalpaiguri) in West Bengal. These have been subsequently validated through a number of on-farm experiments on mustard (Table 23.2), maize (Table 23.3), and different vegetables (Table 23.4) at the farmers' fields located at different mouzas of selective blocks and districts in the State of West Bengal during *rabi* season. Seed yield of mustard (var. Divya 55) has been recorded at the level of 2.32–3.82 t/ha with an average of about 3.0 t/ha (Table 23.2),

TABLE 23.2

Field Performance of Micropot-Raised Mustard Seedlings (var. Dibya 55) in Selective Blocks of West Bengal During *Rabi* Season

Block (District)	Mouza	Date of Sowing	Date of Transplanting	Date of Harvesting	Plant Height (cm)	Branches/ Plant	Siliqua/ Plant	1000-Seed Weight (g)	Av. Yield (t/ha)
2017–18									
Haripal (Hooghly)[a]	Bhagabatipur	16.10.2017	07.11.2017	17.02.2018	210	19	2100	6.4	2.63
	Durgapur	15.10.2017	07.11.2017	16.02.2018	216	20	2100	6.3	2.70
	Kinkarbati	14.10.2017	05.11.2017	15.02.2018	207	21	2150	6.4	2.59
	Paschim Gopinathpur	19.10.2017	10.11.2017	20.02.2018	216	21	2000	7.0	2.66
	Purba Gopinathpur	12.10.2017	03.11.2017	14.02.2018	216	21	2200	7.0	2.70
2018–19									
Haripal (Hooghly)[a]	Chowtara	19.10.2018	12.11.2018	21.02.2019	216	21	2200	6.4	2.66
	Durgapur	18.10.2018	10.11.2018	17.02.2019	216	21	2150	7.0	2.70
	Kinkarbati	12.10.2018	05.11.2018	14.02.2019	210	21	2200	7.0	2.66
	Paschim Gopinathpur	17.10.2018	11.11.2018	18.02.2019	216	21	2150	7.0	2.72
	Purba Gopinathpur	14.10.2018	06.11.2018	17.02.2019	207	21	2150	6.4	2.74
Jalpaiguri Sadar (Jalpaiguri)[b]	Kharija Berubari	04.09.2018	24.09.2018	15.01.2019	219	19	480	6.0	2.55
Malbazar (Jalpaiguri)[b]	Dakshin Chengmari	15.09.2018	05.10.2018	05.02.2019	210	16	510	5.5	2.32
Cooch Behar-I (Cooch Behar)[c]	Howargari	30.10.2018	22.11.2018	25.02.2019	165	38	1145	4.2	3.77
	Khalishaguri	02.11.2018	25.11.2018	27.02.2019	146	33	923	3.8	3.49
	Nawabganj Balashi	01.11.2018	26.11.2018	04.03.2019	149	36	1032	4.0	3.71
2019–20									
Haripal (Hooghly)[a]	Chowtara	17.10.2019	08.11.2019	16.02.2020	216	21	2050	6.3	2.70
	Kaikala	15.10.2019	07.11.2019	14.02.2020	216	20	2100	6.4	2.68
	Durgapur	19.10.2019	12.11.2019	17.02.2020	210	21	2200	6.3	2.74
	Kinkarbati	16.10.2019	09.11.2019	14.02.2020	216	22	2050	6.4	2.70
	Paschim Gopinathpur	14.10.2019	06.11.2019	15.02.2020	210	21	2200	6.3	2.66
Jalpaiguri Sadar (Jalpaiguri)[b]	Kharija Berubari	06.09.2019	26.09.2019	19.01.2020	195	16	495	6.2	2.50
Malbazar (Jalpaiguri)[b]	Dakshin Chengmari	12.09.2019	01.10.2019	20.01.2020	198	14	520	5.5	2.40
Cooch Behar-I (Cooch Behar)[c]	Howargari	06.11.2019	30.11.2019	11.03.2020	152	37	1011	4.4	3.78
	Chatrachekadara	02.11.2019	28.11.2019	06.03.2020	148	36	1005	4.0	3.74
Cooch Behar-II (Cooch Behar)[c]	Gopalpur	04.10.2019	30.10.2019	15.02.2020	160	38	1100	3.0	3.11
	Dhangdhingguri	24.10.2019	10.11.2019	29.02.2020	165	40	1156	4.0	3.41
	Kaljani	25.10.2019	18.11.2019	01.03.2020	150	37	1192	5.0	3.82

No. of plants/m^2: 4 (Cooch Behar and Hooghly), 10–15 (Jalpaiguri).

[a] Office of the Assistant Director of Agriculture, Haripal Block, Hooghly.

[b] Borlaug Vision Society, Darjeeling; Agricultural Technology Management Agency-Block Technology Team, Jalpaiguri Sadar and Malbazar Blocks, Jalpaiguri.

[c] Office of the Assistant Director of Agriculture (Subject Matter), Sadar Cooch Behar; Office of the Assistant Director of Agriculture (Training), Cooch Behar; Farm Information & Advisory Centre, Cooch Behar-I Block, Cooch Behar, West Bengal.

TABLE 23.3

Field Performance of Micropot-Raised Maize Seedlings (Yuvraj Gold in Jalpaiguri and Kalimpong; DKC 9081 in Cooch Behar) in Selective Blocks of West Bengal During *Rabi* Season

Block (District)	Mouza	Date of Sowing	Date of Transplanting	Date of Harvesting	Plant Height (cm)	Cobs/ Plant	Grains/ cob	1000-Grain Weight (g)	Av. Yield (t/ha)
2018–19									
Cooch Behar-I (Cooch Behar)[a]	Howargari	30.11.2018	24.12.2018	03.05.2019	241	1.4	441	224	9.65
	Nawabganj Balashi	25.11.2018	21.12.2018	28.04.2019	234	1.3	434	256	9.96
2019–20									
Jalpaiguri Sadar (Jalpaiguri)[b]	Radhabari,	29.09.2019	17.10.2019	10.02.2020	226	2.0	350	300	8.63
Kalimpong-II (Kalimpong)[c]	Upper Menchu	18.09.2019	09.10.2019	19.02.2020	220	1.0	295	256	6.54

Spacing: 20″ × 12″.

[a] Office of the Assistant Director of Agriculture (Subject Matter), Sadar Cooch Behar; Office of the Assistant Director of Agriculture (Training), Cooch Behar; Farm Information & Advisory Centre, Cooch Behar-I Block, Cooch Behar.

[b] Borlaug Vision Society, Darjeeling and Agricultural Technology Management Agency-Block Technology Team, Jalpaiguri Sadar Block, Jalpaiguri, West Bengal.

[c] Borlaug Vision Society, Darjeeling and Agricultural Technology Management Agency-Block Technology Team, Jalpaiguri Sadar Block, Jalpaiguri; Borlaug Vision Society, Darjeeling and Upper Menchu Farmers' Club, Kalimpong-II, Kalimpong, West Bengal.

TABLE 23.4

Field Performance of Micropot-Raised Vegetable Seedlings in Selective Blocks of West Bengal During *Rabi* Season

Block (District)	Mouza	Vegetable Crop	Hybrid	Date of Sowing	Date of Transplanting	Date of Harvesting	Plant Height (cm)	Av. Yield (t/ha)
2018–19								
Jalpaiguri Sadar (Jalpaiguri)[a]	Malkani	Purple cauliflower	Valentina	03.12.2018	24.12.2019	25.02.2020	57	31.88
2019–20								
Jalpaiguri Sadar (Jalpaiguri)[a]	Malkani	Purple cauliflower	Valentina	20.11.2019	15.12.2019	16.02.2020	55	45.00
		Yellow cauliflower	Carotina	15.11.2019	07.12.2019	06.02.2020	49	46.50
		Tomato	US 1505	22.10.2019	09.11.2019	25.02.2020	91	39.00
	Dakshin Chengmari	Tomato	Suparna	14.11.2019	01.12.2019	13.02.2020	114	24.38
		Broccoli	Sakata Saki	15.12.2019	06.01.2020	19.03.2020	84	40.50
Kalimpong-II (Kalimpong)[b]	Upper Menchu	Purple cauliflower	Valentina	18.10.2019	09.11.2019	08.02.2020	51	31.00

No. of plants/m^2: 6.

[a] Borlaug Vision Society, Darjeeling; Malkani Farmers' Club, Jalpaiguri Sadar; Dakshin Chengmari Sabuj Farmers' Club, Jalpaiguri Sadar; Agricultural Technology Management Agency-Block Technology Team, Jalpaiguri Sadar, Jalpaiguri.

[b] Borlaug Vision Society, Darjeeling; Upper Menchue Farmers' Club, Kalimpong-II, Kalimpong, West Bengal.

whereas it is as high as 2.0 t/ha with the same variety raised under conventional method. Patra et al. (2019a) also reported an early harvest of sunflower (by at least three weeks) with seed yield of more than 2.5 t/ha in the district of South 24 Parganas during *rabi* season. Such yield advantages in mustard and sunflower are due to timely or sometimes early sowing, production of healthier seedlings, and better crop establishment with reduced pest and disease incidence (Patra and Bhowmick, 2019). Likewise, maize hybrids (Yuvraj Gold and DKC 9081) have been found to exhibit higher yield range of about 8.63–9.96 t/ha in these two districts of the State (Table 23.3). Farmers in the North Bengal Districts have been inclined to adopt such an innovative method for raising their desired crop in a remunerative way during the precious window of "pre-*rabi*/autumn". The method is highly suitable for raising healthy seedlings of high-value vegetables and flower crops. A number of vegetables have been evaluated, registering the yield levels of about 24.38–46.50 t/ha in the districts of Jalpaiguri and Kalimpong (Table 23.4). As experienced by the participatory farmers across different locations in the State, the method proved to be highly productive, profit-maximizing, and climate-resilient too (Patra, 2019, 2020; Patra and Bhowmick, 2019; Patra et al., 2019b, 2020, 2021).

23.5 Conclusions

Micropot method of nursery establishment as well as crop production offers multifarious advantages, including timely or even early sowing, reduced seed rate, raising healthier seedlings, easy crop establishment, hands-down weed management, shortened crop duration in the main field, scope of increasing cropping intensity, less incidence of diseases due to noninjury in roots during transplanting etc. The use of secondary and micronutrients would also have a crucial role in making qualitative improvement (oil in oilseeds; protein in pulses and maize) in crop produce. There is a huge scope of farm mechanization that can be explored and made accessible to smallholder farmers as well as rural youths not only for improved plant health management with higher production, but also for making the farming business as a profitable enterprise with greater climatic resilience.

REFERENCES

Patra S. R. 2019. Climate-smart agriculture towards improving livelihood security and farmers' income in West Bengal. Souvenir. National Seminar on "Sustainable Resource Management for Enhancing Farm Income, Nutritional Security and Livelihood Improvement", Feb. 01-03, 2019, Department of Agronomy, Palli Siksha Bhavana (Institute of Agriculture), Visva-Bharati, Sriniketan, West Bengal, India. pp. 37–38.

Patra S. R. 2020. Striding toward food, nutrition and livelihood security. Souvenir. 18th Reunion Meet, Palli Siksha Bhavana (Institute of Agriculture), January 11-12, 2020, Visva-Bharati, Sriniketan, West Bengal, India. pp. 1–4.

Patra S. R. and Bhowmick M. K. 2019. An innovative climate-resilient method of growing crops in micropots. Compendium of Abstracts. National Seminar on "Sustainable Resource Management for Enhancing Farm Income, Nutritional Security and Livelihood Improvement", Feb. 01-03, 2019, Department of Agronomy, Palli Siksha Bhavana (Institute of Agriculture), Visva-Bharati, Sriniketan, West Bengal, India. pp. 21.

Patra S. R., Bhowmick, M. K. and Kar, S. 2020. Raising micropot nursery and crop establishment in main field for improving oilseed productivity. J. Oilseeds Res. **37** (Special Issue): 165–166.

Patra, S. R., Pathak, A. and Bhowmick, M. K. 2019a. Paschimbangey micropot padhyatitey sarishar chas, in Bengali (Micropot method of mustard cultivation in West Bengal). Saar Samachar (Fertilizer Association of India, Eastern Region) **56**(4): 40–45.

Patra, S. R., Bhowmick, M. K. and Sarkar, S. 2019b. Micronutrients for food and nutrition security. Souvenir. National Seminar on "Agro-Chemical Inputs and Its Extension Approaches Towards Food-Security and Bio-Safety: Prospects and Challenges", November 15–16, 2019, organized by the State Agricultural Management and Extension Training Institute (SAMETI)-West Bengal, Ramakrishna

Mission Ashrama, Narendrapur, Kolkata in collaboration with the Integrated Rural Development and Management Faculty Centre, Ramakrishna Mission Vivekananda Educational and Research Institute (RKMVERI), Narendrapur, Kolkata & Sasya Shyamala Krishi Vigyan Kendra, RKMVERI, Arapanch, Sonarpur, Kolkata at SAMETI-West Bengal, Ramakrishna Mission Ashrama, Narendrapur, Kolkata, West Bengal, India. pp. 34–37.

Patra, S. R., Pathak, A., Ahmed, M., Chatterjee, R., Talukder, B. and Bhowmick, M. K. 2021. Micropot nursery establishment for higher crop productivity and greater climatic resilience. SATSA Mukhapatra-Annual Technical Issue **25**: 89–98.

24

Farmers' Innovations in Smallholdings: The Sustainable Transition in Agriculture of West Bengal

Riti Chatterjee[1], Pravat Utpal Acharjee[2], Suddhasuchi Das[3], Amit Baran Sharangi[4], and Sankar Kumar Acharya[1]
[1]*Department of Agricultural Extension, Faculty of Agriculture, Bidhan Chandra Krishi Viswavidyalaya, Nadia, West Bengal, India*
[2]*Department of Agricultural Chemistry and Soil Science, Faculty of Agriculture, Bidhan Chandra Krishi Viswavidyalaya, Nadia, West Bengal, India*
[3]*Malda Krishi Vigyan Kendra, Uttar Banga Krishi Viswavidyalaya, Malda, West Bengal, India*
[4]*Department of PSMA, Faculty of Horticulture, Bidhan Chandra Krishi Viswavidyalaya, Nadia, West Bengal, India*

CONTENTS

DOI: 10.1201/9781003164968-27

24.1 Introduction

Smallholding agriculture is gaining importance every day for raising agriculture growth, maintaining food security, and sustaining livelihoods in India, where Indian agriculture is the home of 99 million small and marginal farmers (almost 80% of the total farms) with average holding from 1.37 to 2.3 ha. Hence, the future of sustainable growth of agriculture and food security in India depends on those small and marginal farmers. However, their share in operated area is near to 44% which implies prevalence of land inequalities in India. Though, the role of smallholders in development and poverty reduction is well documented (Lipton 2006). Smallholdings play necessary role in raising agricultural development and poverty reduction. Smallholdings conjointly face new challenges on the mixing important chains, easement and globalization effects, market volatility and alternative risks and vulnerability, an adaptation of global climate change, etc. (Thapa and Gaiha 2011). Recent 'world-wide processes of farm change – commercialisation of increasing proportions of input and output: institutional developments such as super markets; privatization of key aspects of technical progress, and of output and process grades and standards – now indicate large farm focus' (Lipton 2006). There also are high returns from investments in agricultural research and development, rural roads, and alternative infrastructure and knowledge dimension.

24.2 Threats, Opportunities, and the Roles of Smallholding Agriculture in India

24.2.1 Role of the Smallholding Agriculture

24.2.1.1 Structure of Land Holdings

India is a land of small farmers. As per Agricultural Census 2000–2001, there have been an estimated 98 million small and marginal holdings out of around a 120 million total land households within the country. Marginal and small farmers accounted for around 81% of total operational holdings in 2002–2003 as compared to about 62% in 1960–1961 (Chand et al. 2011).

24.2.1.2 Access to Irrigation

The availability and access to irrigation facility has increased for all categories of farmers. It is the highest for marginal farmers followed by small farmers. The percentage of area under irrigation for the small farmers increased from 40 (1980–1981) to 51 (2000–2001) (NCEUS 2008).

24.2.1.3 Access to Fertilizers and Area under High-Yielding Varieties (HYV)

Fertilizer availability per hectare is inversely related to farm size for both irrigated and unirrigated areas. This is also true in the case of unirrigated areas.

Similarly, the percentage of area under high-yielding varieties is also inversely related to farm size. The coverage was above 50% for marginal, small, and semi-medium in the case of unirrigated areas; however, it is only 30% for large farmers (Dev 2012).

24.2.1.4 Cropping Intensity

Multiple cropping index (MCI) seems to be higher in the case of marginal and small farmers than that of medium and large farmers. For marginal farmers, average cropping intensity ranges from 134 to 139 and in the case of large farmers; it is 116 to 121%. Hence, differences across farm sizes persist over time (Dev 2012).

24.2.1.5 Cropping Patterns

There are four dimensions regarding the status of prevailing cropping patterns: (a) small and marginal farmers allocate greater portion of their cultivable land into high-value crops like fruits, vegetables, or any other cash crops; (b) small and marginal farmers going to transfer their land as to get higher comparative advantage in growing vegetables than fruits because of quicker returns with shorter gestation period; (c) small and marginal farmers distribute higher proportion of rice and wheat than other farmers; (d) small and marginal farmers allocate fewer of their lands to pulses and oilseeds (Birthal et al. 2011).

24.2.1.6 Farm Size, Fragmentation, Output, and Productivity

Marginal and small farmers generally contribute higher output as compared to their share in area. Their share was 46.1% in holding possessed; however, they contribute 51.2% to the total agricultural output of India. The contribution of small and marginal farmers to output is very high, i.e. 86% in West Bengal. In those eastern states of India, the share of both area and output is greater for small and marginal farmers. These farmers contribute around 70% to the total production of vegetables, 55% to fruits with a share of 44% in land area (Birthal et al. 2011).

24.3 Livelihoods of the Small and Marginal Farmers in Indian States

Out of total 120 million farm holdings in India, 98 million are small and marginal farmers. Thus, the sustainability of those farmers is critical for livelihoods in rural India and for the entire nation. NSS (2003) data delineates that the monthly consumption of marginal farmers was Rs. 2,482 and monthly income was Rs. 1,659. It shows that small and marginal farmers run in minus-savings of Rs. 655 and Rs. 823, respectively. Studies also found that the poverty level of smallholding farmers is higher than other ones (NCEUS 2008).

24.4 Issues and Challenges Faced by Smallholders

Smallholding agriculture sector in India still is facing many issues and threats (NSS 2003). There are few prominent as well as critical factors behind the constraints of those farming community: imperfect markets; little access to credit markets; efficiency in human resource base; little access to required and timely extension services; lesser access to 'public goods'; and greater negative externalities from depleting natural resources (NCEUS 2008).

24.4.1 Role of Women Farmers

Women farmers contribute almost 83% of total workforce making agriculture sector increasingly feminized as men are migrating to both urban and rural nonfarm sectors. They are mainly engaged in land preparation, seed selection and seed production, sowing, application of manure, fertilizer, and pesticides, intercultural operations like weeding, transplanting, threshing, harvesting, and value addition as well as in animal husbandry and dairying, fish processing, collection of forest produces (NTFPs), backyard poultry, collection of fuelwood, fodder, and other products for family needs (Dev 2012).

24.4.2 Land Issues

24.4.2.1 Land and Tenancy Security

Small and marginal farmers own and cultivate some a few bighas of cultivable lands, sometimes it becomes a limiting factor for getting all resources. Unregistered cultivators, women farmers as well as tenants face difficulties in accessing institutional credit and other government facilities available to the title farmers with land titles (Jain 2007).

24.4.2.2 Low Level of Formal Education and Skills

Literacy levels are quite lower in the case of smallholding farmers compared to medium and large farmers in India. However, farmers should pose at least a reasonable level of education to create the awareness for information on agricultural technology because improper education limits public dissemination of knowledge and its appropriate application. NSS farmers' survey also found that awareness about biofertilizers, minimum support price (MSP) is associated with farmers' level of education, but it is quite lower for marginal and small farmers (Dev 2012).

24.4.2.3 Globalization Challenges

Increasing globalization has added some more challenges to the pathway of smallholding agriculture. Policies related to heavy subsidies in developed countries shower negative effects on smallholding farmers of developing countries. Thus, if small farmers are not under subsidies, globalization will be advantageous for large farms. For farmers, the single most adverse effect may be the dangerous combination of low prices and output volatility, also in international markets (Dev 2012).

24.4.2.4 Climate Change Impact

Climate change is becoming one of the top challenges for agriculture, food security, and rural livelihoods for millions of people in India. Stresses will be more on smallholding farmers. The brunt of climate change is going to adversely affect farmers' livelihood. To implement climate change responsive and pro-poor policies, there is an urgent need to focus on small farmers through adaptation and mitigation suitable in small farmers' own condition (Dev 2012).

24.4.2.5 Water Problems

In Indian states, smallholding agriculture depends more on groundwater compared to large farmers who have more access on canal water. However, it is on discussion that groundwater is depleting in many parts of India. Thus, sustainable water management is going to be a critical factor toward sustenance of small and marginal farmers also.

24.4.2.6 Diversification

Diversification has already entered in Indian diets, and sometimes away from food grains to high-value products like meat or milk products, vegetables, and fruits. Rapid increase in middle class because of unprecedented urbanization, increasing per-capita income, and increased engagement of women in urban jobs and above all the globalizations have been immensely responsible for the diet diversification in India, including its rural parts (Dev 2012).

24.5 Opportunities for Smallholding Agriculture

Many technological and institutional innovations are coming up nowadays as enabling mechanism to help marginal and small farmers raising agricultural productivity and increase incomes through diversification and high-value agriculture.

24.6 Research and Extension

A paradigm shift is a need of an hour in agriculture sector across Indian states. Due to high fluctuation and variations in agroclimatic conditions in unfavorable areas, research ought to be highly location-specific with greater participation of farmers.

Progress in postharvest technology must be considered in new agendas to promote value addition through establishment of small-scale agro-processing industries. Involvement of private sector in agricultural research, extension, and marketing is to be ensured. However, there is a possibility of private sector participation can be limited to only profitable crops and enterprises occupied by resource-rich farmers in well-endowed regions. However, private sector is indifferent in research for soil and water management, rainfed agriculture, cropping systems, ecological conservation, and long-term sustainability. Here, public sector research has to intervene the problems being faced by the resource-poor farmers (Gaurav and Mishra 2011).

24.7 Technological Innovations

Agricultural technologies are 'scale neutral' but not 'resource neutral' (Singh et al. 2002). Smallholder-oriented research and extension should give importance to cost reduction approach without reduction in crop yields. Thus, the need for adopting the methods of an evergreen revolution is very urgent at the present days. As Swaminathan (2010) says, organic farming is the first pathway. Though, productive organic farming needs strong research support, mainly in the areas of soil fertility replenishment and plant protection, and continuous market support. And the second one is green agriculture. In this context, ecologically resilient practices, viz. conservation agricultural practices, integrated pest management, integrated nutrient management, and natural resource conservation are to be disseminated among farm stakeholders (Swaminathan 2010).

24.7.1 Zero Tillage

Zero tillage or conservation tillage along with residue management and proper fertilizer use can be the best strategy to conserve soil moisture, restricting soil erosion, augment water infiltration, increase carbon sequestration or storage, minimize nutrient runoff, crop diversification, and sustainability of yields. In 2005, in the rice–wheat belt of the Indo-Gangetic plain of India, farmers adopted zero tillage on almost 1.6 million hectares; by 2008, it increased to 20–25% of the wheat in two Indian states (Haryana and Punjab) and is in increasing stage every year (World Bank 2010).

24.7.2 Information Technology

Information technology is one of the magic bullets in the pathway of improving agribusiness and incomes of Indian small farmers, where Indian private companies and NGOs are global leaders in providing information to farmers, as a spinoff from India's meteoric rise as a world leader in ICTs.

24.7.3 E-Choupal

The Indian Tobacco Company (ITC) launched e-Choupal in 2000 in Madhya Pradesh. It has already reached more than 3.5 million farmers growing a range of crops – soybean, coffee, wheat, rice, pulses, and shrimp in over 31,000 villages through 5,372 kiosks across seven states, viz. Madhya Pradesh, Karnataka, Andhra Pradesh, Uttar Pradesh, Maharashtra, Rajasthan, and Kerala (Mondal 2018).

24.8 Institutional Innovations

Smallholding agriculture faces many constraints. However, innovative institutional models are emerging, and they will surely create new opportunities for small and marginal Indian farmers. Institution's innovations are mainly related to (i) land and water management, (ii) gender inclusiveness, (iii) group or cooperative approach for agri-inputs and marketing end and to enhance productivity, farm sustainability, and incomes of smallholding agriculture.

24.8.1 Institutions for Sustainable Land and Water Management

One of the major areas of concern is irrigation, as groundwater table is depleting. Say, if previously a farmer has to use 25 ft. pipe of water lifting for irrigation, now they are to use 35 ft., groundwater lifting for cultivation of boro paddy is already stopped in many districts of West Bengal, increased cost of irrigation and all of these are affecting small and marginal farmers to the greatest extent. Hence, utmost reforms are needed in prioritizing public investment, raising profitability of groundwater exploitation and conserving groundwater resources, rational costing of irrigation water and electricity, engagement of user farmers in the management of irrigation systems, and groundwater markets have to be equitable (Rao 2005). Shah et al. (2009) found from their study that groundwater can be exploited in a big way in eastern region of India. Hence, watershed development and community water conservation are needed as water management strategy. Besides, assets created under the MGNREGA program are there to help in improving land and water management in rural India.

24.8.2 Women's Participation through Group Approach

Women's cooperatives, women's producer groups, and other forms of group initiatives like Self-Help Group (SHG) have to be promoted to overcome the problems faced by the small and uneconomic land holdings, for the equitable transfer of agricultural technology and other inputs, for timely marketing of produce (Agarwal 2010). Deccan Development Society, an NGO is empowering landless women to get access to various government schemes on land, through purchase and lease (Krishnaraj 2006).

24.8.3 Sustainable Agricultural Approaches

Community Managed Sustainable Agriculture (CMSA) initiated by the Society for Elimination of Rural Poverty (SERP) in Andhra Pradesh in 2004 has the mandate to eradicate poverty and to improve livelihoods of the poor section of the society and ensuring better household nutrition. This initiative aimed to address the major causes of agriculture imbalance and made farmers aware of adopting sustainable agricultural practices immediately.

24.8.4 Institutional Approaches for Marketing of Produces in Smallholdings

For the small and marginal farmers, timely marketing of their products is the most important problem in comparison to credit and extension. Recently, there has been some form of contracts for some of the agricultural crops like tomato, chilies, potato, onions, cotton, wheat, basmati rice, groundnut, flowers, and so on. A silent revolution persists in institutions regarding non-cereal foods, emerging production–market linkages in the food supply chain, open-market transactions, agricultural cooperatives, and contract farming (Joshi and Gulati 2003) viz., SHG model, cooperative model, small producer cooperatives, and contract farming. Some of the successful cases of direct marketing are *Apni Mandi* in Punjab, *Rytu Bazars* in Andhra Pradesh, and *Dairy cooperatives* in Gujarat.

24.8.5 Supermarkets

Emerging supermarkets and value chains can provide a good market share to the small and marginal farmers. And their presence in retail trade is rapidly expanding and helping emerging economies. Supermarkets presently hold retail share of 50–60% in South America, East Asia (China excluded), and South Africa; 30–50% in Mexico, Central America, and maximum parts of Southeast Asia. However, in China, India, and Vietnam supermarkets still hold a low and variable position (2–20%), with an annual growth of 30–50% (Reardon and Minten 2011).

24.9 Lessons from Indian Experience

Indian experience on small-scale agriculture provides some essential lessons for the developing countries.

24.9.1 Green Revolution and Its Impact on Small Farms

Mid-1960s is the onset of India's Green Revolution which paved a new way in Indian agriculture. High-yielding varieties were introduced with recommended dosages of fertilizers in water assured areas. Firstly, medium and big farmers in irrigated areas got benefit from the new technology, whereas small farmers also benefited from Green Revolution due to government support through giving access to required services.

24.9.2 Food Grains Management Policy

Indian food grains management policy comprises three components, i.e. minimum support prices and procurement, maintenance of buffer stock, and public distribution system (PDS). Food grain prices increased almost 10% in India as compared to global food price increase of 80–90%. Here, government of India holds responsibility for food grain management through insulation from the hike of global food grain prices. Small and marginal farmers also came under these policies.

24.9.3 Development of Dairy Cooperatives

Dairy cooperatives in India set examples in the world. Indian dairy sector is predominantly run by smallholders. Here, the processors do not have much choice; they have to take milk from smallholder producers (Birthal et al. 2008).

National Dairy Development Board covers 1, 40, 227 village-level societies and 14 million farm families are engaged along with 4 million women, and daily procurement of 22 million liters of milk along with Amul (Joshi and Gulati 2003). Finally, they have generated millions of employments-days for the rural poor and enhanced their socioeconomic condition.

24.9.4 Group Approach

The SHG movement engaging both rural and urban women and SHG-bank linkage program has spread all over the country. The actual power of the SHG-bank linkage model (SBL model) is in the economies of scale created by SHG Federations (having 150–200 SHGs each), e.g. in bulk purchase of agri-inputs like seeds, fertilizers, and marketing of outputs, viz. crops, vegetables, or milk.

24.9.5 Institutional Innovations

As discussed before, in India, there are many institutional innovations in the fields of natural resource management, efficient input procurement, and produce marketing for marginal and small farmers. Other countries can learn from these experiences.

24.9.6 Rights-Based Approaches

India has established Right to Information, 100 days of guaranteed unskilled employment under National Rural Employment Guarantee Act, and Right to Education. It is also going for Right to Food by introducing National Food Security Act. Rights approach puts pressures on both of central and state governments to cater the required services to citizens.

24.9.7 Learning from the Global Initiatives

India learned from China on agricultural transformation, education, micro and macro policies, infrastructural set-ups, business climate, equitable distribution of public asset, and rural nonfarm development approaches. Besides, Latin American Countries like Brazil offered lessons on agriculture research like Agriculture research corporation EMBRAPA and Zero hunger programs.

24.10 Inspiring Stories of Smallholding Farmers from West Bengal

24.10.1 Journey of Commerce Graduate to Become an Organic Farmer

A 30-year-old young commerce graduate from West Bengal, Sk. Jahirul Mondal, is pursuing organic farming at his own farm at Notun Goara Village of Nadia district. Being fully different from what other graduates do after completion of their graduation, Jahirul turned toward agriculture.

Total area under cultivation is 7 ac. There are three enterprises at the farm, (a) crop husbandry: beans (3 types), onion, coriander, cabbage, spinach, basilla, lettuce, brinjal, pointed gourd, bottle gourd, bitter gourd, ribbed gourd, pumpkin (it is grown in sequential pattern in arch, it helps to harvest pumpkin in frequent interval), cucumber, chili, red Amaranthus, radish, lettuce, etc.; (b) animal husbandry: goat raising, poultry, etc.; (c) fishery: several species of Indian carp (Plates 24.1 and 24.2). Also, he has planted

Plate 24.1

PLATE 24.1 How does the farm look? (From Riti Chatterjee, 2018, during the case study.)

PLATE 24.2 Fish pond. (From Riti Chatterjee, 2018, during the case study.)

different trees at the bank of pond. In case of nutrient management, the farmer prepares compost using crop residues from his farm within a pit. After the composting, it is mixed with water in the tank (tank of paints) and their either sprayed or fertigation is done.

Management practices being followed at the farm:

Mulching to save soil

Crop	Perceived Benefits	Limitations
Eggplant (Plate 24.3)	Decrease in weed infestation, irrigation in 15-day intervals, and water conservation up to 50%.	Labor cost is higher, material cost is also high.
Cucumber (Plate 24.4)	Decrease in weed infestation, irrigation in 15-day intervals, and water conservation up to 50%.	
Leafy vegetables (Plate 24.5)	Aid in temperature regulation during the winter months.	
Onion (Plate 24.6)	Decrease weeds infestation	
Pointed gourd (Plates 24.7 and 24.8)	Better coloration of pointed gourd, lesser vegetative growth.	If the leaves of pointed gourd come into contact with the plastic mulch, the leaves get burnt. That is why scaffold is made to facilitate proper aeration.

Yield Variability: (a) While overall production is lower than conventional farming, taste and quality of produce were much better. He gets *higher market price.* (b) Sprouts of onion were harvested and sold in the market. *Market price* of onion sprout is higher than the onion itself.

In the case of pointed gourd, the size of the gourd is bigger, so is the yield.

Marketing of organic produce from his farm: the farmer created a WhatsApp group and added 200 wiling households. Those households order vegetables as per their requirement through this group. Besides, he also supplies his produce to the nearby supermarkets and wholesale markets.

Income and livelihood: (a) five workers are currently employed. (b) He has a *fixed selling price* for the produce throughout the year.

Achievements: he received assistance from several government programs, e.g. Agricultural Technology Management Agency (ATMA) and Prime Minister Krishi Sinchayee Yojana (PMKSY).

24.10.2 Some other Significant Success Stories on Horticultural Crops

24.10.2.1 Coconut Sap (Neera) Production

Coconut (*Cocos nucifera*) cultivation is gradually becoming not profitable due to several causes. To make it profitable, an emphasis is being given on production of neera and its value-added products like sugar, jaggery, neera-based milk made sweets, etc. It can be consumed even by diabetic patients because of its low-glycemic index. As value-added product neera jaggery, neera sugar and neera-based milk made sweets are also being prepared and marketed with higher demand in West Bengal. Considering

PLATE 24.3, 24.4, 24.5, 24.6, 24.7, AND 24.8 Eggplant with plastic mulch, cucumber with mulching, leafy vegetables with straw mulching, onion with live mulch as coriander, pointed gourd with scaffold + mulching, and pumpkin raising. (From Riti Chatterjee, 2018, during the case study.)

the above fact, the scope for tapping of coco sap neera may also be explored in the coconut growing northeastern states which may reflect on the state GSDP by uplifting the economic status of the farmers (Plates 24.9–24.11).

24.10.2.2 Innovative Value Addition in Terms of Space through Strawberry Cultivation

Strawberry is predominantly temperate fruit. Cultivation of strawberry in hot and dry climate of Bankura district, West Bengal is an uncommon practice. As an alternate crop with adequate profit margin, strawberry cultivation took place in a small piece of land (4 Cottah) with four varieties, Kamruja, Interdown, Sweet Charlee, and Tioguae of Maharashtra. 'Interdown' was found best for this region in terms of quality, quantity, and postharvest life. Cultivation practices included maintaining 2 × 1 ft. spacing (Length and width) with land preparation by providing four number of tillage. Cow dung manure @ 15 t/ac along with N 30 kg + 40 kgP205 + K20 30 kg + neem cake 100 kg + bone meal 100 kg, and horn meal 100 kg was thoroughly mixed with soil. After one month of planting, 15 kg N/ac was added in the soil.

PLATE 24.9, 24.10, 24.11 Coconut sap (neera) collection, strawberry var. interdown, precooling unit.

Another 15 kg N was applied just before flowering and foliar spay of 10-26-26 @ 5 g/L of water at an interval of 15 days. The bed was kept moist by providing irrigation and straw mulching. Harvesting of fruits started from January onwards till the end of March. It was found that on an average 500 g of fruits per plant can be obtained making the yield of 60 q/ac.

Practical utility of the Innovation: farmers who are searching for alternate crop in place of paddy and potato, straw berry as a cash crop have well been recognized. This technology is very simple to adopt by the local farmers, though cost of cultivation is more than the other crops. However, profit margin is so lucrative, farmers can very well venture toward cultivation of strawberry. A number of outlets like Big Bazaar, Reliance Mart, etc., are approaching the farmers for buyback arrangements which are expected to ensure steady return from this crop.

24.10.2.3 Indigenous Precooling Methodology of Seed Potato before Storage

Seed potatoes are generally kept open in a shed for a few days before storing in cold storage. Potatoes are not covered either with paddy straw or any other material. The practice leads to deterioration of quality in seed potato. In this method, seed potatoes are heaped up to a height of 3 ft. in an open shed and then covered with paddy straw for 15 days. Seed tubers are then processed properly to make it soil free. Cleaned potatoes are then treated with 3% boric acid and mancozeb before storing in cold storage during evening or night.

Potato is the most important cash crop of Hooghly and the district's economy largely depends on its production and marketing. But the nonavailability of good quality potato seed (tuber) often forces the farmers either to go for costlier seed tuber of other states or use their deteriorated seed material for cultivation. Potato seeds are generally stored in a very cool chamber at 2.5–3.0°C. If seeds are directly or with a very little care sent to the cold storage for storing, it may choke or its quality may reduce for low temperature. But in this innovative method, potato skin gets hardened as well as preconditioning of the seeds takes place which help seed potatoes adjusting the sudden low temperature in cold storage. And the seed quality is also maintained for next year. This methodology of precooling before storage of seed potato is sustainable, eco-friendly, economically viable, and widely acceptable.

24.10.2.4 Low-Cost Turmeric Grinder – A Workable Design

It is a new type of turmeric grinder operated through 5 HP motor having six blades with speed of 600 rpm. The blades slice the dry turmeric into power-like tiny pieces. The grinder has two chambers – one for slicing turmeric and another for accumulating turmeric powder. A speed-adjustable blower is fitted between the chambers for transfer of turmeric powder from one chamber to another and put the small pieces of turmeric into the grinding chamber again. Practical utility of the Innovation: North 24 Pgs. district of West Bengal is known for its potential in relation to production and marketing of spices powder, turmeric being one of the most important one. Grinding of turmeric at household level generally is done by 'Atta Chakki' which generates substantial amount of heat. It deteriorates the quality of powder and sometimes large particles are mixed with powder due to defective or broken sieve. The developed grinder has helped in improving the quality of powder, producing more quantity, and reducing drudgery in grinding operation (Plates 24.12–24.14).

24.10.2.5 Aloe Vera Gel Extractor

The innovator has developed an effective multipurpose unit capable of pulverizing, steaming, and extraction of gel for herbal applications. With this device, the innovator uses the specially designed pressure-cooking chamber to extract the essence from *Aloe vera*.

24.10.2.6 Banana Bunch Cover

Banana bunch cover is a physical protection method which will improve the visual quality of fruit by promoting skin coloration and reducing blemishes but can also change the microenvironment for

Plate 24.12 Plate 24.13 Plate 24.14

PLATE 24.12, 24.13, AND 24.14 Low-cost turmeric grinder, banana bunch cover. (From Suddhasuchi Das, 2018, during the case study.)

fruit development, which can have several beneficial effects on internal fruit quality. Bunch cover can also reduce the incidence of disease, insect pest and/or mechanical damage, sunburn of the skin, fruit cracking, agrochemical residues on the fruit, and bird damage. Bunch covering is laborious, and its benefit cost ratio must be investigated in order to promote adoption of the method in much of the World.

24.10.2.7 Another Innovation

A herbal plant growth promoter, which is effective in protecting the plants from a broad spectrum of pests apart from providing necessary nutrition, has been developed. It is named as 'Kamaal' meaning wonderful, due to its performance. It is effective in field crops as well as in vegetable crops. Year-round coriander production is a profitable venture for the farmer for leaf and seed purpose. Coriander leaf has high demand in local markets. So, farmers are started to grow year-round coriander production for their financial benefits. Farmers are started to grow ginger production in jute or nylon bags. It's also a new innovation to proper utilization of vacant place in different orchard through planting as an intercrop. By this way, farmers can get extra income from orchard. Nowadays, farmers are preparing organic or natural color from marigold, it's also a new innovation and profitable venture for the farmers. Aloe vera gel, cosmetics, etc. are prepared from different medicinal plants and agarbatti, insect repellent oil are prepared from different aromatic plants (Plates 24.15 and 24.16).

Plate 24.15 Plate 24.16

PLATE 24.15 AND 24.16 Ginger cultivation in bag. (From Suddhasuchi Das, 2018, during the case study.)

24.11 Conclusion

To conclude, agriculture is entitled as a state subject in the Indian constitution. Thus, states are to play active roles along with central government for attaining the three goals of growth, inclusions, and sustainability. Besides, the forward and backward linkage effects of Indian agriculture indicate the growing income from the non-agriculture sector. Some of the commercial and cash crops (both agricultural and horticultural) got significant potential for augmenting exports bringing faster development of agro-based industries. Thus, agriculture sector not only contributes to overall national economic growth at the same time generates sustainable livelihood but also reduces poverty and ensures food security as the most inclusive growth sector. Hence, agriculture transformation must be more holistic in terms of rural transformation and urban linkages for sustainable transformation and farmers' welfare.

REFERENCES

Agarwal, B. 2010. Rethinking agricultural production collectivities, *Economic and Political Weekly* 45(9):64–78.

Birthal, P. S., Jha, A. K., Tiongco, M. M. and Narrod C. 2008. Improving farm to market linkages through contract farming: A case study of small holder dairying in India, IFPRI discussion paper 00814. Washington, DC.

Birthal, P. S., Joshi, P. K. and Narayanan, A. V. 2011. Agricultural Diversification in India: Trends, contribution to growth and small farmer participation, ICRISAT, mimeo.

Chand, R., Lakshmi Prasanna, P. A. and Singh, A. 2011. Farm size and productivity: Understanding the strengths of smallholders and improving their livelihoods, *Economic and Political Weekly* 46: 26–27.

Dev, S. M. 2012. Small Farmers in India: Challenges and Opportunities, presented at Emerging Economies Research Dialogue, Beijing, China, 14–15 November 2011 organized by ICRIER.

Gaurav, S. and Mishra, S. 2011. Size Class and returns to Cultivation in India: A Cold Case Reopened, IGIDR working paper No.WP2011-27, Mumbai.

Jain, R. C. A. 2007. Regulation and Dispute Settlement in Contract Farming in India.

Joshi, P. K. and Gulati, A. 2003. From Plate to Plough: Agricultural Diversification in India, Paper presented at the Dragon and Elephant: A Comparative Study of Economic and Agricultural Reforms in China and India, New Delhi, India, March 25–26.

Krishnaraj, M. 2006. Food security, agrarian crisis and rural livelihoods: Implications for women, *Economic and Political Weekly* 41(52): 5376–5388. doi: 10.2307/4419084.

Lipton, M. 2006. Can small farmers survive, prosper, or be the key channel to cut mass poverty, *Journal of Agricultural and Development Economics* 3(1): 58–85.

Mondal, S. 2018. Text book of agricultural extension with global innovations, Kalyani Publishers, New Delhi, ISBN: 978-93-272-2877-9.

NCEUS. 2008. A Special Programme for Marginal and Small Farmers, A Report prepared by the National Commission for Enterprises in the Unorganized Sector, NCEUS, New Delhi.

NSS. 2003. Access to modern technology for farming, Report No. 499.

Rao, C. H. Hanumatha. 2005. Agriculture, food security, poverty and environment, Oxford University Press, New Delhi.

Reardon, T. and Minten, B. 2011. Surprised by supermarkets: Diffusion of modern food retail in India, *Journal of Agribusiness in Developing and Emerging Economies* 1(2): 134–161.

Shah, T., Gulati, A., Hemant P., Shreedhar, G. and Jain, R. C. 2009. Secret of Gujarat Agrarian Miracle after 2000, *Economic and Political Weekly* 44(52).

Singh, M. and Dwivedi, R. N. 2002. Farmers preferences on tree/crop species and livestock feeding – A study through PRA approach, *Progressive Agriculture* 2(2): 135–137.

Swaminathan, M. S. 2010. From green to evergreen revolution, Indian agriculture: performance and challenges. Academic Foundation, New Delhi, 410.

Thapa, G. and Gaiha, R. 2011. Smallholder farming in Asia and the Pacific: Challenges and Opportunities, paper presented at the Conference on new directions for small holder agriculture, 24–25 January 2011, Rome, IFAD.

World Bank. 2010. World development report 2010: development and climate change. World Bank, Washington, DC. https://openknowledge.worldbank.org/handle/10986/4387 License: CC BY 3.0 IGO.

25

Promoting Gender Equality in the Context of Agriculture and Natural Resource Management: Opportunities, Challenges, and Management Policies in Indian Mid-Himalayas

Kushagra Joshi[1], **Arunava Pattanayak**[1,2], **Renu Jethi**[1], **and Vijay Singh Meena**[1,3]
[1]*ICAR-Vivekananda Parvatiya Krishi Anusandhan Sansthan, Almora, Uttarakhand, India*
[2]*ICAR-Indian Institute of Agricultural Biotechnology, Ranchi, Jharkhand, India*
[3]*Borlaug Institute for South Asia (BISA), CIMMYT, Pusa, Samastipur, Bihar, India*

CONTENTS

25.1 Introduction

In most of the hill states of India, women are the main rural workforce, due to their major involvement in agriculture, animal husbandry, fodder and fuel-wood collection, and household activities. Among the three hill states in North-western Himalayas, women are contributing ~90% of the total work in agriculture and animal husbandry (National Sample Survey Organization 2009–2010). The structure of employment in particularly in rural areas (and that too in hill regions), is predominated by low yielding employment in agriculture. This has dramatically dislocated in the village communities. Since, the hills are constrained in the development of large-scale industrialization, and due to infrastructure constraints,

DOI: 10.1201/9781003164968-28

the development of the service sector is also limited, the growth and development of the agriculture sector remains the prime focus in hills.

The migration of men has resulted in women being left behind with heavy workload on the farm (Jain 2010; Maharjan *et al.* 2012) and they also become the de facto heads of the households. As women cannot move as autonomous economic migrants because of social, cultural, and economic constraints, they remain at home as dependents. In the remittance-based economy of hill regions, farmwomen are key driver of growth and development in the absence of their husbands/male counterparts. Migration by the male population has also given rise to various social and psychological problems among the hill women (Dighe 2008). The work burden of females left behind almost doubled. For improving the status of hill agriculture, the status of farmwomen of these regions needs to be addressed. The empowerment status of women needs to be examined and then implications for policy can be drawn out. The present chapter is an attempt to bring out the gender issues in hill agriculture and comes up with insights for policymakers, planners, extension personnel, and other stakeholders so that the plight of the farmwomen left behind as de facto farmers in hills can be addressed and hill agriculture, resting on shoulders of women, can be sustained.

25.1.1 The Context

Why is it needed to study agriculture and women, particularly in hills in developing countries like India? Basically, three opinions rationalize a research thrust on women in agriculture. First, agriculture continues to outweigh in the economies of developing countries. According to the economic survey, the sector share the Gross Domestic Product (GDP) was ~17.4% in 2015–2016. Census (2011) indicates that an estimated ~62% of the 1300 million Indian populations is rural and dependent on agriculture. The number of farming households is 159.6 million. Second, women are half the world's population, receive one-tenth of the world's income, account for two-third of the world's working hours, and own only one-hundredth of the world's prosperity (ILO 2011). The sustainable development goals (SDG 5) call for achieving gender equality. In Indian context, the customs and traditions play important role. The roots of gender role segregation lie in cultural beliefs that are responsible for the overburden of women. The gendered pattern of distribution of work in the households is ingrained within local culture and tradition. Women, in the region, are expected to take on sole responsibility for managing their families' nutritional and other needs-in addition to tending crops and animals.

Third, women are overwhelmingly affected by the current agricultural system. As primary caregivers, they suffer from food insecurity and malnutrition mostly due to their maternal instinct to feed their children and family first before themselves. Difficult living conditions in hills such as fragmented fields, steep slopes, and women's continuous toil in the fields for 14–15 hours consistently, coupled with household chores, make their workload overwhelming. Most women in hills suffer from lower back pain due to carrying heavy loads over long distances; they also suffer from various skin problems due to long exposure to sun and are also exposed to several health hazards.

This article presents the context that despite major share in agricultural work force; the women of hill regions of Indian Himalayas face a gamut of issues, which needs attention by the policymakers for sustaining the agriculture and food security. Though studies highlight their role as farmers and caregivers, their role as managers of natural resources is still trivialized. The chapter analyzes women's status in hill agriculture, examining closely questions such as: Who is involved in agricultural processes and why? How can agricultural processes be more effectively engendered? Engendering not only requires mainstreaming in agriculture alone but also at household and community level of which women are a part.

The article follows the conceptual framework that due to certain socioeconomic, demographic, and cultural factors, women in hills are bound to perform various instrumental roles along with their caregiving roles (Figure 25.1). This could be seen as a shift in the gender norms, but it poses some adverse consequences for women due to lack of enabling environment. The consequences of male outmigration and dual role as caregivers and cultivators on their health are also neglected, thus leaving women's occupational issues unaddressed. We hypothesize that intervention and policies that reduce the gender gap in resource availability and control will lead to achieve development outcomes related to food security, nutrition, occupational health, and other aspects of well-being by providing enabling environment to women by gender sensitization and cultural as well as policy reforms.

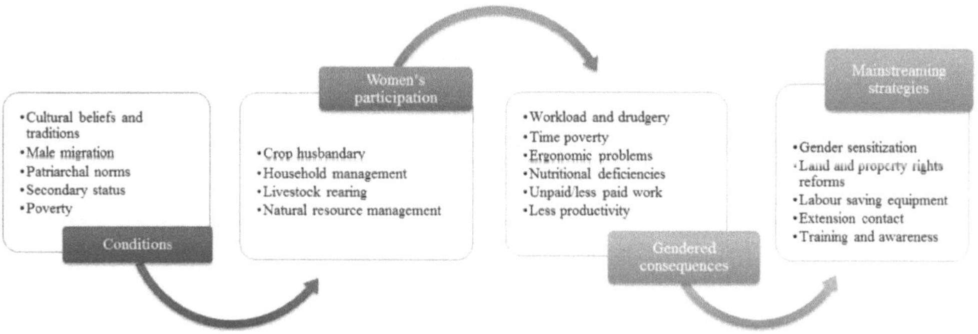

FIGURE 25.1 Conditions and consequences of women's participation in agriculture and mainstreaming concerns.

25.2 Materials and Methods

The paper attempts to document women's involvement in hill agriculture in north-western Himalayas at both macro and micro level. It uses a mixed method approach drawing conclusion based on both the secondary and primary data. The main sources of secondary data are the publications of government agencies like the Directorate of Agriculture and Economics and Statistics, Government of Uttarakhand, Census of India, National Family and Health Survey, and other scientific publications.

25.3 Results and Discussion

25.3.1 Share of Women in the Agricultural Workforce

Work participation rate (WPR) is an important indicator of development showing the proportion of working population to total population in an economy. It is generally believed that the higher WPR is an indication of well-being of population. Gender gap in WPR is an important measure, which helps to assess the progress made by a social group in social and economic platform. Available literature denotes that the work force participation in most hill regions of North-Western Himalayas is differentiated by gender. In all the three hill states, the percentage share of women in agriculture has increased. In Uttarakhand, the work-force participation rate of population is higher than the national level, particularly due to significant contribution of women population in different economic activities. Uttarakhand hill region has highly gender biased work structure in the rural areas where women tremendously work in agricultural occupations while their male counterparts work in highly paid non-farm occupations (Awasthi & Nathan 2015). The workforce participation rate for female in agriculture as cultivators is ~64% against ~29% for males (Census 2011). Male (~11%) and female (~9%) accounted themselves as agricultural laborers (Tables 25.1a and b).

The interesting fact is that women outnumbered their male counterparts as cultivators. This change can be accounted to male migration from the hills to prosperous areas with higher wages, leaving women to accept low paid casual work in agriculture. Women's participation as cultivators is more which indicates that women are taking care of their own small farms for securing family's consumption needs. Relatively, less percentage of women is working as agricultural laborers, where they are working for wages in other's fields. Here, men outnumber women in paid work as agricultural laborers. Fontana and Paciello (2010) also stated that wage work and non-wage work have a sexual division of labor which further widens the gender gap in agriculture and allied activities.

25.3.2 Division of Labor in Crop Farming

Available data in crop farming in hills shows women shouldering more responsibilities as farmers in comparison to men. Women participate in almost all the agriculture operations while the participation

TABLE 25.1A

Share Contribution of Women and Men in Agricultural in North-Western Himalayan States (Labour Bureau 2012–13)

		Cultivators		Agricultural Laborers	
Years	**States**	**Male (%)**	**Female (%)**	**Male (%)**	**Female (%)**
2001	Uttarakhand	34.29	77.84	9.54	6.08
	Jammu and Kashmir	37.50	54.70	7.10	5.20
	Himachal Pradesh	49.50	85.80	3.30	2.90
	India	31.06	32.93	20.85	38.87
		Cultivators		**Agricultural Laborers**	
		Male (%)	**Female (%)**	**Male (%)**	**Female (%)**
2011	Uttarakhand	28.82	64.00	11.23	8.84
	Jammu and Kashmir	23.96	42.55	12.97	11.83
	Himachal Pradesh	44.35	76.24	5.04	4.75
	India	24.92	24.01	18.56	55.21

TABLE 25.1B

Relative Change in Percentage Share of Men and Women in Agriculture in North-Western Himalayan States in 2011 (Base Year 2001)

	Cultivators		Agricultural Laborers	
States	**Male (%)**	**Female (%)**	**Male (%)**	**Female (%)**
Uttarakhand	–0.16	–0.18	0.18	0.45
Jammu and Kashmir	0.36	–0.22	0.83	1.28
Himachal Pradesh	–0.10	–0.11	0.53	0.64
India	–0.20	–0.27	–0.11	0.42

of male members is confined to a limited number of operations. In field survey in mid hills, segregation in crop husbandry was observed (Table 25.2). It was evident that women dominate in labor-intensive and monotonous tasks. Women are regarded as submissive and docile, having greater dexterity for tasks that require care and patience and are done for long hours, manually. Weeding is such a task which requires care and patience. Thinning and replanting of plants and transplanting of rice also require unique dexterity in handling the young seedlings which is assumed as a trait of women. Harvesting with sickle is also a

TABLE 25.2

Women's Role in Crop Husbandry in Hill Region Based on Focused Group Discussion

Activity	Women Dominant	Jointly with Men
Land preparation		✓
Sowing		✓
Seed treatment	✓	
Thinning and gap filling	✓	
Weeding	✓	
Harvesting	✓	
Threshing		✓
Loading and unloading	✓	
Storage	✓	
Chasing away wild animals		✓

women dominant activity as it also requires patience and careful selection of matured stalks. All agricultural work, except plowing, was done by women either alone or jointly with men. However, increasingly women are taking up plowing also, especially in female-headed households, single women homes, and among nuclear families (USNPSS 2005).

25.3.3 Women as Managers of Natural Resources

Women play a crucial role in management of natural resources. Although women are the natural resource managers at local level, they participate very less or little in decision-making process at intra-household or inter-household level (Figure 25.2). There are many instances in mountainous terrains where women are striving toward natural resource management and biodiversity conservation. They manage natural resources by their own way but lack the scientific rationale and practices. If appropriate directions in form of trainings be provided to them on the values, management, and sustainability of natural resources, it can provide opportunities to women as alternative sources of livelihood (Gupta *et al.* 2014).

Women trek for longer hours and distances, using more energy, to ensure family sustenance. In hills, women collect flowers particularly Buransh (*Rhododendron* sp.) from forests which has great medicinal value and has high demand in food processing industries. They collect it in the month of April and sell it to the nearby processing units. Fodder collection is a main household activity in hills and women visit the forests once or sometimes even twice a day to collect fodder (Dhyani & Maikhuri 2012). In field survey, it was found that all women (100%) collected fodder whole year except in the months when rainfall and

FIGURE 25.2 Photographs (a–h) show the various stages crop farming practices from land preparation to threshing by hill farm women dominated and jointly with men.

TABLE 25.3

Participation of Women in Collecting Minor Forest Produce (MFPs)

Collection of Minor Forest Produce (MFPs)	Women Involved (%)
Fuel wood	100
Fodder	70.5
Flowers and leaves	51.6
Medicinal plants	2.5

snowfall occur (Table 25.3). About 70% were involved in collecting fuel wood. About half of the women were involved in collecting flowers and leaves from forests as minor forest produce, which is seasonal. Women also collect pine needles during summer for spreading it on fields and use as litter for animals. Apart from agricultural activities, women are predominantly involved in collecting nontimber forest produce which also requires trekking and considerable time.

Women are also the conservers of biodiversity. They act as custodians of seeds and remain actively involved in its selection, multiplication, and management. In Garhwal hills of Uttarakhand, women's role in Chipko Movement is an immortal example of women's role as custodian of natural resource. In this movement under the leadership of Gaura Devi, a woman of hill, 27 women hugged the trees, embracing the trees to prevent their being cut down. Women of Chipko Movement achieved a 60–80% tree survival rate that contributed to reducing landslides and provided fuel and fodder (Joshi 1981).

Another instance to quote is the save our seeds movement, or *Beej Bachao Andolan*, which begun in the Central Himalayan region of Garhwal in Uttarakhand in the late 1980s to save the traditional seeds of the hills. As a consequence, it preserved in situ a rich variety of traditional seeds, ensuring food security and the well-being of both the people and the land (Kotamraju 2009). Women, in their primary role in seed conservation, struggled collectively to conserve the local seed. These movements solicit the crucial role played by women in hills as custodians and preservers of natural resources.

Though these movements contributed to conservation of forests but it also created conflicts of powers. This calls for a need to add gender dimension to sustainable development policies for natural resource management in the Himalayas so that the contribution of women in resource conservation can be acknowledged. As women are directly involved in managing, utilizing, and conserving natural resources, they must be involved in decisions that affect their vulnerabilities and needs.

25.3.4 Time Poverty and Occupational Drudgery

Majority of agricultural activities, which are full of drudgery, have not been supported by the mechanical advantages of tools and appliances. While doing agricultural operations the farmers adopt many unnatural postures like bending, stretching of different body parts that lead to increase in cardio vascular stresses. Among the crop production task, most of the activities are moderately to very heavy to perform based on workload and energy expenditure (Table 25.4). In hills of Uttarakhand, cutting of grasses for

TABLE 25.4

Workload of Women in Managing Management on Daily Basis

Collection of Natural Resource	Distance Travelled (km)	Time (Hours/day)	Frequency (Trips/day)
To fetch water for irrigation and drinking purpose	1.71	1.45	1.36
To collect fodder	4.70	4.13	1.00
To collect firewood	5.80	3.87	1.64 (seasonal)

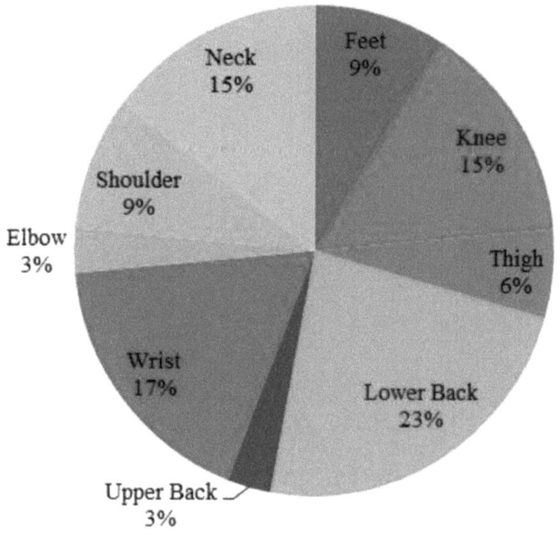

FIGURE 25.3 Percentage of hill farmwomen suffering from chronic pain and body discomfort.

preservation are the activities solely performed by women. These activities are carried by women in peak season resulting in occupational fatigue due to climate, work environment, and traditional tools used for carrying out the activity.

Future crops like millets are the main pseudo-cereal crops of hills. Post-harvest operations in these crops are done manually which is a tedious and lengthy process (Joshi & Moharana 2015). The homestead work often affects their health, in particular because of the hardships of collection activities and the air pollution they are exposed to when working indoors. In the hills, the dependence of most households on firewood, for cooking and heating is well documented. They even take the risk of climbing up on trees to ensure lighting of *chulhas* at home (Duflo *et al.* 2008; Sidh & Basu 2011; Nayak *et al.* 2013).

Fodder cutting and chaff cutting is one of the hazardous daily activity carried by hill farm women. In a field survey on occupational health hazards of hill farm women, manual chaff cutting was analyzed as highly risk prone which requires immediate ergonomic intervention due to the posture adopted, the repetitiveness involved and the hazardous chaffing tool used (Joshi *et al.* 2018). Field survey in mid hills of Uttarakhand (Figure 25.3) showed that women reported severe pain in lower back (~80%) followed by wrist/hands (~60%), knees and neck (~50% each), and ankles and shoulders (~30% each) since a year. These tasks not only demand considerable time and energy but also are sources of drudgery for hill farmers, which are not yet precisely categorized and quantified. Studies reveal that majority of the females in the working age group in Uttarakhand suffer from various health problems which further aggravated by the poor health facilities for the women in these regions and lack of awareness on occupational health techniques (Kandari 2013). There remains a need to create awareness among farmwomen about the health hazards and risks posed by agricultural activities, if not done in a proper technique, and their prevention.

25.3.5 Hill Farm Women and Nutritional Inadequacy

Not specifying major contributions by women to economic performance and development has important consequences. At household level, the failure by the male heads of household to recognize the economic value of these functions results in the low social status of the women and the neglect of her health, nutrition, and social well-being. This effect is institutionalized and has been incorporated in the traditional gender role of women. A woman's health affects the household economy also as a woman with poor health will be less productive as work force. The productivity will be more if their health remains better (Gupta *et al.* 2014).

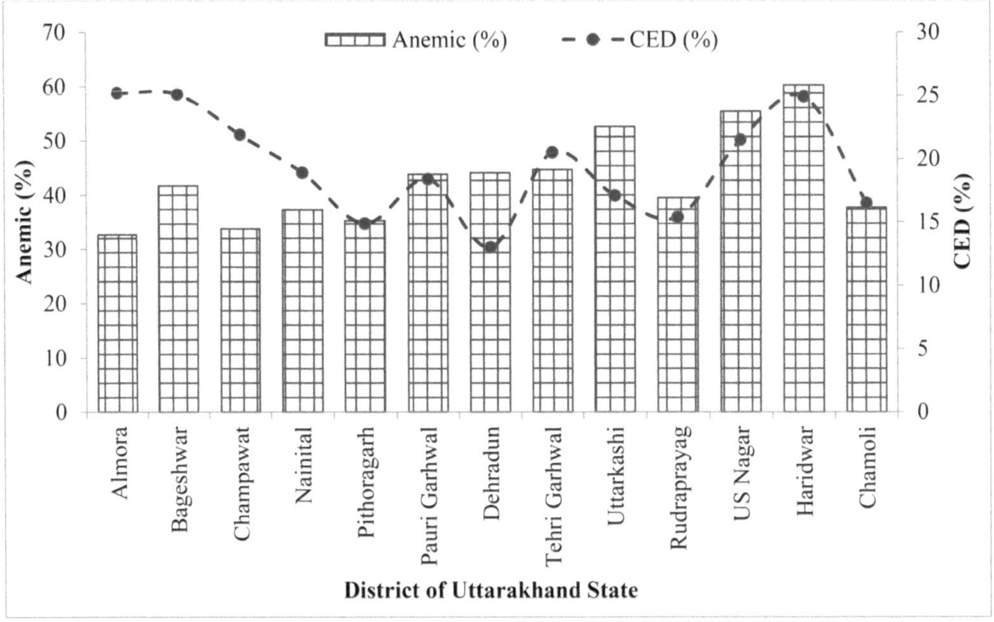

FIGURE 25.4 Prevalence of chronic energy deficiency and anemia among women (NFHS-4 2015–16).

In Uttarakhand, the malnutrition was found more widespread in both the rural areas of Kumaon and Garhwal region. The incidence of anemia was higher in Garhwal region namely in Uttarkashi district while in Kumaon, more cases were reported in Bageshwar (Figure 25.4). In a study conducted on farm women of Kumaon region, 45% subjects were found as Chronic Energy Deficient (Joshi & Moharana 2015). As energy deficiency impairs the work performance, the working potential of such women is likely to be reduced (Figure 25.5).

The prevalence of anemia indicates the micro-nutrient deficiency in the diets of women which calls for agricultural and dietary interventions in these areas. Improvement in health and well-being of women through better nutrition may contribute to reducing their burdens, emotionally and financially,

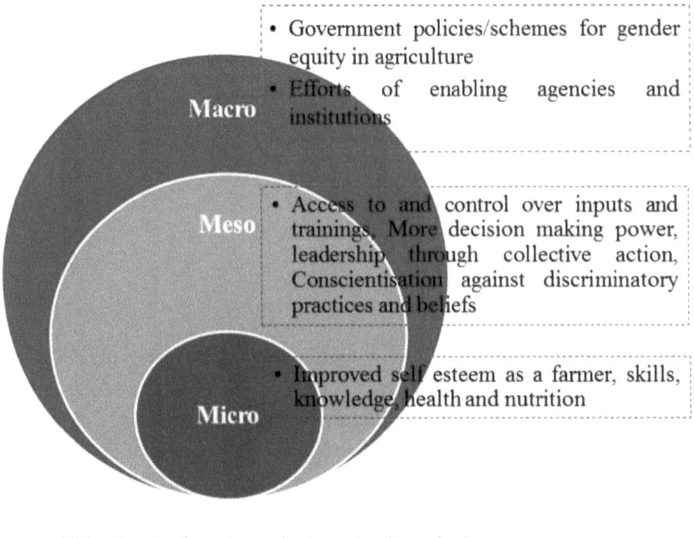

FIGURE 25.5 Conceptualizing levels of gender mainstreaming in agriculture.

and alleviating time constraints. Gained time and resources which become available in this way can be used for productive activities or participating in educational, health, or social engagements, from which women and their families can benefit.

25.3.6 Suggested Strategies for Mainstreaming Hill Farm Women in Agriculture and Natural Resource Management

Although various government programs have targeted vulnerable sections of the population, further concerted effort is required to bring about tangible change in the situation. This study suggests that women in hills are involved in management of farming and natural resources; unfortunately, they cannot effectively contribute to the management of resources in which they do not directly own or control. Strategies aimed at mainstreaming women can be applied both to the study region and beyond, since these strategies have specific gender policy implications. Some of the recommendations for effective planning and implementation of women specific and gender friendly policies and programs can be:

25.3.7 Acknowledging Women as Conservers of Natural Resources

It is necessary to highlight the gender differentiated practices and knowledge of women and men in their relations with biodiversity resources and to recognize that women and men can contribute differently to the conservation of biological diversity. For example, in the Himalayan region, the creation of protected areas has caused several conflicts. A conflict started between the women and the forest personnel while collecting wood and other non-timber products and many admitted that they started stealing biomass from the protected areas. The *Binsar Wildlife Sanctuary* realized these problems and involved women in eco-development planning. Once recognized, women began to take pride in planning for the protected areas.

25.3.8 Enhancing Women's Reach to Finance and Credit

Unequal access to land rights has major ramifications for women's access to credit, as land use rights certificates are generally required by banks as collateral. To increase women's access to financial resources provisions should be made for ownership of land by women farmers. In the beginning, joint pattas (entitlements) may be issued to both men and women of the household so that the women can also be considered as beneficiary of social and financial schemes as de jure farmers. In hill economy, where out migration of men very common, land rights for women farmers is a much-required initiative. One of the efforts of Government has been initiated where 'Kisan credit cards (Farmer credit cards)' have been provided to farmers on name of the land owner. It could be made more women friendly if the credit cards are issued on the name of women of the household. Access of women to rural credit shall be facilitated through simple documentation procedures where proof of residence should suffice to get a loan. This mere step of allotting some ownership to the women can be a stepping stone for women empowerment.

25.3.9 Creating Platform for Women's Participation

It has been observed that women participate more actively and confidently in productive activities besides homestead when organized in groups. In studies by Mathrani and Periodi (2007) and Sharma and Sudarshan (2010), participants in women's collectives and networks in India reported increased influence over decision-making in the household and community; in Dhaka, Bangladesh, women participating in local advocacy networks have reported improved mobility outside the household and influence over community affairs (Banks 2008); in Kenya Namibia (Crone 2010), women have risen through leadership roles in local networks of community caregivers to sit on government decision-making bodies, and in the Niger Delta, women's groups are active in traditional governance and in protests on environmental and social issues (Ikelegbe 2005). Women farmers shall be held at key positions in such organizations so that the strategic needs of women farmers shall get a representation at local, national as well as international level.

25.3.10 Women to be Acknowledged as Farmers by Extension Service Providers

Extension needs across gender vary considerably as often by virtue of ascribed gender roles. Often extension service providers don't acknowledge women as farmers, thereby excluding women as participants in agricultural training programs. To bridge this extension gap, a gender-sensitive extension approach should be devised and implemented. Women should be recruited as extension agents and male extension workers should be trained on gender issues and mainstreaming techniques. Such gender sensitized extension agents will proactively involve farm women in various training programs so that innovations, knowledge, and technology trickle down to women farmers at grassroots level.

25.3.11 Strong Policy Measures for Occupational Health and Safety in Agriculture

Occupational safety and health (OSH) is still nonexistent for unorganized sector including agriculture. The Insecticides Act, 1968 and the Dangerous Machines (Regulation) Act, 1983 of India are the two legislations presently available to specific aspects of farming. Time has come when policymakers should manage their attitude toward occupational health and recognize that occupational health improvement is a necessary condition for socioeconomic development. Training and awareness programs on OSH should be made an essential part of all R&D work in agriculture. Close collaboration of agricultural ministry with labor ministry and ILO is required for alleviating occupational health hazards among farming population. Separate funding should be made by the state government for arranging large-scale demonstration programs of technologies for farmwomen in hill areas.

25.3.12 Promoting Gender Friendly Technologies

Gender-sensitive/specific agricultural technology development is still a distant dream in majority of developing countries. Many ICAR institutes are catering the needs of farmwomen by developing and refining gender friendly technologies but rarely these technologies are multiplied by the manufacturers either due to less demand because of ignorance or less prioritization of women dominant agricultural tasks for mechanization. Economically viable and tailor-made drudgery reducing technologies as per the physical strength of women need to be introduced. A proactive approach called 'participatory ergonomic intervention' can be adopted to bring change in work method by involving women farmers at all stages *viz.* identifying the problem, technology design, and evaluation.

25.3.13 Nutritional Empowerment of Farmwomen and Families as a Whole

The emphasis of national agricultural research policies is usually placed on increasing agricultural production with the application of science and technology. However, policymakers must give importance to nutrition and family health issues in agricultural research programs as nutrition and productivity are related. Farming system shall be planned based on nutritional needs, local, and seasonal availability with sufficient variety in the crop mix to provide a balanced diet to the population. In hills, more emphasis should be given on production of locally available nutritious food crops like millets, *bhatt* (black soybean), maize, and their distribution must be ensured through public distribution system. Quality protein maize (QPM) interventions should be promoted to meet the protein needs of women and their families. Crops, which women predominantly cultivate need to be promoted. Strategies such as food crop diversification, development of nutrition gardens, tending small livestock, and backyard fishery and poultry can be promoted to diversify their food basket.

25.4 Conclusions and Future Prospective

Improvement in status of hill farmwomen requires a process of empowerment. Empowerment is about people – both men and women – taking control of their own lives: setting their own issues, acquiring skills as better workers, improving their competencies, seeking solutions to problems, and develop autonomy.

In the hills, the economy is remittance based and agriculture is being feminized. Time has come when the women farmers must be empowered so that they can have their own voice, could sought various alternatives, make wise choices for the betterment of themselves, their families, and hill ecosystem as a whole. The change in societal attitude is essential for ensuring their empowerment. Unless rural farm-women are not given economic benefits along with social and political opportunities, the women empowerment will be a distant dream. Keeping in purview the importance of these women in the economy and their significant contribution as work force, which is rapidly increasing, it is very important to develop such policies, which could give highest priority to the health of the female work force in this region. An integrated approach is required for minimizing the gendered consequences of women's involvement and the factors responsible for these consequences.

REFERENCES

Awasthi, I. C. & D. Nathan. 2015. Poverty and gender analysis of Uttarakhand: Some insights from the field. Working Paper 2016, Giri Institute of Development Studies.

Banks, N. 2008. A tale of two wards: political participation and the urban poor in Dhaka city. *Environment and Urbanization* **20**: 361–376.

Census 2011. Register general of India, New Delhi: Ministry of Statistics and Program Implementation.

Crone, T. 2010. Transforming the national AIDS response advancing women's leadership and participation, New York, NY: UNIFEM and ATHENA.

Dhyani, S. & R. K. Maikhuri. 2012. Fodder banks can reduce women drudgery and anthropogenic pressure from forests of western Himalaya. *Current Sciences* **103(7)**: 763.

Dighe, A. 2008. Women's empowerment at the local level (WELL)-a study undertaken in the state of Uttarakhand. Commonwealth of learning, Vancouver. http://hdl.handle.net/11599/263

Duflo, E., M. Greenstone & R. Hanna. 2008. Cooking stoves, indoor air pollution and respiratory health in rural Orissa. *Economic and Political Weekly* **9**: 71–76.

Fontana, M. & C. Paciello. 2010. Gender dimensions of rural and agricultural employment: differential pathways out of poverty: a global perspective. In Gender dimensions of agricultural and rural employment: differentiated pathways out of poverty. Rome: FAO/IFAD/ILO.

Gupta, U. C., P. U. Verma & H. A. Solanki. 2014. Role of ethnic women in biodiversity conservation. *International Journal of Research and Development in Pharmacy and Life Sciences* **3**: 855–858.

Ikelegbe, A. 2005. Engendering civil society: oil, women groups and resource conflicts in the Niger Delta Region of Nigeria. *The Journal of Modern African Studies* **43(2)**: 241–270.

ILO. 2011. International Labour Organization Report VI (1): Safety and health in agriculture. Geneva: ILO.

Jain, A. 2010. Labour migration and remittances in Uttarakhand. Kathmandu: ICIMOD.

Joshi, G. 1981. Forest policy and tribal development. *Social Action* **31**: 446–468.

Joshi, K. & G. Moharana. 2015. Policies and programmes for women empowerment. Gender mainstreaming in integrated watershed management programme. ICAR-CIWA 62–70.

Joshi, K., B. M. Pandey, R. K. Khulbe & A. Pattanayak. 2018. Women's drudgery and maize sheller intervention: a case of tribes of Jaunsar region of Uttarakhand. *Indian Journal of Hill Farming* Special Issue: 96–100.

Kandari, P. 2013. Migration pattern and the increasing participation of females in the economy of hill rural areas: A study of Pauri district in Uttarakhand. *IOSR Journal of Humanities and Social Science* **17**: 27–33.

Kotamraju, P. 2009. Beej Bachao Andolan: The analysis of a movement: the unpublished thesis. Ahmedabad: Mudra Institute of Communications.

Labour Bureau. 2012–13. Statistical profile on women labour. Chandigarh/Shimla: Labour Bureau Ministry of Labour and Employment, Government of India.

Maharjan, A., B. Siegfried & K. Beatrice. 2012. Do rural women who stay behind benefit from male out-migration? A case study in the hills of Nepal". *Gender, Technology and Development* **16**: 95–123.

Mathrani, V. & V. Periodi. 2007. The Sangha Mané: The translation of an internal need into a physical space. *Indian Journal of Gender Studies* **13**: 317–349.

Nayak, J., S. P. Singh & G. Moharana. 2013. Occupational health hazards of farmwomen. Technical Bulletin 24. Bhubaneswar: ICAR-DRWA, 1–32.

NFHS-4. 2015–16. National family health survey (NFHS-4). Uttarakhand, Mumbai: IIPS.

Sharma, D. & R. Sudarshan. 2010. Towards a politics of collective empowerment: learning empowerment: learning from hill women in rural Uttarakhand, India. *IDS Bulletin* **41(5)**: 43–51.

Sidh, S. N. & S. Basu. 2011. Women's contribution to household food and economic security: a study in the Garhwal. *Mountain Research and Development* **31**: 102–111.

USNPSS 2005. Beyond practical gender needs: women in north-eastern and hill states: Uttaranchal,' national research programme on growth and human development. UNDP-Planning Commission, Government of India (mimeo).

26

Making Rice-Farming System More Climate Resilient and Nutrition Sensitive: Heritage of Kurichiya Tribe Community of Western Ghats

N. Anil Kumar, Merlin Lopus, Raveendran Telapurath, and Vipin das
Community Agrobiodiversity Centre, M. S. Swaminathan Research Foundation, Wayanad, India

CONTENTS

26.1 Introduction

Rice Farming System (RFS) is the second-largest food production system in the world, which supports food, nutrition, and livelihoods to nearly 80% of the people of the developing region (FAO 2013). In South Asia, the intensive RFS is confined to Bangladesh, West Bengal, and a few smaller areas of Tamil Nadu and Kerala states of India, and southern Sri Lanka. This system supports some 130 million people (Dixon et al. 2001). Traditionally, RFS landscapes in south Asia are maintained by creative landscape designs networked with water channels, raised bunds, bushes, fish ponds, intermittent small huts and broad canopy shade-giving trees, as well as coconut tree and banana-based agro-forests along with the borderlands. Through such an integrated farming and intensive management practices, rice paddies improve land and soil health and conserve rich biodiversity – fish, insects, frogs, reptiles, birds, medicinal plants, edible plants, forage grasses, and so on. The system also provides critical ecosystem services like the production of water and support to the nutrient cycles, aesthetic and recreational benefits. Further, rice is the only crop that adapted to grow in a wide range of habitats, altitudes, weather conditions, and soil types and has an impressive rate of genetic variation.

Even though it required an increase of 114 Mt of rice in South Asia by 2036 to feed the growing population, the RFS experiences a rapid decline or oversimplification in its integrity due to shortage of irrigation, land use changes, reduction in agriculture labor and increased cost of agricultural inputs (Suzanne et al. 2012; Chauhan et al. 2017). Simplified or a degraded RFS is characterized by a low diversity of plant species and crop varieties, for instance, just 70 species only are reported from an area of 1440 million ha of presently cultivated land in the world (Altieri, 1999). A meta-analysis,

DOI: 10.1201/9781003164968-29

covering 66 independent studies, reported the species richness, including the predatory insects (as well as birds and spiders) significantly higher in integrated farms, whereas the richness of pest species and non-predatory insects turned up higher in the farms that are simplified and use high external chemical inputs (Bengtsson et al. 2005).

Asian Rice (*Oryza sativa*), despite being a self-pollinated species, is the only crop estimated with over 100,000 morphologically distinct allelic variations (Chang 1985). The Asian region is considered as the center of origin and diversification of *O. sativa* as evidenced there are 104,427 accessions of rice germplasm conserved at the national gene bank of National Bureau of Plant Genetic Resources (NBPGR), New Delhi, India as of April 2015. But, the paradox is that despite the genetic richness and high adaptability of the rice crop, the Asia-Pacific region experiencing significant food insecurity and undernourishment issues. The Asia-Pacific region holds 60% of the world's population and has a hungry population of more than 550 million (FAO 2014). In terms of absolute numbers of stunted children, the south Asian RFS is the fifth-worst affected system in the world (Dixon et al. 2001). The COVID-19 pandemic crisis has accelerated this situation and is expected 270 million people more will be added to the hitherto estimated 821 million food-insecure populations of the world, mostly from the Asia-Pacific and African regions (WFP and Oxfam 2020).

Added to the accelerated COVID-19 induced hunger crisis, there will be a substantial reduction in agricultural productivity and nutritional quality of many food crops, including rice in the coming years for the major reason of climate-induced extreme weather conditions and emerging pests and diseases (IPCC 2019). Without an ecosystem-based adaptation approach, the projection is up to an 8% reduction in the mean yield of all crops across south Asia by 2050 due to changing pests and diseases and even greater losses caused by extreme weather conditions (Knox et al. 2012). Promoting an effective RFS-based adaptation (RFSbA) approach can address the risks related to loss of agriculture labor and livelihoods, input-intensive production, climate change, and undernutrition. RFSbA combines ecological, social, and economic principles through the integration of diverse crops (millets, legumes, banana) as well as fish, cattle, poultry, tree, and soil biodiversity in the RFS. An integrated and intensified RFS offers huge potential to enhance food production manifold and to provide adequate food, nutrition, and employment to several millions of families who are pushed into starvation due to the COVID pandemic.

Despite a multitude of challenges, marginal and smallholder farmers in some interior pockets of Kerala and Tamil Nadu continue maintaining the intensity and heterogeneity of RFS and keeping diverse assemblage of crops and breeds in the system. Restoring the integrated RFS practices becomes critically important in the era of uncertainty in improving health, increasing food production, and keeping the soil fertility and productivity.

In this chapter, we describe the methods and results of such an effort taken in partnership with an indigenous community in the Wayanad district of Kerala called Kurichiya in sustaining their RFS.

26.2 Wayanad District – Home of Kerala's Heritage Rice Farming System

Wayanad – a steep mountainous plateau in the Kerala part of the Western Ghats is one of the 115 'aspirational districts' (otherwise economically backward districts) of India. The district is located from 11° 26′ 28″ – 11° 58′ 22″ N latitude and 75° 46′38″ – 76° 26′11″ E longitude covering an area of 2136 sq. km and an average altitude of around 750 m (Figure 26.1). Wayanad with its vast forest cover, luxuriant rainfall, extensive rice paddies, and a sizable percentage of tribal community keeps dynamic both biodiversity and cultural diversity of the state. The biodiversity richness of the district is evidenced by over 2100 flowering plants with as many 600 rare, endemic, and threatened plant species of Western Ghats, among which around 180 species are as food, 244 for medicinal uses, and 76 plants as Minor Forest Produces (Anil Kumar 2018). Agrobiodiversity is also equally rich in the district where the rice varieties are at the forefront presenting impressive genetic diversity of landraces suited to various land types and agroclimatic peculiarities. The name Wayanad itself is considered to have derived from the preeminence of the region for vast stretches of Rice paddies (The name 'Wayanad' is emerged from these two names 'vayal' and 'nadu').

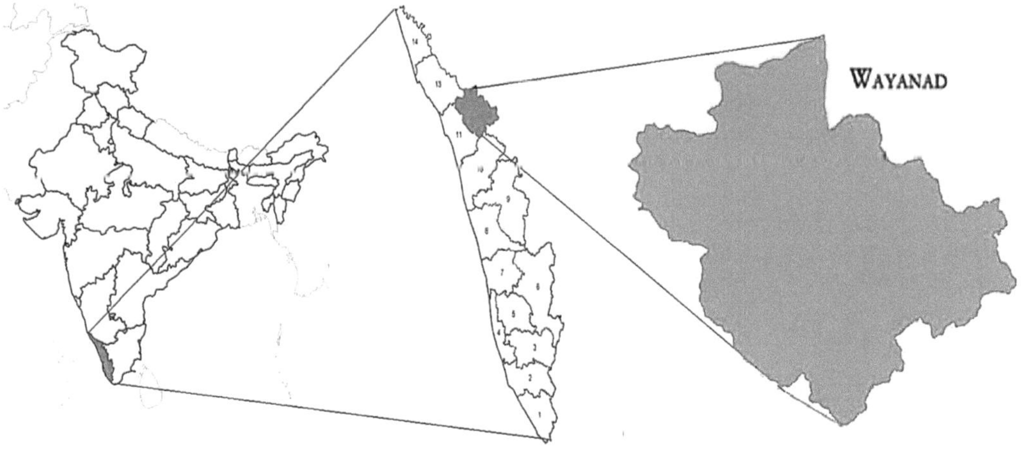

FIGURE 26.1 Location of Wayanad district.

Wayanad district has traditionally followed two seasons of rice cultivation; June-December (*Nanja* rice season) and January-May (*Punja* rice season). It is based on the season of the year, physical nature of the terrain, water availability, soil type, and climate; the varieties are selected for cultivation. A study conducted by M S Swaminathan Research Foundation (MSSRF) in 2011 showed that there were more than 75 traditional rice varieties cultivated in the Wayanad district. Some of the varieties are believed to have evolved in this place and some came during the immigration of people from the plains. But this has reduced to 30–35 from the 1980s and 1990s and out of which only less than 20 are now being cultivated in the district. Local farmers hold seeds of several exclusive traditional rice varieties, some of them, like *Njavara,* or *Navara,* which have unique medicinal properties. Apart from this, the varieties such as *Chennellu, Gandhakasala* and *Jeerakasala, Mullan Channa,* and *Veliyan* are some speciality rice varieties unique to the district. Kerala's legacy of traditional RFS and speciality rice cultivation has now largely confined to this district. In the district Kurichiya along with Kuruma are the custodians of around 20 traditional rice varieties, each having unique characteristics, like flood/drought resistance, pest/disease resistance, medicinal and aromatic properties.

26.3 Kurichiya Rice Farming System: Profile and Practices

Kurichiya, though a minor community in size with roughly 30,000 people (of the 151,443 tribal community population in Wayanad district of Kerala) is one of the largest joint families ever reported in anthropological literature (Kumaran 1996). They keep large land holding, big herds of cattle, and all adults of the joint family engage in the cultivation of diverse kinds of food crops. The Family farming approach ensures the cultural, culinary, and curative needs of the family and is considered a sustainable agriculture system (Anil Kumar and Vedavalli 2019). The food production landscape of Kurichiya habitations has the typical characteristics of the south Asian Wetland RFS.

26.3.1 Biodiversity Conservation Practices

Kurichiya maintains their RFS with integrated production fields, the main wetland rice field that is maintained usually at three levels based on the contour gradient as raised upland, middle land, and the bottom low lying extent called Kunduvayal, Kunivayal and Koravuvayal, respectively. A unique form of rice cultivation practiced by Kurichiya in the past in the flood line that bordering the Koravuvayal was the ValichaKrishi. In this method, long-duration flood-tolerant rice varieties like Chenthadi and

Chettu Veliyan (7–11 months) that grow above flood line were selected. In this cultivation practice, the intervention is limited to only the seed broadcasting and the crop would be maintained more in the form of a population of wild rice. The benefit of this mode of rice cultivation is that the crop would grow as perennials, and even if the tillers are damaged and the harvest is done; the plant develops new tillers and survive many years and yield sufficient grains. This form of cultivation was followed as a backup mechanism to cope with the emergency conditions that occur due to flooding of the normal rice fields.

The land above the main rice fields is maintained as garden land for the production of other food crops, mainly fruits like mango, jack, banana (5–10 varieties) and vegetables, tuber crops like yams, spices such as turmeric and ginger, and medicinal plants. Kurichiya women maintain small vegetable gardens on different parts of the commonly owned land and are often proud of their contribution to the common stock of vegetables for the next year (Suma 2014). The cultivated diversity of vegetables, rice, and other cereals and tubers in Kurichiya homestead is still heisted among other communities in the region. The garden land merges with the upper hill slopes that are kept usually under coffee plantation intermixed with a large number of diverse wild trees of the past forest to keep shade and as support trunks for growing black pepper vines. These areas were also part of the once practiced slash burn cultivation sites by their earlier generations to grow their staple food of millets and hill rice called Karuthan. When the coffee crop introduction became popular during the early 1900, such areas were converted for coffee cultivation, but without leaving the forest characteristics of the land.

No external chemical inputs are applied in the RFS. They are very specific to eat pure and toxin-free food. Due to this purity in cultivation, one can see the abundance of birds, insects, fish, and other forms of diversity in their farm fields, which in turn helps to reduce pest attacks. The RFS in Wayanad is associated with 16 species of birds rendering invaluable services to rice crop by controlling 10 species of pests of rice (Vishnudas 2007). The men and women of Kurichiya have profound knowledge about knowing the services of the birds and other beneficial organisms associated with rice fields. They easily differentiate farm-friendly creatures and pests and avoid hunting water birds as they increase the fertility of the field by depositing guano and feed on the pests. Rich species diversity of Egret, Heron, Ibis, Cormorant, numerous other birds and fishes are consciously managed by them at their fields. Similarly, many types of dragon and damselflies, crustaceans, and insects, including various spiders, also live in their farmlands. The rice fields are identified with three different types of crabs and each crab sole acts as a water storage unit that can hold 2–10 l of water (Unnikrishnan Nair 2020).

Kurichiya owns all of the 20 varieties (mostly landraces) known in cultivation in Wayanad and maintains the seeds through the two seasons of rice cultivation (see Box 26.1 for details of these varieties, including the use and perception about the value of each variety). All the specific rituals of Kurichiya are linked with different rice varieties based on duration and distinct characteristics. All of the rice-related rituals are observed in the first season (Nanja) and the varietal conservation has linked with culture. This is one of the motives that Kurichiya cultivating traditional rice varieties during the rainy season Nanja (Suma 2014).

There are several native plants associated with the Kurichiya food, healing, handicraft, weapons, and implement making tradition. More than 50 species of inland fish and numerous plant species seen in the paddy fields are part of the Kurichiya diet. The game meat from the neighboring forests was an integral part of Kurichiya culture and food basket until the enactment of the Indian wildlife protection act in 1972. The cultivated diversity of vegetables, rice, and other cereals and tubers in Kurichiya homestead is still heisted among other communities in the region (Suma 2014). We have listed some 110 items from their food basket (barring rice), which was intact till recently (see Table 26.1 to have a glimpse).

A nutrition analysis conducted in ten traditional rice varieties viz., *Adukkan, Chennellu, Chenthadi, Chomala, Gandhakasala, Jeerakasala, Kalladiaryan, Marathondi, MullanKaima* and *Kodu Veliyan by* MSSRF showed that the range of protein content of the ten rice varieties was found to be greater than the desired range (7%–8%) (Juliano 1985). The fiber content for the majority of the varieties was higher than the range reported by previous studies in other rice varieties (Diako et al. 2011; Verma and Srivastav 2017).

TABLE 26.1

The Food Diversity of Kurichiya Community[a]

Foods Produced/Managed in the RFS				
Fruits	**Leafy Greens and Nuts**	**Roots and Tubers**	**Fish**	**Game Meat (Now Banned)**
1. Palakaya	1. Mathanila	1. Chena	1. Cherumeen	1. Kooman
2. Kodampuli	2. Payarela	2. Katuchena	2. Puluvala	2. Moonga
3. Kotakaya(Pannikuru)	3. Mulluullacheera	3. Vellakachil	3. Potuvala	3. Pravu
4. Keerikaya	4. Katuthalu	4. Kayamakachil	4. Chakkamullan	4. Mayil
5. Challorum kaya	5. Thavi	5. Inchikachil	5. Thotumullan	5. Pokana
6. Ayanichakka	6. Thava	6. Mullankachil	6. Venmeen	6. Kakapokana
7. Njaval	7. Thakara	7. Kuzhikachil	7. Kaichule	7. Kolakozhi
8. Elanjikaya	8. Mudungachappu	8. Kaduvakayyan	8. Maran	8. Kannancheri
9. Edalakaya	9. Pressure cheera	9. Cherukachil	9. Valanjil	9. Katimota
10. Puli	10. Kochucheera	10. Neendi	10. Aaral	10. Pannipullu
11. Ambazham	11. Ponnamkanni	11. Noora	11. Vannalan	11. Necharipullu
12. Panalpazham	12. Mulakila	12. Nara	12. Mavoolu	12. Vatyanpullu
13. Kollinjaval	13. Murikila	13. Kavala	13. Mushi	13. Kudukkapullu
14. Chadachikaya	14. Valapayar	14. Venni	14. Mechom	14. Manjayan
15. Anelakayi	15. Elavanila	15. Chakarachempu	15. Karimeche	15. Koroolu
16. Cheri		16. Kavala	16. Vaala	16. Udumpu
17. Anguttikaya		17. Vayalchempu	17. Kata	17. Kooran
18. Nellika	*Nuts*	18. Kandichempu	18. Kallelmuti	
19. Thalumpazham	1. Chakkakuru	19. Kuzhichempu	19. Kanuman	
20. Velikuru	2. Anjilikuru	20. Malaramanchempu	20. Konjan	
21. Sivanunnikaya	3. Thanikuru	21. Makalepoti	21. Kannanparal	
22. Avalchunda	4. Katujathika	22. Pindalan	22. Valiyaparal	
23. Bamboo shoot (kallamula)	5. Pulinkuru	23. Eyanchempu	23. Thodan	
24. Katumanga			24. Keeru	
25. Kattuchakka			25. Nunayanvaala	

[a] The depicted names are in the local language of *Kurichiya*.

BOX 26.1 RICE VARIETIES NOW IN CULTIVATION BY KURICHIYA

Veliyan (MannuVeliyan): Drought and flood tolerant, source of high-calorie energy, used in brewing home liquor and the burned husk is most preferred for homemade tooth powder

Chettuveliyan: Flood resistant, comparatively high yield, bold and red-colored grain, nutritious, and tasty rice, it gives a feeling of fullness when consumed, resistant to various biotic and abiotic stresses, high fodder yield as well as grain yield

Palveliyan: Highly preferred for rice gruel ('Kanji'), white kernel
Thondi: Tasty rice, red kernel
Palthondi: Highly preferred for rice gruel, white kernel
Marathondi: Red and stiff rice
Chennellu: Holy and Medicinal rice, used as a cure for stomach ulcers, vomiting, etc.
Kaima: Scented rice, preferred for preparing breakfast dishes and ghee rice
Urunikaima: Scented, preferred for preparing breakfast dishes
Mullankaima: Scented, used on special occasions in the family

Poothadikaima: Scented with a strong aroma, preferred for preparing beaten rice
Gandhakasala: Scented, preferred for Biriyani and Payasam on special occasions in the family
Jeerakasala: Scented, preferred for Biriyani and Payasam on special occasions in the family
Mullanpuncha: Drought resistant
ThonnuranThondi: Short duration, traditionally treated as famine crop, harvested on the emergency during the scarce periods
Kalladiyaryan: Highly drought resistant. Suitable for valleys and terrains
Onavattan: Tasty rice, an introduced variety
Chempathi: Scented rice
Chomala: Highly tasty rice, white kernel, preferred to prepare break-fast dishes during special occasions
Chenthadi: Flood tolerant, tasty grains

26.3.2 Water Conservation Practices

Since they follow a joint family system, their land is under collective ownership that avoids fragmentation and results in maintaining large rice fields, good tree canopy cover on the hillsides assures water availability throughout the year. As the natural vegetation and swamps left undisturbed and non-degraded, they remain as water reservoirs all the seasons (Suma 2014). Keeping head ponds (Thalakulam) along the downline of such marshes where it is fed by innumerable water channels from the mountains is a common method practiced by them for water conservation. Kurichiya people are also experts in making PanamKeni a traditional water conservation method for drinking and cooking purposes using the trunk of a Palm (*Caryota urens* L.), locally known as *Aanapana*, *Choondapana*, and *Yakshipana*. The lowermost stem portion of the Pana is made into wooden cylinders after soaking in water for a long period so that the inner core gets degraded and the firm outer layer remains. After the processing, the stem is immersed deep into some corner of the rice fields where there is a good groundwater spring. *Panamkeni* is considered sacred by them and ladies in their menstrual period are not allowed to collect water from it (Unnikrishnan Nair 2016). The life span of a *Panamkeni* is around 40–60 years depending upon the maturity of the palm used (Sujana and Joseph 2015). These two practices along with keeping rice fields that are the reservoirs of rainwater turned out to be the secret of plentiful water availability to them and for the people around their habitations even during the hottest months.

Despite this richness in the RFS of Kurichiya, the low productivity and low profitability of these rice varieties and economic pressure compelling them to leave rice cultivation as other rice farmers from non-tribe community shifted in favor of more profitable crop banana. During the last decade, the area under rice in Wayanad drastically has decreased from 21,770 to 7300 ha. This shift from paddy to banana has been impacting a cascading effect on the rice farming ecosystem; firstly, on rainwater storage and aquifer recharging, then the livelihood options of another tribe community Paniya, who largely depends on this system for their employment and income, and finally on the environment, which paves the way to dump with larger quantities of highly toxic pesticides to ensure high yield from banana cultivation.

26.4 Declining Rice Production in Wayanad District

The area under rice cultivation in Wayanad is decreasing rapidly. During 1985–1986, nearly 30,000 ha area were under rice cultivation in the district (Armando et al. 2014); by 2009–2010, it had shrunken to 10,576 ha and in 2018–2019, it further slashed to 7300 ha. Similarly, the production was decreased from

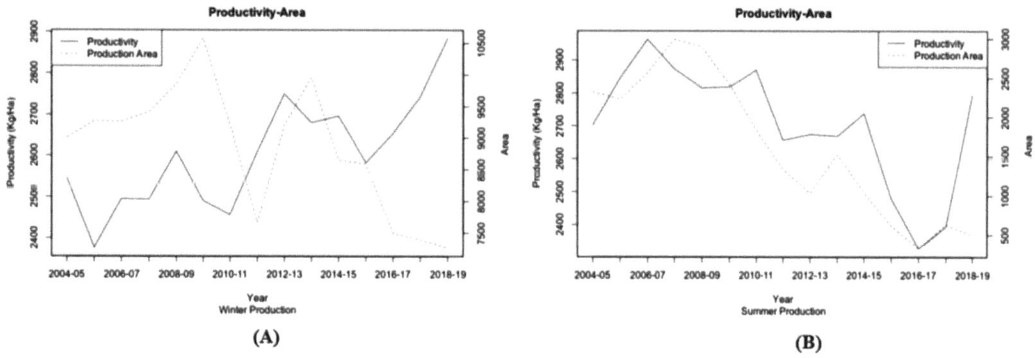

FIGURE 26.2 (A) Graph showing productivity and area during winter season (Nancha) over the period 2004–2005 to 2018–2019, (B) graph showing productivity and area during summer season (Puncha) over the period 2004–2005 to 2018–2019. (Adapted from Merlin et al. 2020.)

29,206 t (2004–2005) to 22,340 t (2018–2019), whereas the productivity was observed to be increased for the past few years (Figure 26.2). The data on rice productivity (kg/ha) and area of rice cultivation (ha) were analyzed for a period of 13 years from 2004–2005 to 2018–2019. The graph of the winter season crop (*Nancha*) depicted a gradual increase in rice productivity from 2010 to 2011 and the highest productivity was in the year 2018–2019. Similarly, a gradual decrease in rice cultivating area was observed from 2013–2014 to 2018–2019, whereas the graph for the summer season depicted a gradual decrease in both productivity and area size except for the year 2018–2019. The highest rice productivity was in 2018–2019 for both seasons.

In general, the rice farmers of Wayanad commonly use high-yielding varieties (HYVs), except one or two high-income value traditional varieties, whereas the common traditional varieties are cultivated only by the tribal groups like Kurichiya. The HYVs like *Athira, Uma, Kanchana* and IR20 dominate in the cultivation and replacing very fast the traditional rice varieties. The reason for the increased productivity even the area under production has declined is the replacement of traditional rice varieties with HYV. Figure 26.3 depicts the trend of decline in the usage of traditional rice variety over HYV.

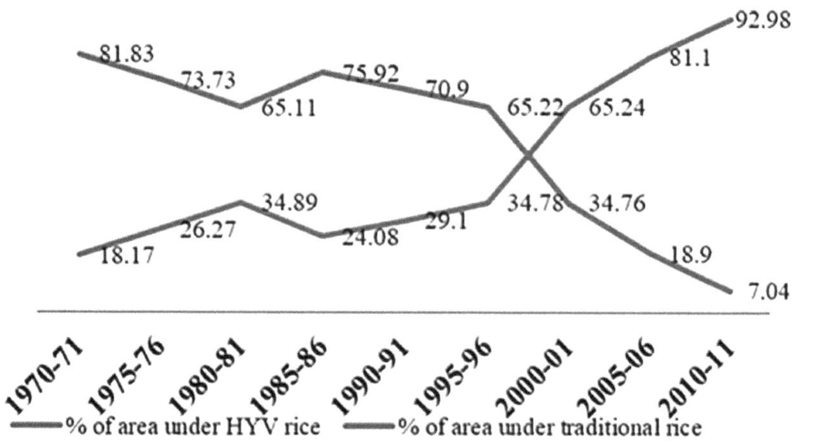

FIGURE 26.3 Area under HYV and traditional rice varieties in Kerala since 1970–1971. (Modified from Girigan and Manjula 2018)

26.5 Trend Analysis of Rainfall, Temperature, and Irrigated Area for Rice in Wayanad

The previous climate analysis studies of Wayanad show that there is a decreasing trend of the amount of rainfall and an increase in variability of precipitation (Rajendran et al. 2012; Gopakumar 2011; Aype and Rajan 2005). But the trend is changing slightly in the case of rainfall as two consecutive years 2018 and 2019 witnessed flood in the district. The seasonal pattern of rainfall in each month has been visualized using a box plot (Figure 26.4) created by taking average rainfall (mm) for a period of 30 years, each month from 1991 to 2019. The monthly precipitation depicts an increase in its variability from 1991 to 2019. Similarly, analysis of temperature through anomalies of linear trend portrays an increasing trend of maximum and minimum temperatures at Wayanad district (Figure 26.5).

The increasing trend of the irrigated area during the winter season (Nancha) was reported in previous studies. Reduction in the water retention capacity of landscapes (Anil Kumar et al. 2012), increased

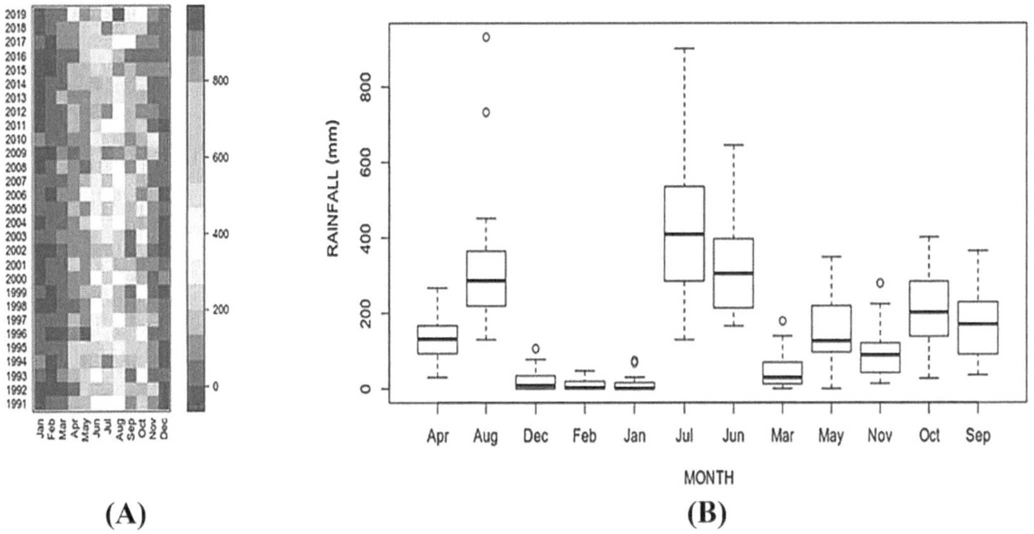

FIGURE 26.4 (A) Monthly precipitation at Wayanad (mm/month), (B) box plot of seasonal effect of rainfall in Wayanad (bold line – median, q1 – maxT, q2 – minT, o – outlayers). (Adapted from Merlin et al. 2020.)

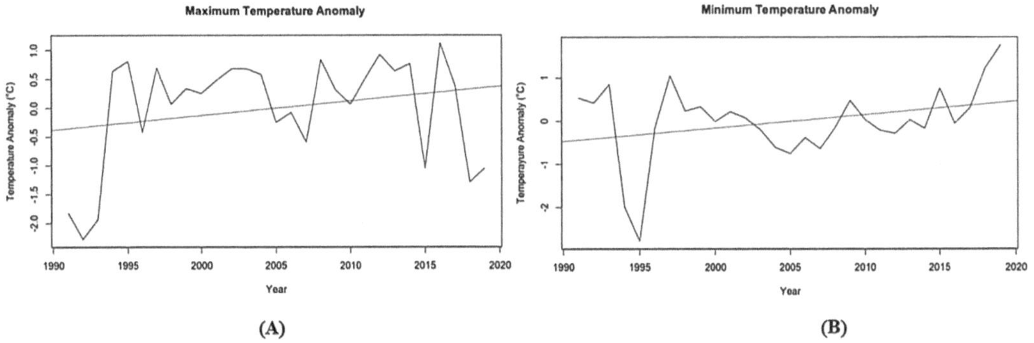

FIGURE 26.5 Graphical representation of trend analysis of (A) maximum and (B) minimum temperatures at Wayanad district from 1991 to 2019. (Adapted from Merlin et al. 2020.).

runoff, abandoned irrigation projects, and reduction of groundwater recharge have made farmers either transform their paddy fields to other crops or to alternate means to irrigate rice fields (Kumar et al. 2012). In the case of HYV during the winter season (*Nanja*), rice cultivation in an irrigated area in 2004–2005 was 4462 ha and in unirrigated area, it was 1950 ha. By 2009–2010, there was a gradual increase in irrigated area and a gradual decrease in unirrigated area, and it was 7661 ha and 615 ha, respectively, for the period. In the year 2017–2018, there were no high-yielding rice varieties cultivated in unirrigated area and the cultivation in the irrigated area was limited to 6237 ha. Similar fluctuations were seen in traditional varieties cultivated in the winter season. In 2017–2018, traditional varieties were cultivated only in irrigated area. In the case of summer season (*Punja*) rice cultivation, the usually cultivated HYV is limited to the irrigated area. The cultivation of traditional varieties in the summer season (*Punja*) was negligible (Merlin et al. 2020).

26.6 Protection of Kurichiya Rice Farming System Heritage

Since 2004, MSSRF has partnered with Kurichiya community men and women for seed purification and multiplication that resulted in the production of quality seeds of few speciality rice varieties. Traditional rice varieties like *Veliyan, Gandhakasala, Chennellu, and Navara* were tried in the first phase of this intervention. Participatory mode of seed purification wherein scientists work to strengthen farmers' informal research and development system was adopted and it was found to be the best approach to conserve these varieties. System of Rice Intensification (SRI) method of cultivation was introduced in the district to boost the yield of the traditional rice varieties. In 2011, participatory purification methods were adopted for the selection and purification of seeds sourcing the expertise of lead farmers. Purification techniques like rouging of weeds and removing off-types were performed and the purified seeds were brought into the distribution chain, every year. Extensive numbers of training were organized on purification techniques, seed and grain management, mechanization for helping the community in their efforts to conserve the speciality varieties.

These efforts in conservation and cultivation of traditional varieties of rice have resulted in recognizing the Kurichiya and Kuruma tribal communities of Wayanad, with the Second Plant Genome Savior Community Recognition in 2008 by the Protection of Plant Varieties and Farmers' Rights Authority of Govt. of India. Subsequently, in 2010–2011, Wayanad District Tribal Development Action Council, a community-based organization comprises primarily Kurichiya and Kuruma tribal communities and facilitated by MSSRF received the Plant Genome Savior Award of Ten Lakhs rupees and Certificate. The Centre had also played a lead role to form another grass root organization called SEED CARE for the conservation of traditional crop germplasm and facilitated the applications for 27 Farmers' Paddy Varieties under the Protection of Plant Varieties and Farmers' Rights Authority of Govt. of India. This has resulted in 16 of them were declared as Farmers' Varieties in 2013. Steps such as seed fair, farmer, and breeder interface were also initiated that facilitate access to these varieties (that are conserved ex situ in MSSRF gene bank), from the part of formal seed system actors.

The SEED CARE forum spearheads the processes of community mobilization, awareness generation in Plant Genetic Resources Management, including quality seed production of traditional paddy varieties. There are at present 105 households, spread across 10 selected seed villages in 74.8 ha in the district engaged in the traditional rice seed production. Every village has taken up the cultivation of 2–8 traditional varieties of paddy. The farmers of the seed villages can sell off their excess produce every year at a rate that is much higher than the rate in the local market. During the year 2018–2019, the production of 2.15 t of seeds of traditional rice varieties was reported.

The farmers also into conserve and cultivate many other food crops associated with the RFS. MSSRF has recorded 30–40 cultivated varieties of yams and taros from such farms in Wayanad and adjoining regions. All these varieties are now maintained at MSSRF's Community Agrobiodiversity Centre's Field Gene Bank in Puthurvayal village of the district. Protein analysis of these varieties showed that *Inchikachil* (*Dioscorea alata*) had the highest protein (14.52%) and *Neelakachil* had the lowest protein

(3.67%). 38.72% of *Thoonankachil* (*D. alata*) was the dry matter, which was the highest among the samples. *Nanakizhangu* (*Dioscorea esculenta*) had the lowest with only 15.59%. Many of the cultivated varieties of yams are orphan crops, as there is little attention to their improvement.

An important aspect of yams is that they have a very low water footprint, which means they can be grown in extreme marginal environments. They are traditionally high-resilient crops, which matters to counteract the deleterious effect of climate change and have the potential to help the poor and marginal farmers to adapt to the vulnerabilities of climate. Also as the volume of food (tuber) is very large with long shelf life compared to any other food crops in the region, and correspondingly the availability of food per person at household is very high. The aroids are used for both their corms and culms, which form an important source of readily available vegetable.

Due to changes in the climate and unpredicted floods and drought, now there is a growing trend to move toward traditional varieties that are more hardy and tolerant to climate vagaries. There is an increased demand for the seeds of such varieties, provided the quality of the seeds is maintained. To overcome the issue of contamination/mixing of seeds at farmers' fields, 2 ha of paddy land is being consistently maintained by the MSSRF research team for purification of the selected ten varieties such as *Chennellu, Chenthadi, Chomala, Jeerakasala, Gandhakasala, Mullankaima, Thondi, Adukkan, Veliyan,* and *Kalladiyaryan.*

26.7 Conclusion and Way Forward

Kurichiya's agriculture heritage reflects the tangible and intangible traditions and knowledge systems embodied and sustained by a community over centuries in farming and food production. Such food and agricultural production systems managed with a whole of family approach have conserved the local food specialities and diversity of crop varieties, animal breeds, and many forms of biodiversity. Despite the fast erosion of cultural traditions, the value of agricultural heritage continues to be potted in the heart of their culture. But, how can such heritage be realized within the context of the fast-growing economic needs and aspirations is the biggest development challenge? The agriculture heritage practices of this community are unlikely to continue for a long unless there are enabling conditions and incentives for local communities to promote their quality of life and play a greater role in incentivizing their conservation efforts.

As an enabling mechanism to protect the world agriculture heritage, the UN has launched an initiative in 2010 called Globally Important Agriculture Heritage System (GIAHS). As of 2020, this multi-donor program hosted by FAO has identified and designated 62 such heritage sites across 22 countries. These sites are exceptionally diverse landscapes with rich biodiversity compatible and climate-resilient agricultural practices. From Asia and the Pacific, 40 GIAHS sites were designated so far in 8 countries with 15 in China, 11 in Japan, 5 in the Republic of Korea, 3 in India, 1 each in Bangladesh and Sri Lanka. The three Indian heritage systems are the Below Sea-Level Farming system of Kuttanad (Kerala), Tribal Agriculture system of Koraput (Odisha), and the Saffron cultivation system of Jammu and Kashmir. In India, there are many more potential GIAHS sites to qualify from each state. India is a home for more than 2000 ethnic groups; 415 living languages, different agroecological regions, and hundreds of villages where everyday life of people revolves around agriculture-, forestry- or fishery-based livelihoods which are better placed for sustainable promotion of world 'agricultural' heritage.

Section 37 of the Biodiversity Act of India also advocates the central and state governments to identify and declare bio-cultural diversity heritage sites. Section 36 (2) of the Act gives power to the Central Government to give directives to the concerned state government to take immediate ameliorative measures if they are convinced, any areas rich in biological diversity, biological resources and their habitats is being threatened by overuse, abuse, or neglect (BD Act 2002 & Rules 2004). These provisions along with the GIAHS mechanisms as well as other associated commitments are to be made use of for identifying and making the agriculture heritage sites fully functional and promote happy livelihoods around agriculture. The Wayanad RFS comprise rice paddies, canals, wild bushes, sacred groves, and individual tree species and an enormous number of agrobiodiversity nurtured by the indigenous tribal communities like Kurichiya is an ideal system for such kind of recognition.

REFERENCES

Altieri, M. A. 1999. The ecological role of biodiversity in agroecosystems, *Agriculture, Ecosystems & Environment*, 74: 19–31.

Anil Kumar, N. 2018. Aspiring to Rise on Rice: Case of Rice Genetic Diversity of Wayanad District of Kerala. (https://www.linkedin.com/pulse/aspiring-rise-rice-case-genetic-diversity-wayanad-district-anil-kumar)

Anil Kumar, N. and Vedavalli, L. 2014. Conservation of family farming heritage, background. In Paper prepared for the Asia Pacific Regional Consultation on "Role of Family Farming in the 21st Century: Achieving the Zero Hunger Challenges by 2025", 7–10 August, MSSRF, Chennai.

Armando, G., Nidhi, N., Shradha, S., and Devi Prasad, K. V. 2014. Examining the Influence of Climate Change Impacts on Agrobiodiversity, Land Use Change and Communities. An Empirical Experience from Wayanad-Kerala, India. 10.13140/RG.2.1.1050.0326.

Aype, T. P. and Rajan C. K. 2005. South-West Monsoon Rainfall of Kerala and Its Variability, Document by Cochin University of Science and Technology, Kochi.

Bengtsson, S., Nagy, Z., Skare, S. et al. 2005. Extensive piano practicing has regionally specific effects on white matter development. *Nature Neuroscience*, 8: 1148–1150.

Chang, T. T. 1985. Crop history and genetic conservation: rice – a case study. *Iowa State Journal of Research*, 59: 425–456.

Chauhan, B. S., Jabran, K., and Mahajan, G. 2017. Rice Production Worldwide, Springer International Publishing AG, Springer, Cham.

Diako, C., Sakyi-Dawson, E., Amoa, B., Saalia, F., and Manful, J. 2011. Cooking characteristics and variations in nutrient content of some new scented rice varieties in Ghana. *Annals. Food Science and Technology*, 12: P5.

Dixon, J., Gulliver, A., and Gibbon, D. 2001. Farming Systems and Poverty: Improving Farmers' Livelihoods in a Changing World, FAO and World Bank, Rome.

FAO. 2013. Food Systems for Better Nutrition. http://www.fao.org/publications/sofa/2013/en/.

FAO. 2014. Innovation in Family Farming. http://www.fao.org/publications/sofa/2014/en/.

Gopakumar, C. S. 2011. Impacts of Climate Variability on Agriculture in Kerala. MPhil thesis, Department of Atmospheric Sciences, Cochin University of Science and Technology. India.

Girigan, Gopi, and Manjula, M. 2018. Speciality Rice Biodiversity of Kerala: Need for Incentivising Conservation in the Era of Changing Climate. *Current Science*. 114. 997–1006. 10.18520/cs/v114/i05/997-1006.

IPCC. 2019. Summary for policymakers. In Climate Change and Land an IPCC Special Report on Climate Change, Desertification, Land Degradation, Sustainable Land Management, Food Security, and Greenhouse Gas Fluxes in Terrestrial Ecosystems, https://www.ipcc.ch/srccl/chapter/summary-for-policymakers/

Juliano, B. O. 1985 (ed) (1985). *Rice: Chemistry and Technology* (Vol. 69). St Paul, MN: American Association of Cereal Chemists.

Knox, J., Hess, T., Daccache, A., & Wheeler, T. 2012. Climate change impacts on crop productivity in Africa and South Asia. *Environmental Research Letters*, 7: 034032.

Anil Kumar, N., Gopi, G., and Prajeesh, P. 2012. Genetic erosion and degradation of ecosystem services of wetland rice fields. In S. Lockie and D. Carpenter. (Eds.), A Case Study from Western Ghats, India, *Agriculture, Biodiversity and Markets*, London: Earthscan, 137–53.

Kumaran, V. 1996. Kurichyarude Jeevithavum Samskaaravum, Current Books, Kottayam.

Merlin, L., Aghosh, C. H., Aleena, T., Parvathy, M., Kuriakose, J. 2020. Impacts of Climate Change on Rice Productivity of Wayanad District, Kerala, International Webinar – DOCTRINA 11, 05 to 07 June, Sir Syed College, Taliparamba.

Rajendran, K., Kitoh, A., Srinivasan, J., Mizuta, R., and Raghavan, K. 2012. Monsoon circulation interaction with Western Ghats orography under changing climate. Projection by a 20-km mesh AGCM. *Theoretical and Applied Climatology*, 110. 10.1007/s00704-012-0690-

Sujana, K. A and Joseph, J. 2015. Panamkutty – a traditional water harvesting system in Wayanad district, Kerala, *ENVIS Newsletter*, 20(1), P6.

Suma, T. R. 2014. Case Study on Kurichyas of Wayanad: One of the Largest Family Farmers of the World.

Suzanne, K. R., Nadine, A., and Binamira, J. S. 2012. Rice in Southeast Asia: Facing Risks and Vulnerabilities to Respond to Climate Change. Food and Agriculture Organization of the United Nations, Organisation for Economic Co-operation and Development, Rome, 295–314.

The Biological Diversity Act. 2002 and Biological Diversity Rules 2004. http://nbaindia.org/uploaded/act/BDACT_ENG.pdf.

Unnikrishnan Nair, G. S. 2016. Traditional wisdom in harvesting water. *Traditional and Folk Practices*, 04: 50–53.

Unnikrishnan Nair, G. S. 2020. Guardians of the Grain. (https://idsffk.in/2017/06/02/guardians-of-the-grain-guardians-of-the-grain/)

Verma, D. K. and Srivastav, P. P. 2017. Proximate composition, mineral content and fatty acids analyses of aromatic and non-aromatic Indian rice. *Rice Science*, 24: 21–31.

Vishnudas, C. R. 2007. Silent Services of Winged Beauties in Agriculture LEISA, Dec. 13–14.

WFP and Oxfam. 2020. The Hunger Virus: How Covid-19 is Fuelling Hunger in a Hungry World. Oxfam Media Briefing. (https://www.oxfam.org/en/research/hunger-virus-how-covid-19-fuelling-hunger-hungry-world)

27

Agricultural Innovation and Technology Adoption among Small-Scale Producers in Developing Countries: Is Biotechnology a Sustainable Alternative?

Alejandro Barragán-Ocaña, Adalberto de Hoyos-Bermea, Katya Amparo Luna-López, and Tamara Arizbe Virgilio-Virgilio
Instituto Politécnico Nacional (National Polytechnic Institute), Mexico City, Mexico

CONTENTS

27.1 Introduction to Agricultural Technology and Biotechnology

The limited availability and scarcity of natural resources are barriers to food production, especially among small-scale producers in developing economies, which makes agricultural practices for conservation an alternative to achieve sustainable agricultural progress (Ashoori et al., 2017). Collaboration among agricultural producers is also relevant as a source of new knowledge (e.g., agricultural knowledge, sharing information related to the activity, or support mechanisms to optimize their activities) (Saint Ville et al., 2016). The sustainable intensification of agriculture will require approaches from different areas of knowledge to generate solutions of different kinds taking into account political economy, the corporate sector, and other factors that determine the demand and commercialization of agricultural products (Rockström et al., 2017).

However, the agricultural, geographical, and institutional environments circumscribing the reality of a country can promote or impede technology transfer and adoption; thus, developing countries must work more intensely than developed ones as consequence of the low level of technological adoption frequently associated with economic and human aspects in many places (Obi and Nwakaire, 2014; Mwangi and Kariuki, 2015). In this regard, different agricultural technologies are intended to operationalize and optimize the development of the field in different ways. These include increasing productivity, improving crops, promoting sustainability, and mitigating pollution, among others. Nevertheless, the adoption, transfer, and implementation of these technologies is clearly still a challenge for farmers, especially small-scale farmers.

These technologies include the following: (1) use of wireless sensor networks in smart farming (Taheri et al., 2020); (2) use of climate-smart agriculture (CSA), intended to increase sustainable production and characterized by its environmental friendliness and potential to cope with adverse climate effects

DOI: 10.1201/9781003164968-30

(Khatri-Chhetri et al., 2017); and (3) precision agriculture as a strategy to adapt crops to space and time and, consequently, improve production (quantity and quality), favor the environment, and optimize the use of material and economic resources (Paustian and Theuvsen, 2017).

Among other tools to contribute to agricultural development and mitigate pollution caused by chemical fertilizers are microbial inoculants (biofertilizers), which contribute to the sustainable development of the field by improving the plant's nutritional potential, productivity, and resistance to diseases and abiotic stressors (Igiehon and Babalola, 2017; Olanrewaju et al., 2017). On the other hand, the use of endogenous technology (biofertilizers) for agricultural activities in developing countries has been shown to have a positive effect on the social well-being of farmers (Barragán-Ocaña and del Valle-Rivera, 2016). Thus, the benefits of biofertilizers are associated with improved agricultural productivity and economic savings. However, experiences in developing countries point out the need to improve the quality, efficacy, and product standardization of biofertilizers. It is also necessary to work on elements associated with laws and regulations, disseminate information about these technologies, raise awareness about their use, and develop related infrastructure, supplies, human resources, labeling, and technological developments taking into account different application contexts (geographically and environmentally) (Naveed et al., 2015).

Biopesticides are also an option; some of these technological developments are of microbial origin, others are products derived from microorganisms or obtained from natural sources, including transgenic biopesticides. Therefore, since they are considered as an alternative to the use of chemical pesticides, their purpose is to produce residue-free food and to function as an ecologically friendly option, albeit with multiple requirements concerning its composition, toxicity, and efficacy, among other features (Gupta and Dikshit, 2010). Composting is an example of a biotechnological technique that can help to recover degraded soils and increase their fertility (soil quality); it contributes to carbon sequestration (in soils), and it represents a sustainable alternative compatible with the bioeconomy. Composting methods depend on different technical criteria, which should also consider economic and environmental issues (Pergolaa et al., 2018; Viaene et al., 2016).

Based on this set of ideas, it is important to point out that there are several cases of successful use of biotechnology in agriculture, such as the following: biofertilizers (increased productivity and/or decreased use of fertilizers): (1) nitrogen-fixing filamentous cyanobacteria in rice crops (Pereira et al., 2009); (2) microdosification of fertilizers (precision), lime, and *Bacillus megaterium* as a biofertilizer in acid soils to increase maize production (Kubheka et al., 2020); and (3) applications using plant-growth-promoting rhizobacteria (PGPR) – *Bacillus* – to optimize crop development in blackberry production (*Rubus glaucus* Benth) (Robledo-Buriticá et al., 2018). Biopesticides (alternative to the use of agrochemicals): (1) entomogenous fungi as a viable mechanism for the biological control of ticks (Kaaya and Hassan, 2000); (2) use of baculovirus to control insects as an integrated pest management strategy (Haase et al., 2015); and (3) aqueous extracts of plants (*Thevetia peruviana* and *Azadirachta indica* seeds) to control cocoa mirids (*Sahlbergella singularis*) (Mboussi et al., 2018).

A third element is represented by composts, which involve the processing and utilization of generated residues and consequently reduce environmental problems; these developments can be used in different agricultural activities, especially to recover soils impoverished by their intensive use due to the growing demand for food. Some of these cases include the following: (1) residues from the palm oil industry (biomass) for the production of compost to be used in agricultural soils (Vakili et al., 2015); (2) inclusion of nitrogen-fixing cellulolytic microorganisms in the composting process of lignocellulosic residues of crops as a sustainable strategy (Harindintwali et al., 2020); and (3) composting of organic waste (food), which due to its massive generation, can help to reintegrate organic matter into soils (carbon restitution) and fertilize soils (Azim et al., 2018).

27.2 Ethical Considerations and Public Policy Recommendations for the Adoption of Biofertilizers, Biopesticides, and Composts

Biotechnology associated with biofertilizers, biopesticides, and composts is an area in which the general population should have an important role: the possibility of universal access to nutritious and healthy foods is a human right promoted by technological capabilities offered by biofertilizers in fields related to food security. The awareness of small-scale producer organizations and communities is essential since their way of life is also affected by the way in which such technologies are developed and applied. Supported by biotechnology, the production of biofertilizers, biopesticides, and composts combines the benefits of agricultural production with a positive impact on issues related to soil degradation, the sustainability of agricultural practices, and the health of people who consume these agricultural products.

These biotechnological innovations have the potential to create solutions to be incorporated into production in the form of processes, inventions, or models. However, the introduction of a technological system based on values such as efficiency, production, and economic efficiency needs to take into account the needs of the societies in which it is applied and possible solutions to local problems. In adopting this technology, it is important to consider that these technological systems are not value-neutral; they introduce a new way of interacting with reality in the adopting communities that may be alien to their established values and representations about the world. This is especially important in the case of small-scale producers, whose organizational forms can be seriously disrupted, and they are a most sensitive issue in the case of indigenous communities, who have historically been forced to adopt forms of production and commercialization unrelated to their reality (Gómez, 2010).

The local knowledge of a situation makes the point of view of these communities valuable for solving specific problems and provides a very rich diagnosis of the technological elements that would be adequate to introduce into a given situation. Therefore, a technology developer has a social responsibility, as well as those who commercialize the technology, who should use a dialogic approach when introducing it so that community members may benefit rather than being imposed a technological system (Olivé, 2007).

The adoption of a technology will have better prospects of success if it is commensurate with the needs of the community, and any possible risks should be openly acknowledged (Genus and Iskandarova, 2018). The assessment of needs should be sensitive to community values, not only in terms of market needs for certain products and available resources. This is achieved by involving the stakeholders in agricultural production, that is, farmers' groups and their communities, technology developers, academia and authorities, but it would be important for consumer groups to participate in the discussion, so that solutions are harmonious and thoroughly consensual. Ample and democratic participation in innovation activities results in respect for the values and actors of the societies where biotechnological systems are developed and deployed and allows for local forms of life to voluntarily and actively integrate them. An introduction of biotechnologies that takes into account only technological components and capabilities and production effectiveness can disrupt the social and economic life of rural and indigenous peoples.

Substantial participation of all stakeholders can be complicated when it comes to small-scale producers, but it can be assumed that the planning and development of viable technologies together with a participating citizenry will have better prospects of resulting in technological solutions that are socially viable and sustainable for communities, preserving social and natural resources for future generations (Thompson et al., 2020). Caring for the diversity of plant and animal species of an ecosystem is another of the elements common to different participants, who could have a shared interest in preserving them and therefore adding harmonious values to the environment, considering not only how much the land produces and the extent to which it can be exploited. The use of biofertilizers, composts, and biopesticides can be an interesting response to concerns about biosecurity around the use and implementation of these technologies. Therefore, these communities cannot be expected to cope with the local risks, such as those associated with water pollution and soil degradation, while others have access to the benefits of production based on these technologies (Mouter et al., 2018).

On the other hand, one of the driving forces behind the adoption of new knowledge by small-scale agricultural producers are innovation policies, understood as the actions implemented by government authorities to promote the use of scientific and technological resources to address a given specific problem or take advantage of a technology-based business opportunity that results in economic and social benefits. The set of policy instruments that can be used to encourage research, development, or the adoption of innovative production techniques includes, in addition to financial support programs, legal and planning programs, as proposed in the following paragraphs:

1. *Training, education, and new skills programs:* They contribute to the creation of capacities to generate, adopt, and assimilate new technologies; these skills are key to consolidating technology development and implementation plans in a given territory. They are the most commonly implemented instruments, including promoting the development and use of new biofertilizers (Naveed et al., 2015).
2. *Innovation plans:* This type of instrument involves outlining agendas to define objectives and integrate strategies to promote innovation by regionalizing production and defining priority crops, reconverting production, modernizing production techniques, and fighting pollution and the effects of climate change.
3. *Legal framework:* This is the set of laws, regulations, standards, norms, and certifications that regulate the execution of innovation activities, providing certainty to producers and consumers about the desirable performance and quality parameters of biotechnological production processes and products.
4. *Creation of new organizations:* These programs aim to encourage the creation of new research centers or companies, or even new research divisions in technological areas of interest, as well as intermediary organizations to access investors with the purpose of bringing biotechnology developments to the market. They can also promote new service centers for common use (laboratories, pilot plants), as well as the integration of public distribution systems to facilitate the distribution and supply of raw materials from the agricultural sector (Lele and Goswami, 2017).
5. *Cluster creation and strengthening programs:* They seek to create synergies among companies and research centers around clearly defined technology transfer objectives located in a defined space where public-private partnerships are created to join efforts for innovation (Stadler and Chauvet, 2018).
6. *Subsidies to research and development (R&D):* These financial support instruments often have low interest rates or are nonrepayable to stimulate new R&D varieties or to adopt new technologies. They often involve funds concurrently contributed by the beneficiaries, which encourages them to invest in innovation projects.
7. *Tax incentives:* Beneficiaries can deduce a percentage of the income tax corresponding to technological development projects. They involve the total financing of the project by companies, which contributes to the long-awaited consolidation of the R&D culture.
8. *Value chain and supply chain integration programs:* They seek to optimally articulate actors involved in R&D, production, and commercial activities in order to consolidate the assimilation of new technological knowledge among all members of the production chain.

Defining an idoneous set of policy instruments for small-scale agricultural producers requires considering the degree of novelty of the technological knowledge to be generated or assimilated, as well as the complexity of the production reconversion, since the disruption of traditional production practices correlates with the degree of resistance to the implementation of public policies. Therefore, community participation is essential; small-scale farmers must be involved in the definition of the plans and programs to be carried out to minimize the resistance to change caused by the introduction of new and unknown technological alternatives (Van Damme et al., 2014).

27.3 Methodology

The proposed methodology was based on a network analysis. First, we mapped the knowledge around the issue of agricultural technology. This was carried out using a co-occurrence study based on data obtained from the bibliographic analysis carried out in the Scopus database (2020) using keywords (authors) and the total counting method. To present the most recent data, the search was limited to a period of time that included the most recent information; a set condition was to use the year 2019 as the final search date for results including up to 2,000 records; this was established because the database allows only citation information to be exported when the search yields a larger amount of results (greater than 2,000 and up to 20,000 documents). Based on these restrictions, we searched documents for a three-year period (2017–2019) using the term "agricultural technology" in the following fields: (1) Article title; (2) Abstract; and (3) Keywords. A total of 1,489 records were obtained based on the following search criteria:

(TITLE-ABS-KEY ("agricultural technology") AND (LIMIT-TO (PUBYEAR, 2019) OR LIMIT-TO (PUBYEAR, 2018) OR LIMIT-TO (PUBYEAR, 2017)))

Subsequently, to continue with the analysis of technologies related to agricultural biotechnology, the search was extended to the terms "biofertilizer," "biopesticide," and "compost" within the same fields (article title, abstract, keywords) but for a five-year period, except in the case of compost, because the number of documents was too large and the exercise had to be limited to 2019. Therefore, the search criteria were established as described in Table 27.1.

Subsequently, once the four databases were integrated, network studies were carried out using co-occurrence analysis based on the following elements: (1) Type of analysis: Co-occurrence; (2) Unit of analysis: Author keywords; and (3) Counting method: Full counting. Additionally, the exercise was limited to the analysis of terms with an occurrence (frequency) of at least five words out of the total number of keywords identified. This means that each element or node appearing in each network map did so at least five times in the search. Based on this information, the study was conducted aiming at two goals: the first was to identify the total number of occurrences, their links, and the total strength of the link for each of the four network maps (overall result); the second was to analyze the number of occurrences (including their links and their total link strengths [TLSs]) between the most representative nodes associated with all the resulting clusters of each map.

TABLE 27.1

Search for Agricultural Biotechnological Inventions

No.	Term	Time interval	Query	Documents
1	Biofertilizer	2015-2019	(TITLE-ABS-KEY(biofertilizer) AND (LIMIT-TO (PUBYEAR,2019) OR LIMIT-TO (PUBYEAR,2018) OR LIMIT-TO (PUBYEAR,2017) OR LIMIT-TO (PUBYEAR,2016) OR LIMIT-TO (PUBYEAR,2015)))	1757
2	Biopesticide	2015-2019	(TITLE-ABS-KEY(biopesticide) AND (LIMIT-TO (PUBYEAR,2019) OR LIMIT-TO (PUBYEAR,2018) OR LIMIT-TO (PUBYEAR,2017) OR LIMIT-TO (PUBYEAR,2016) OR LIMIT-TO (PUBYEAR,2015)))	1811
3	Compost	2019	(TITLE-ABS-KEY(compost) AND (LIMIT-TO (PUBYEAR,2019)))	1635

Source: Authors' elaboration based on Scopus (2020).

27.4 Analysis of Results

The first case presents a network map for agricultural technology as obtained from the co-occurrence analysis using data obtained from the Scopus database. Based on this information, a total of 4,906 keywords were identified, of which only 108 met the minimum five occurrences criterion (only 107 of these had links with other terms). The obtained network map consists of 14 clusters with a total of 507 links and a TLS of 640. Thus, in the case of the first four clusters, the most important node (based on highest number of occurrences [O]; we also include data on links [L] and TLS) were: Cluster 1 (red): Precision agriculture – O: 33, L: 18, TLS: 26. Cluster 2 (green): Sub-Saharan Africa – O: 17, L: 20, TLS: 25. Cluster 3 (dark blue): Food security – O: 29, L: 35, TLS: 44. Cluster 4 (lemon yellow): Maize – O: 18, L: 15, TLS: 18.

In the case of clusters 5–14, the results were: Cluster 5 (dark purple): Technology – O: 24, L: 31, TLS: 44. Cluster 6 (light blue): Technology adoption – O: 40, L: 35, TLS: 50. Cluster 7 (orange): Yield – O: 18, L: 11, TLS: 17. Cluster 8 (brown): Agricultural technology – O: 38, L: 25, TLS: 33. Cluster 9 (fuchsia): Organic farming – O: 9, L: 5, TLS: 5. Cluster 10 (pink): Climate change – O: 37, L: 31, TLS: 50. Cluster 11 (light green): Tanzania – O: 14, L: 17, TLS: 21. Cluster 12 (gray): Adoption – O: 30, L: 29, TLS: 49. Cluster 13 (olive green): Fertilizer – O: 9, L: 12, TLS: 13. Cluster 14 (light purple): Agriculture – O: 74, L: 52, TLS: 92 (see Figure 27.1).

This analysis highlights the presence of different analysis elements such as countries or regions, different agricultural inputs and technologies, important crops, and other topics associated with food security, technological adoption, sustainability, productivity, and climate change, among others. However, within agricultural technologies, biotechnology showed a low number of occurrences in cluster 3 as

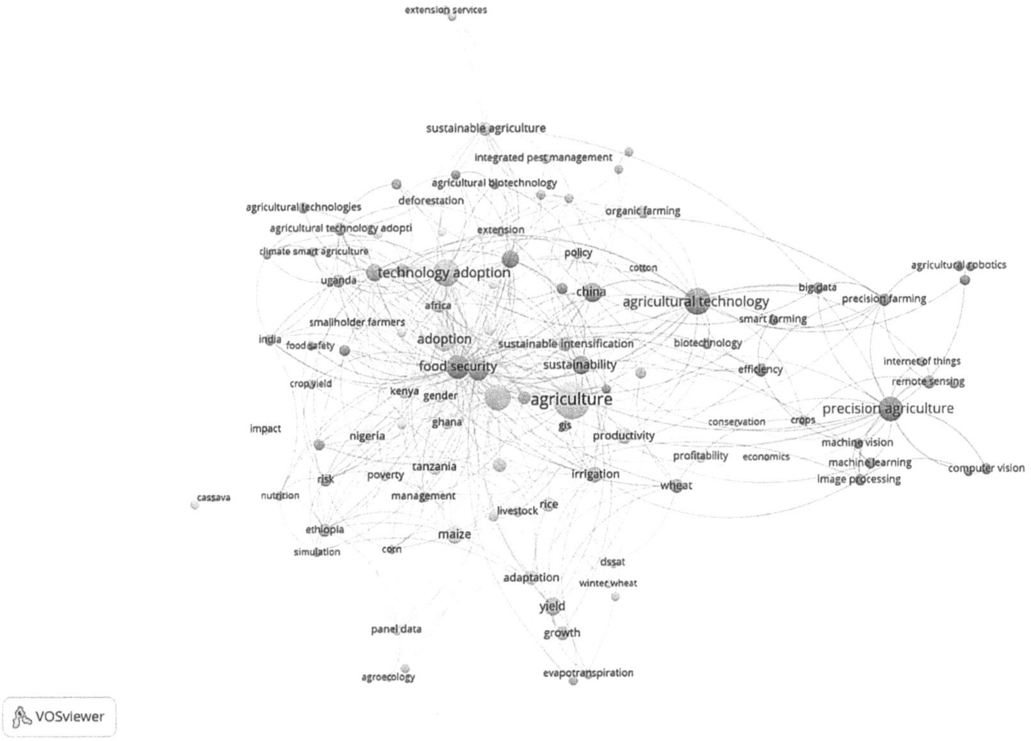

FIGURE 27.1 Co-occurrence analysis of keywords (author) – total count (agricultural technology). (Source: Authors' elaboration based on Scopus, 2020 and VOSviewer, 2020.)

"agricultural biotechnology" (O: 6, L: 4, TLS: 4) and, in cluster 8, as "biotechnology" (O: 7, L: 4, TLS: 7) – see supplementary material 1 (Barragán-Ocaña et al., 2021): all terms are shown in lower case in the annexes because VOSviewer unifies nominal sentences and, for example, capitals are converted to lowercase. This reflects its limited presence and links with other nodes, at least in studies related to agricultural technology. This highlights the need to continue addressing the use of biotechnology as a group of agricultural technologies, as well as other issues concerning technological acceptance and adoption by agricultural producers, productivity and biosecurity themes, as well as other relevant issues in the context of these developments, especially if these products are aimed at small-scale agricultural producers.

Thus, when analyzing the previously mentioned three biotechnological alternatives for agricultural use in connection with the term "biofertilizer" in the following map, we identified a total of 4,259 keywords; 239 of them occurred at least five times. Thus, the map consisted of 10 clusters with 2,177 links and a TLS of 3,332 – see supplementary material 2 (Barragán-Ocaña et al., 2021). Thus, the most important results for the first five clusters were as follows: Cluster 1 (red): Biogas – O: 49, L: 43, TLS: 98. Cluster 2 (green): Biofertilizers – O: 115, L: 100, TLS: 208. Cluster 3 (dark blue): Yield – O: 58, L: 74, TLS: 141. Cluster 4 (lemon yellow): Biofertilizer – O: 458, L: 209, TLS: 784. Cluster 5 (purple): Phosphate solubilization – O: 23, L: 37, TLS: 63. Results for the remaining five clusters were as follows: Cluster 6 (light blue): Plant growth promotion – O: 39, L: 44, TLS: 75. Cluster 7 (orange): Azotobacter – O: 32, L: 38, TLS: 69. Cluster 8 (brown): Bio-fertilizer – O: 34, L: 48, TLS: 64. Cluster 9 (fuchsia): Wheat – O: 31, L: 52, TLS: 78. Cluster 10 (pink): Compost – O: 23, L: 29, TLS: 40 (see Figure 27.2).

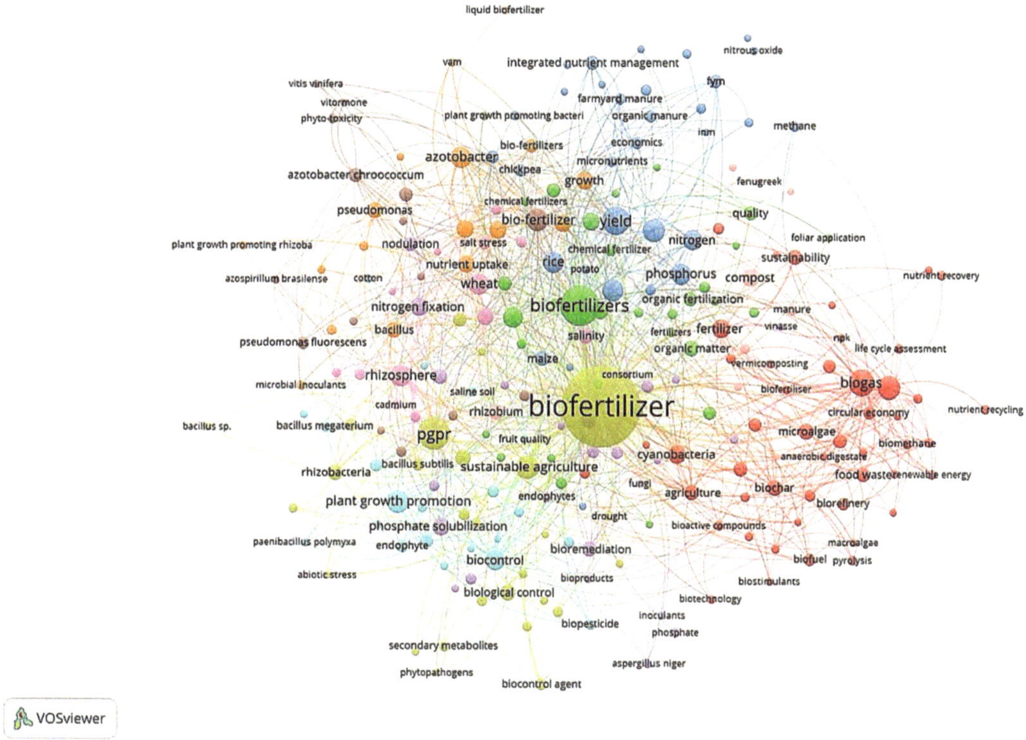

FIGURE 27.2 Co-occurrence analysis of keywords (author) – total count (biofertilizer). (Source: Authors' elaboration based on Scopus, 2020 and VOSviewer, 2020.)

For the case of the term biopesticide, results show that the analysis was based on the identification of 236 key words out of a total of 4,747. As a result, we identified 12 clusters with 1,859 links and a TLS of 2,781 – see supplementary material 3 (Barragán-Ocaña et al., 2021). Thus, the most relevant results for each cluster were: Cluster 1 (red): Azadirachtin – O: 29, L: 25, TLS: 40. Cluster 2 (dark green): Biopesticide – O: 256, L: 151, TLS: 358. Cluster 3 (dark blue): Biopesticides – O: 257, L: 163, TLS: 451. Cluster 4 (lemon yellow): Pesticides – O: 29, L: 36, TLS: 47. Cluster 5 (purple): Biocontrol – O: 74, L: 73, TLS: 142. Cluster results 6 through 11 were expressed as follows: Cluster 6 (light blue): Biological control – O: 115, L: 100, TLS: 211. Cluster 7 (orange): Pest management – O: 26, L: 34, TLS: 45. Cluster 8 (brown): Insecticidal activity – O: 23, L: 19, TLS: 21. Cluster 9 (fuchsia): *Bacillus thuringiensis* – O: 86, L: 65, TLS: 117. Cluster 10 (pink): Fungi – O: 13, L: 20, TLS: 36. Cluster 11 (light green): Bioinsecticides – O: 10, L: 8, TLS: 9. Cluster 12 (gray): Insecticide – O: 21, L: 26, TLS: 53 (see Figure 27.3).

Finally, in the case of the term "compost," the co-occurrence analysis included 194 key words out of 4,449, which resulted in 12 clusters with 1,295 links and a TLS of 1,854 – see supplementary material 4 (Barragán-Ocaña et al., 2021). As in the previous cases, the most relevant results for each cluster were as follows: Cluster 1 (red): Compost – O: 300, L: 150, TLS: 421. Cluster 2 (green): Organic waste – O: 22, L: 33, TLS: 48. Cluster 3 (dark blue): Heavy metals – O: 33, L: 32, TLS: 49. Cluster 4 (lemon yellow): Composting – O: 139, L: 98, TLS: 200. Cluster 5 (purple): Co-composting – O: 18, L: 25, TLS: 29. Cluster 6 through 12: Cluster 6 (light blue): Yield – O: 28, L: 36, TLS: 56. Cluster 7 (orange): Biochar – O: 73, L: 72, TLS: 138. Cluster 8 (brown): Vermicompost – O: 45, L: 51, TLS: 70. Cluster 9 (fuchsia): Enzyme activity – O: 10, L: 17, TLS: 19. Cluster 10 (pink): Plant growth – O: 11, L: 25, TLS: 31. Cluster 11 (light green): Nitrogen – O: 19, L: 23, TLS: 37. Cluster 12 (light blue): Bacterial community – O: 13, L: 13, TLS: 25 (see Figure 27.4).

Many terms associated with each of these three technologies stand out from the analysis. In general, it can be observed that each co-occurrence study produced good results in terms of the identified items,

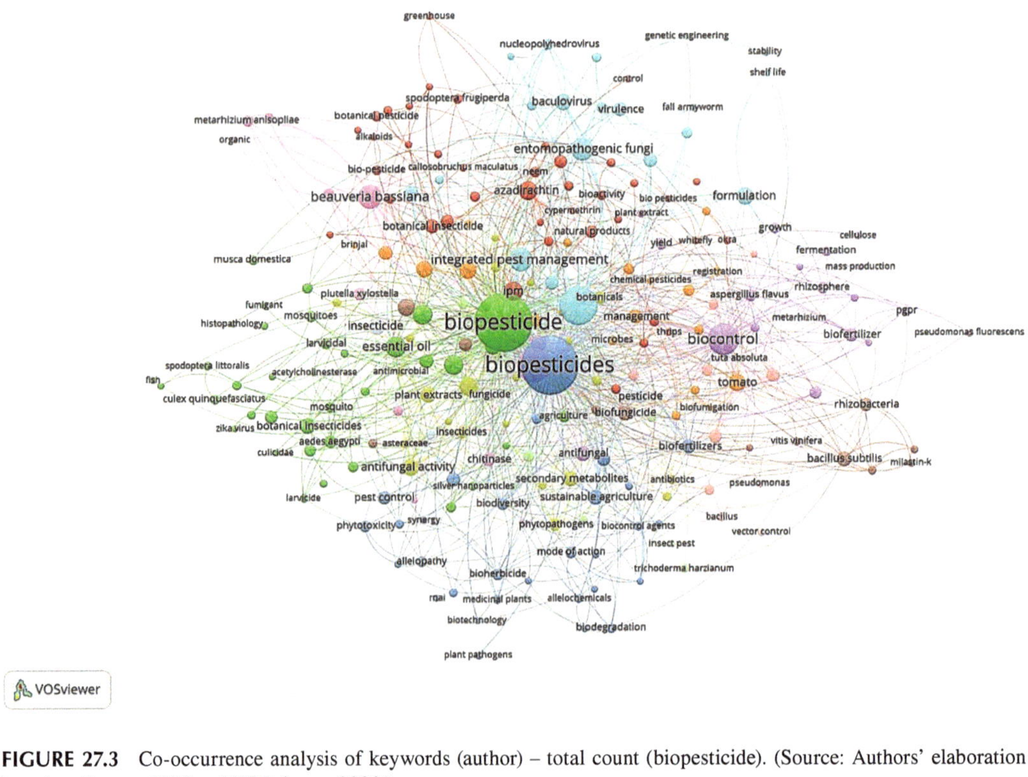

FIGURE 27.3 Co-occurrence analysis of keywords (author) – total count (biopesticide). (Source: Authors' elaboration based on Scopus, 2020 and VOSviewer, 2020.)

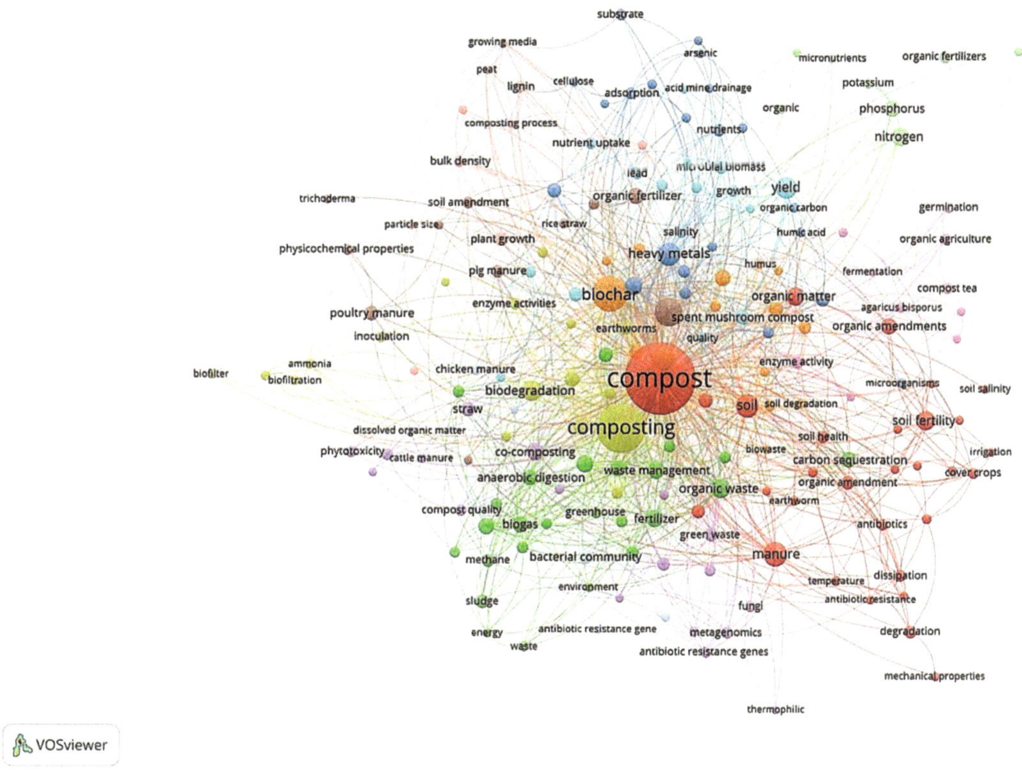

FIGURE 27.4 Co-occurrence analysis of keywords (author) – total count (compost). (Source: Authors' elaboration based on Scopus, 2020 and VOSviewer, 2020.)

their associated links, and TLS, which reflects the widespread interest of basic research in the study of biofertilizers, biopesticides, and composts; these topics are addressed very frequently, and they are profusely linked with other nodes. The emergence of additional themes that are beginning to generate important associations can also be observed. Among the results, it was possible to identify an important number of topics associated with processes and technologies, multiple microorganisms, agricultural problems and their management, nutrients, and even the presence of greenhouse gases conventionally derived from agricultural activities. A number of nodes focused on areas of interest such as productivity and sustainability were also identified; these are undoubtedly two of the most important challenges in the face of climate change and increasing global demand for food.

However, it is also true that the use of this type of technologies presents important technical and economic challenges related to its application and development, both if applied at an intensive scale and if used by small-scale producers. Therefore, agricultural innovation and sustainability are two necessary elements in the activities and lives of small-scale agricultural producers in developing countries. These technological innovations are intended to facilitate their day-to-day work, but without neglecting economic, social, and environmental development. In this chapter, we have described a series of agricultural biotechnology alternatives, but in order to properly materialize these alternatives, viable options of biofertilizers, biopesticides, and composts have to be evaluated in each context. Such an exercise should be carried out using different approaches and disciplines and to guarantee the technical and economic viability of developing, transferring, or adopting these technologies and promoting and investing in R&D and infrastructure to improve all stakeholders' capabilities to participate and operate, always prioritizing the implementation of public policies aiming at a wide social, cultural, and ethical impact (see Figure 27.5).

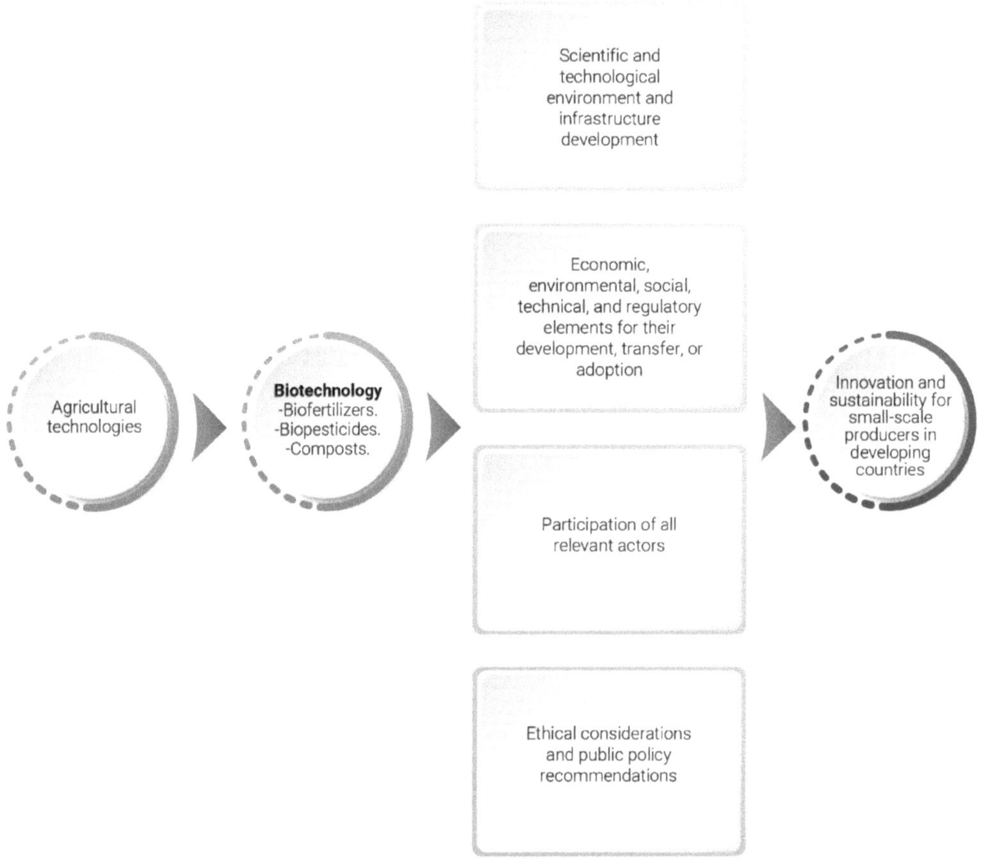

FIGURE 27.5 Biotechnology as an alternative for sustainable small-scale agricultural management. (**Elaborated by the authors.**)

27.5 Conclusions

As science and technology advance, available options for the agricultural sector increase. The effects of globalization, population growth, and climate change present new challenges to food production and environmental care. Consequently, the use of technology to respond to the many problems of the agricultural sector becomes essential. Challenges exacerbate in developing countries, particularly for small-scale agricultural producers. This is due to food insecurity and poverty, which result in inequality, uncertainty, and insecurity conditions, which in turn restrict their access to the inputs and technologies necessary to carry out their activities. It is then a priority for these economies to continue developing policies, institutions, regulations, and all other elements intended to provide certainty to this group of producers, which is highly relevant for the development of the rural sector.

Socially sensitive innovation, which addresses the values of local communities around the development and introduction of biotechnology, can significantly benefit the agricultural sector and generate profits and benefits for all stakeholders, such as technology-producing companies, small-scale farmers, and consumers, under an adequate governance framework allowing for sustainable production to be achieved while addressing the needs and lifestyles of the communities where new technologies are introduced. Therefore, a local diagnosis must be based on identifying the needs of the communities and not only focus on the products demanded by the market; in this regard, local diagnosis is an essential element to determine a starting point to implement the necessary strategies to promote sustainable development in the rural sector and have a favorable impact on the lives of producers and their families.

On the other hand, it is also necessary to discuss the risks associated with the development of a technology, but especially the way in which these risks are addressed. It is not convenient to force the incorporation of technological development based only on productivity, production, and market demand criteria. Although technical and economic viability is undoubtedly a central issue, consideration should also be given to local resources, community interests, and relevant technologies capable of exerting a positive social and environmental impact. This study presents a set of public policy instruments related to capacity-building, innovation plans, attention to the legal framework, R&D subsidies, and fiscal incentives, among others, that may contribute to the generation of specific guidelines according to the context and specific case, which is clearly of considerable relevance in the context of biofertilizers, biopesticides, and composts.

This exploratory approach provides positive evidence on the relevance of biotechnology in agricultural activities carried out by small producers in developing countries based on the use of the previously analyzed technologies. It is important to recognize how these can be integrated, for instance, into production schemes based on organic farming, or how waste generation problems can become a solution to recover overexploited soils. The range of scientific and technological options offered by these technologies is ample and can help to address different agricultural problems and mitigate pollution caused by the use of agrochemicals. The intention should not only be creating capabilities for technological adoption but also having the ability to evaluate technologies from different countries and create endogenous technology for the reality of each location. However, as it is with any type of biotechnology, it is necessary to guarantee biosecurity and production, create regulations, develop infrastructure and human resources, achieve technical and economic viability, and address ethical issues, always involving all stakeholders.

Acknowledgments

We wish to acknowledge the support provided by the National Polytechnic Institute (Instituto Politécnico Nacional) and the Secretariat for Research and Postgraduate Studies (Secretaría de Investigación y Posgrado), grant numbers 20210383, 20210515, and 20210755.

REFERENCES

Ashoori, D., Allahyari, M. S., and C. A. Damalas. 2017. Adoption of conservation farming practices for sustainable rice production among small-scale paddy farmers in northern Iran. Paddy and Water Environment 15: 237–248.

Azim, K., Soudi, B., Boukhari, S., Perissol, C., Roussos, S., and A. I. Thami. 2018. Composting parameters and compost quality: A literature review. Organic Agriculture 8: 141–158.

Barragán-Ocaña, A. and M. C. del Valle-Rivera. 2016. Rural development and environmental protection through the use of biofertilizers in agriculture: An alternative for underdeveloped countries? Technology in Society 46: 90–99.

Barragán-Ocaña, A., De Hoyos Bermea, A., Luna López, K. A., and T. A. Virgilio Virgilio, 2021. Supplementary materials on agricultural technology, biofertilizers, biopesticides and compost. Available in: https://repositoriodigital.ipn.mx/handle/123456789/26382 (uploaded on 7 July 2021).

Genus, A. and M. Iskandarova. 2018. Responsible innovation: its institutionalisation and a critique. *Technological Forecasting and Social Change* 128: 1–9.

Gómez, M. 2010. El devenir de las identidades mexicanas en tiempos de innovación biotecnológica in D. E. García, I. Cepeda, S. A. Shea (eds.) *Visiones de México: 1810–2010. Devenires de una identidad cultural.* México: Porrúa-UNESCO-ITESM: 67–85.

Gupta, S., and A. K. Dikshit. 2010. Biopesticides: An ecofriendly approach for pest control. Journal of Biopesticides 3: 186–188.

Haase, S., Sciocco-Cap, A., and V. Romanowski. 2015. Baculovirus insecticides in Latin America: Historical overview, current status and future perspectives. Viruses 7: 2230–2267.

Harindintwali, J. D., Zhou, J., and X. Yu. 2020. Lignocellulosic crop residue composting by cellulolytic nitrogen-fixing bacteria: A novel tool for environmental sustainability. Science of the Total Environment 715: 136912.

Igiehon, N. O. and O. O. Babalola. 2017. Biofertilizers and sustainable agriculture: Exploring arbuscular mycorrhizal fungi. Applied Microbiology and Biotechnology 101: 4871–4881.

Kaaya, G. P and S. Hassan. 2000. Entomogenous fungi as promising biopesticides for tick control. Experimental an Applied Acarology 24: 913–926.

Khatri-Chhetri, A., Aggarwal, P. K., Joshi, P. K., and S. Vyas. 2017. Farmers' prioritization of climate-smart agriculture (CSA) technologies. Agricultural Systems 151: 184–191.

Kubheka, B. P., Laing, M.D., and K. S. Yobo. 2020. Combinations of a biofertilizer with micro-dosed chemical fertilizers increased yield of maize in a high acid saturated soil. Rhizosphere 13: 100189.

Lele, U. and S. Goswami. 2017. The fourth industrial revolution, agricultural and rural innovation, and implications for public policy and investments: A case of India. Agricultural Economics 48: 87–100.

Mboussi, S. B., Ambang, Z., Kakam, S., and L. B. Beilhe. 2018. Control of cocoa mirids using aqueous extracts of *Thevetia peruviana* and *Azadirachta indica*. Cogent Food Agriculture 4: 1430470.

Mouter, N., de Geest, A., and N. Doorn. 2018. A values-based approach to energy controversies: Value-sensitive design applied to the Groningen gas controversy in the Netherlands. Energy Policy 122: 639–648.

Mwangi M. N. and S. Kariuki. 2015. Factors determining adoption of new agricultural technology by smallholder farmers in developing countries. Journal of Economics and Sustainable Development 6: 208–216.

Naveed, M., Mehboob, I., Shaker, M. A., Hussain, M. B., and M. Farooq. 2015. Biofertilizers in Pakistan: Initiatives and limitations. International Journal of Agriculture and Biology 17: 411–420.

Obi O. F. and J. N. Nwakaire. 2014. Repositioning agricultural development in Africa through appropriate technology transfer. African Journal of Science, Technology, Innovation and Development 6: 45–50.

Olanrewaju, O. S., Glick, B. R., and O. O. Babalola. 2017. Mechanisms of action of plant growth promoting bacteria. World Journal of Microbiology and Biotechnology 33: 1–16.

Olivé, L. 2007. *La ciencia y la tecnología en la sociedad del conocimiento: Ética, política y epistemología.* México: Fondo de Cultura Económica.

Paustian, M., and L. Theuvsen. 2017. Adoption of precision agriculture technologies by German crop farmers. Precision Agriculture 18: 701–716.

Pereira, I., Ortega, R., Barrientos, L., Moya, M., Reyes, G., and V. Kramm. 2009. Development of a biofertilizer based on filamentous nitrogen-fixing cyanobacteria for rice crops in Chile. Journal of Applied Phycology 21: 135–144.

Pergolaa, M., Persiania, A., Palesea, A. M. et al. 2018. Composting: The way for a sustainable agriculture. Applied Soil Ecology 123: 744–750.

Robledo-Buriticá, J., Aristizábal-Loaiza, J. C., Ceballos-Aguirre, N., and T. Cabra-Cendales. 2018. Influence of plant growth-promoting rhizobacteria (PGPR) on blackberry (*Rubus glaucus* Benth. cv. thornless) growth under semi-cover and field conditions. Acta Agronomica 67: 258–263.

Rockström, J., Williams, J., Daily, G. et al. 2017. Sustainable intensification of agriculture for human prosperity and global sustainability. Ambio 46: 4–17.

Saint Ville, A. S., Hickey, G. M., Locher, U., and L. E. Phillip. 2016. Exploring the role of social capital in influencing knowledge flows and innovation in smallholder farming communities in the Caribbean. Food Security 8: 535–549.

Scopus. (2020). Abstract and citation database. Available in: https://www.scopus.com (accessed 20 November 2020).

Stadler, T. and J. Chauvet. 2018. New innovative ecosystems in France to develop the bioeconomy. New Biotechnology 40: 113–118.

Taheri, F., D'Haese, M., Fiems, D., Hosseininia, G. H., and H. Azadi. 2020. Wireless sensor network for small-scale farming systems in southwest: Application of Q-methodology to investigate farmers' perceptions. Computers and Electronics in Agriculture 177: 105682.

Thompson, M., Cochrane, A., and J. Hopma. 2020. Democratising food: The case for a deliberative approach. Review of International Studies 46: 435–455.

Vakili, M., Rafatullah, M., Ibrahim, M., Salamatinia, B., Gholami, Z., and H. Zwain. 2015. A review on composting of oil palm biomass. Environment, Development and Sustainability 17: 691–709.

Van Damme, J., Ansoms, A., and P. Baret. 2014. Agricultural innovation from above and from below: Confrontation and integration or Rwanda's Hills. African Affairs 113: 108–127.

Viaene, J., Van Lancker, J., Vandecasteele B. et al. 2016. Opportunities and barriers to on-farm composting and compost application: A case study from northwestern Europe. Waste Management 48: 181–192.

VOSviewer. 2020. Vosviewer (Software). Version 1.6.15.

Index

Note: Locators in *italics* represent figures and **bold** indicate tables in the text.